S0-ATB-874

THE MYTH OF RACE

Sussman, Robert W., 194
The myth of race : the
troubling persistence o
2014.
33305231675756
cu   10/17/14

THE MYTH OF RACE

The Troubling Persistence of an Unscientific Idea

Robert Wald Sussman

Harvard University Press

Cambridge, Massachusetts

London, England

2014

Copyright © 2014 by the President and Fellows of Harvard College
All rights reserved
Printed in the United States of America

First Printing

Library of Congress Cataloging-in-Publication Data

Sussman, Robert W., 1941–
The myth of race : the troubling persistence of an
unscientific idea / Robert Wald Sussman.
pages cm
Includes bibliographical references and index.
ISBN 978-0-674-41731-1 (hardcover : alk. paper)
1. Race. 2. Racism. I. Title.
HT1521.S83 2014
305.8—dc23   2014011078

I dedicate this book to my wife, Linda;
my two daughters, Katya and Diana;
and my students

Contents

Abbreviations

AA	*American Anthropologist*
AAA	American Anthropological Association
AAAS	American Association for the Advancement of Science
AAPA	American Association of Physical Anthropologists
ABA	American Breeders Association
AES	American Eugenics Society
AICF	American Immigration Control Foundation
AR	*American Renaissance*
CIW	Carnegie Institution of Washington, D.C.
ECUSA	Eugenics Committee of the United States of America
EN	*Eugenical News*
ERA	Eugenics Research Association
ERO	Eugenics Record Office
FAIR	Federation for American Immigration Reform
IAAEE	International Association for Advancement of Ethnology and Eugenics
IFEO	International Federation of Eugenics Organizations
ISM	Institute for the Study of Man
JAMA	*Journal of the American Medical Association*
KWIA	Kaiser Wilhelm Institute for Anthropology
MQ	*Mankind Quarterly*

MTFS	Minnesota Twin Family Study
NAACP	National Association for the Advancement of Colored People
NAS	National Academy of Sciences
NRC	National Research Council
NSDAP	National Socialist German Worker's Party
UNESCO	United Nations Educational, Scientific and Cultural Organization
USM	University of Southern Mississippi

Introduction

In 1950, UNESCO issued a statement asserting that all humans belong to the same species and that "race" is not a biological reality but a myth. This was a summary of the findings of an international panel of anthropologists, geneticists, sociologists, and psychologists. A great deal of evidence had accumulated by that time to support this conclusion, and the scientists involved were those who were conducting research and were most knowledgeable about the topic of human variation. Since that time similar statements have been published by the American Anthropological Association and the American Association of Physical Anthropologists, and an enormous amount of modern scientific data has been gathered to justify this conclusion. Today the vast majority of those involved in research on human variation would agree that biological races do not exist among humans. Among those who study the subject, who use and accept modern scientific techniques and logic, this scientific fact is as valid and true as the fact that the earth is round and revolves around the sun.

Yet as recently as 2010, highly acclaimed journalist Guy Harrison (2010) wrote:

> One day in the 1980s, I sat in the front row in my first undergraduate anthropology class, eager to learn more about this bizarre and fascinating species I was born into. But I got more than I expected that day as I heard for the first time that biological races are not real. After hearing several perfectly sensible reasons why vast biological categories don't work very well, I started to feel betrayed by my society. "Why am I just hearing this now? ... Why didn't somebody tell me this in elementary school?" ... I never should have made it through twelve years of schooling before entering

a university, without ever hearing the important news that most anthropologists reject the concept of biological races. (27, 30)

The Myth of Human Racism

Unfortunately, along with the belief in the reality of biologically based human races, racism still abounds in the United States and Western Europe. How can this be when there is so much scientific evidence against it? Most educated people would accept the facts that the earth is not flat and that it revolves around the sun. However, it is much more difficult for them to accept modern science concerning human variation. Why is this so? It seems that the belief in human races, carrying along with it the prejudice and hatred of "racism," is so embedded in our culture and has been an integral part of our worldview for so long that many of us assume that it just must be true.

Racism is a part of our everyday lives. Where you live, where you go to school, your job, your profession, who you interact with, how people interact with you, your treatment in the healthcare and justice systems are all affected by your race. For the past 500 years people have been taught how to interpret and understand racism. We have been told that there are very specific things that relate to race, such as intelligence, sexual behavior, birth rates, infant care, work ethics and abilities, personal restraint, life span, law-abidingness, aggression, altruism, economic and business practices, family cohesion, and even brain size. We have learned that races are structured in a hierarchical order and that some races are better than others. Even if you are not a racist, your life is affected by this ordered structure. We are born into a racist society.

What many people do not realize is that this racial structure is not based on reality. Anthropologists have shown for many years now that there is no biological reality to human race. There are no major complex behaviors that directly correlate with what might be considered human "racial" characteristics. There is no inherent relationship between intelligence, law-abidingness, or economic practices and race, just as there is no relationship between nose size, height, blood group, or skin color and any set of complex human behaviors. However, over the past 500 years, we have been taught by an informal, mutually reinforcing consortium of intellectuals, politicians, statesmen, business and economic leaders, and their books that human racial biology is real and that certain races are biologically better than others. These teachings have led to major injustices to Jews and non-

Christians during the Spanish Inquisition; to blacks, Native Americans, and others during colonial times; to African Americans during slavery and reconstruction; to Jews and other Europeans during the reign of the Nazis in Germany; and to groups from Latin America and the Middle East, among others, during modern political times.

In this book I am not going to dwell upon all of the scientific information that has been gathered by anthropologists, biologists, geneticists, and other scientists concerning the fact that there are no such things as human biological races. This has been done by many people over the past fifty or so years. What I am going to do is describe the history of our myth of race and racism. As I describe this history, I think that you will be able to understand why many of our leaders and their followers have deluded us into believing these racist fallacies and how they have been perpetuated from the late Middle Ages to the present. Many of our basic policies of race and racism have been developed as a way to keep these leaders and their followers in control of the way we live our modern lives. These leaders often see themselves as the best and the brightest. Much of this history helped establish and maintain the Spanish Inquisition, colonial policies, slavery, Nazism, racial separatism and discrimination, and anti-immigration policies. Although policies related to racism seem to be improving over time, I hope to help clarify why this myth still exists and remains widespread in the United States and throughout Western Europe by describing the history of racism and by exploring how the anthropological concepts of culture and worldview have challenged and disproven the validity of racist views.

Over the past 500 or so years, many intellectuals and their books have created our story of racism. They developed our initial ideas of race in Western society and solidified the attitudes and beliefs that gradually followed under the influence of their economic and political policies. Then, approximately 100 years ago, anthropologist Franz Boas came up with an alternate explanation for why peoples from different areas or living under certain conditions behaved differently from one another. People have divergent life histories, different shared experiences with distinctive ways of relating to these differences. We all have a worldview, and we all share our worldview with others with similar experiences. We have culture. It took many years for Boas and his few followers to develop this idea and pass it on to others. However, over the past fifty or sixty years, anthropologists, biologists, and geneticists have written many articles and books explaining why biological race in humans is nonexistent. At first, scientists attempted to classify human races based on variations in characteristics such as skin color, hair color and form, eye color, facial anatomy, and blood groups. In the recent past, various scientists have divided us into anywhere between three and more than thirty different races, without any success (see Molnar 2006). Most of these hypothetical "races" were developed using assumptions about genetic relationships and distributions among different human populations.

What Is a Race or Subspecies?

In 1942, Ashley Montagu, a student of Franz Boas, claimed that "there are no races, there are only clines." Traits considered to be "racial" are actually distributed independently and depend upon many environmental and behavioral factors. For the most part, each trait has a distinct distribution from other traits, and these traits are rarely determined by a single genetic factor. This type of distribution of a biological trait is referred to as a cline. For example, skin color is related to the amount of solar radiation, and dark skin is found in Africa, India, and Australia. However, many other genetic traits in peoples of these areas are not similar. Furthermore, similar traits such as skin color are convergent; different genes can cause similar morphological and behavioral characteristics. For example, genetic pathways to dark skin are different in Tamil Nadu and in Nigeria. Genetic traits usually do not correlate with one another and are not distributed in the same place or in the same way over time.

Race is supposed to tell us something about our genetic history. Who is related to whom? How did populations evolve over time and how isolated were they in the past? Recent studies have shown us that humans have been migrating since *Homo sapiens* evolved some 200,000 years ago. This migration has not been in one direction but had happened back and forth. Our genes have been mixing since we evolved, and our genetic structure looks more like a complex, intermixed trellis than a simple candelabra (Templeton 1998). It is very difficult to tell what our particular genetic background is over human historic time. We humans are more similar to each other as a group than we are to one another within any particular racial or genetic category. Many anthropological books have been written to explain this phenomenon (for recent examples see Tattersall and DeSalle 2011; Smedley and Smedley 2012; Relethford 2013; Jurmain et al. 2014; and Mukhopadhyay, Henze, and Moses 2014).

Our view of genetics has also changed in recent times. Although many people still believe that genes, or a series of genes, directly determine some of our most complex behavioral or cognitive characteristics, the reality is more complicated. Studies now show that each gene is only a single player in a wondrous, intricate drama involving nonadditive interactions of genes, proteins, hormones, food, and life experiences and learning that interact to affect us on different levels of cognitive and behavioral functions. Each gene has an effect on multiple types of behaviors, and many behaviors are affected by many genes as well as other factors. The assumption that a single gene is causative can lead to unwarranted conclusions and an over-interpretation of any genuine genetic linkage (Berkowitz 1999; Weiss and Buchanan 2009; Charney 2013).

Before beginning this story, however, it is important to understand how scientists define the concept of race. How is race defined in *biological* terms? What do we mean by the term *race* when describing population variation in large mammals such as humans? Do the criteria used in describing these variations hold when we examine human population variation? In biological terms, the concept of race is integrally bound to the process of evolution and the origin of species. It is part of the process of the formation of new species and is related to subspecific differentiation. However, because conditions can change and subspecies can and do merge (Alan Templeton, 2013, personal communication), this process does not necessarily lead to the development of new species. In biology, a species is defined as a population of individuals who are able to mate and have viable offspring; that is, offspring who are also successful in reproducing. The formation of new species usually occurs slowly over a long period of time. For example, many species have a widespread geographic distribution with ranges that include ecologically diverse regions. If these regions are large in relationship to the average distance of migration of individuals within the species, there will be more mating, and thus more exchange of genes, within than between regions. Over very long periods of time (tens of thousands of years), differences would be expected to evolve between distant populations of the same species. Some of these variations would be related to adaptations to ecological differences within the geographic range of the populations, while others might be purely random. Over time, if little or no mating (or genetic exchange) occurs between these distant populations, genetic (and related morphological) differences will increase. Ultimately, over tens of thousands of years of separation, if little or no mating takes place between separate populations, genetic distinctions can become so great that individuals of the different populations could no longer mate and produce viable offspring. The two populations would now be considered two separate species. This is the process of speciation. However, again, none of these criteria require that speciation will ultimately occur.

Since speciation develops very slowly, it is useful to recognize intermediate stages in this process. Populations of a species undergoing differentiation would show genetic and morphological variation due to a buildup of genetic differences but would still be able to breed and have offspring that could successfully reproduce. They would be in various stages of the process of speciation but not yet different species. In biological terminology, it is these populations that are considered "races" or "subspecies" (Williams 1973; Amato and Gatesy 1994; Templeton 1998, 2013). Basically, subspecies within a species are geographically, morphologically, and genetically distinct populations but still maintain the possibility of successful interbreeding (Smith, Chiszar, and Montanucci 1997). Thus, using this biological definition of race, we assume that races or subspecies are populations of a species that have genetic and morphological differences due to barriers to mating. Furthermore, little or no mating (or genetic exchange) between them has persisted for extremely long periods of time, thus giving the individuals within the population a common and separate evolutionary history.

Given advances in molecular genetics, we now have the ability to examine populations of species and subspecies and reconstruct their evolutionary histories in an objective and explicit fashion. In this way, we can determine the validity of the traditional definition of human races "by examining the patterns and amount of genetic diversity found within and among human populations" (Templeton 1998, 633) and by comparing this diversity with other large-bodied mammals that have wide geographic distributions. In other words, we can determine how much populations of a species differ from one another and how these divergences came about.

A commonly used method to quantify the amount of within- to among-group genetic diversity is through examining molecular data, using statistics measuring genetic differences within and between populations of a species. Using this method, biologists have set a minimal threshold for the amount of genetic differentiation that is required to recognize subspecies (Smith, Chiszar, and Montanucci 1997). Compared to other large mammals with wide geographic distributions, human populations do not reach this threshold. In fact, even though humans have the widest distribution, the measure of human genetic diversity (based on sixteen populations from Europe, Africa, Asia, the Americas, and the Australia-Pacific region) falls well below the threshold used to recognize races for other species and is among the lowest value known for large mammalian species (Figure I.1). This is true even if we compare humans to chimpanzees (Templeton 2013).

Using a number of molecular markers, Templeton (1998, 2013), further, has shown that the degree of isolation among human populations that would have been necessary for the formation of biological subspecies or races never occurred during the 200,000 years of modern human evolution. Combined genetic data reveal that from around one million years ago to the last tens of thousands of years, human evolution has been dominated by

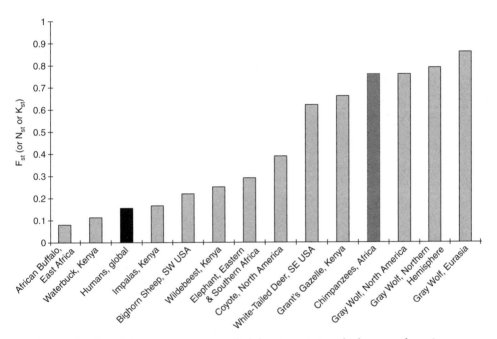

Figure I.1 Geneticists use a measure called the F_{st} statistic, which ranges from 0 (when agenetic diversity within a species is equally shared by all populations with no genetic differences among populations) to 1 (when genetic diversity within a species is found as fixed differences among populations with no diversity within populations). Using this measure, the F_{st} for humans is 0.156 (Barbujani et al. 1997), indicating that most human genetic diversity exists within populations and only a small percent can be used to genetically differentiate between human populations. The standard criterion for a subspecies in large nonhuman mammals under this traditional definition is to have an F_{st} value of 0.30 (Smith, Chiszar, and Montanucci 1997). Thus, the human F_{st} value is much too small to have any biological, taxonomic significance under the traditional definition of subspecies or race in biology. Although subspecies or races do exist among many large mammals, using those biological criteria, races do not exist in humans. (Modified from Templeton 1998).

two evolutionary forces: (1) constant population movement and range expansion; and (2) restrictions on mating between individuals only because of distance. Thus, there is no evidence of fixed, long-term geographic isolation between populations. Other than some rare, temporary isolation events, such as the isolation of the aborigines of Australia, for example, the major human populations have been interconnected by mating opportunities (and thus genetic mixture) during the last 200,000 years (as long as modern humans, *Homo sapiens,* have been around). As summarized by Templeton (1998, 647), who is among the world's most recognized and respected geneticists:

> Because of the extensive evidence for genetic interchange through population movements and recurrent gene flow going back at least hundreds of thousands of years ago, there is only one evolutionary lineage of humanity and there are no subspecies or races. . . . Human evolution and population structure has been and is characterized by many locally differentiated populations coexisting at any given time, but with sufficient contact to make all of humanity a single lineage sharing a common, long-term evolutionary fate.

Thus, given current scientific data, biological races do not exist among modern humans today, and they have never existed in the past. Given such clear scientific evidence as this and the research data of so many other biologists, anthropologists, and geneticists that demonstrate the nonexistence of biological races among humans, how can the "myth" of human races still persist? If races do not exist as a biological reality, why do so many people still believe that they do? In fact, even though biological races do not exist, the concept of race obviously is still a reality, as is racism. These are prevalent and persistent elements of our everyday lives and generally accepted aspects of our culture. Thus, the concept of human races is real. It is not a biological reality, however, but a cultural one. Race is not a part of our biology, but it is definitely a part of our culture. Race and racism are deeply ingrained in our history. In this book, I use the term *scientific racism* to refer to scientists who continue to believe that race is a *biological* reality and categorize human "races" in a hierarchical fashion.

The Goals of This Book

I began teaching courses in human evolution and human variation in the mid-1970s. In 1974, Richard Popkin, a philosopher, wrote a paper on the history of racism in Western Europe and the United States that focused on the period from the Spanish Inquisition of the fifteenth century to the seventeenth century (Popkin [1974] 1983). Popkin traced the germs of modern racism to two major themes, or ideologies: the pre-Adamite and the degenerate views of human racial variation. Others also have recognized these two approaches to racist history and have used similar terms or have substituted the terms *polygenism* and *monogenism,* respectively. Although Popkin did not extend his historical scenario to Darwin's time, the eugenics movement, Nazism, or current concepts and debates about race and racism, some of the basic underlying premises he described can be traced through the nineteenth and twentieth centuries and still exist today. In this volume, I will review Popkin's original theme and show how some ancient perspectives on race have survived and persist in some current views on this topic.

I am not a historian. What I have tried to do here is trace the ideas of monogenism and polygenism that continue to resurface through many sources and to follow this common thread underlying the concept of race and the basis of racism from the fifteenth century to modern times. Furthermore, as I will show, this history of race and racism is a major component in the development of modern anthropology. I have not done a great deal of archival research myself but instead have depended upon the published works of many historians, biographers, and philosophers who have painstakingly searched out many exchanges, letters, and difficult-to-locate original sources. I have tried to synthesize this literature and show how it all comes together into one story, a fairly consistent story outlining the underpinnings of a six-century-old racist ideology. Thus, I hope to illuminate why some of us are still living in the Middle Ages in terms of the subject of race. I will also explore why some very early views of race based on ignorance, emotion, hatred, intolerance, and prejudice have been repeated, almost verbatim, through the ages. They are still being voiced by modern racists despite empirical data and growing scientific evidence through the centuries and decades that contradict this racist ideology. As one result of this exploration I hope that at least some of those caught in the past will understand why their intolerance is as baseless as the idea of a flat earth. For others, I hope that this book will further their understanding of why racism is still so prevalent in our society; why the anthropological concept of culture is so important and the influence of culture on our lives is so profound. Unfortunately, one's cultural background often trumps logic, empirical data, and modern science. In fact, modern science is by its very nature framed by culture (Kuhn 1962; Benson 2011). However, in the end, I hope to show that racism and bigotry are fueled by deep-seated hatred and intolerance of human variation and are not supported by empirical evidence or modern science.

1

Early Racism in Western Europe

Early Christians, Hebrews, and the Greeks allowed out-groups to overcome their alleged inferiority by converting to the "superior" or dominant group, or through the process of assimilating (Longhurst 1964). The Greeks, for example, allowed so-called barbarians to learn to speak, write, think, and live as Greeks. However, in the fifteenth century, the Spanish introduced a new form of racism. In order to squelch the large and rising number of Jews who had been forced to convert to Catholicism and who were gaining status financially and in the church, Old Christians were separated from New Christians, or *conversos*, on biological grounds. Anyone with Jewish ancestry in the previous five generations was considered a New Christian and was subject to a number of restrictions, including an inability to attend college, join certain religious orders, or hold government positions. Certificates of "purity of blood" were issued to non-Jews to prove that an individual was not a member of this "inferior" group.

The Spanish Inquisition

The Spanish Inquisition was established to ensure that those of Jewish ancestry were kept apart and out of the mainstream of society. Although it was mainly directed at Jews, the inquisition also focused on Christianized Muslims and Gypsies and later moved to Asia and America, where it targeted indigenous people (Popkin [1974] 1983; Kamen 1998; Murphy 2012). In Spain, the inquisition was formally established in 1478, although it built on earlier inquisitions in other places. When it moved to Rome in the sixteenth century, although still persecuting Jews, the inquisition expanded its focus to include Protestants, homosexuals, people accused of witchcraft, freethinkers, public intellectuals, and people considered to be quirky or "gadflies" (Murphy 2012).

The inquisitions discriminated against and separated one group from another without allowing any legal means for the discriminated group to assimilate. Unlike earlier inquisitions, the Spanish Inquisition did not focus on religion alone but expanded to include ethnicity or race, introducing the notion of *limpieza de sangre,* or "impurity of blood." "It was about classes of people rather than just categories of belief," author Cullen Murphy notes (2012, 70). Furthermore, it was run by those in political power. It was political: religion, ideology, and race or ethnicity were ruled and defined by the state. Minority or conquered peoples could not change their identities; they could not convert or assimilate into mainstream society. Although these discriminating practices began as a result of economic and political conditions, "scientific" theories justifying this kind of racism began to appear in Spain and Portugal in the fifteenth century, and after the discovery of America, they were expanded to justify similar racist ideas toward Native Americans, Asians, and, later, enslaved Africans. It is interesting to note that Columbus's voyage to America was at the peak of the Inquisition in Spain. It was financed mainly by *conversos,* and there were *conversos* among the ships' crews. In fact, a large number of Jews who had refused to be baptized were leaving Spain at that time (Murphy 2012).

The initial cause of anti-Semitism in Spain and Portugal may have been jealousy of the power, wealth, and influence of some Jews (and others) in early Spanish society. However, it also could be explained and justified by biblical explanations of Jews as the killers of Christ and eternal enemies of Christianity (Cohen 2007). But when the Spaniards and Portuguese began to colonize America, the people they conquered and whose land they were taking had no established role in European society. Prior to this, travelers and explorers saw continuity between neighboring peoples as they traveled slowly through adjacent areas instead of traveling long distances to entirely new regions—basically jumping continents (Brace 2005; Jablonski 2012). New rationalizations had to be made to justify mistreating the peoples Europeans encountered and new theories formed to explain their place in the universe.

As described by Popkin (1973), although numerous explanations were expounded, two major theories emerged, became prominent, and exhibited remarkable staying power: the pre-Adamite and the degenerate theories. These theories first centered on the question of whether Native Americans' origins were traceable to migrations of biblical people that had somehow become degenerate or were not descendants from the biblical world at all but had a separate origin. In this latter theory, American Indians were not descendants of Adam and Eve but had an independent, earlier origin—they were pre-Adamites.

The conquistadores justified their maltreatment of Native Americans by claiming they were subhuman and incapable of having abstract ideas and of running their own world. They also were deemed incapable of morality and unable to become Christian. These views were promulgated by Spanish theorists such as Sepulveda and Oviedo in the early sixteenth century (Popkin [1974] 1983; Brace 2005). In 1512, Montesinos, a preacher in Santo Domingo, opposed the mistreatment of the Indians and insisted that they were human. Bartholemé de Las Casas, who became bishop of Chiapas, became an advocate of this cause and debated Sepulveda and his followers for almost half a century (Hanke 1949; Popkin [1974] 1983; Brace 2005). He claimed that "all people in the world are men . . . all have understanding and volition . . . all take satisfaction in goodness and [feel] pleasure with happy and delicious things, all regret and abhor evil" (quoted in Popkin [1974] 1983, 129).

The first professor of philosophy in the New World, Alonso de la Vera Cruz, argued in his first and only course at the University of Mexico that Spaniards did not have the right to subjugate the Indians, and Pope Paul III, in 1537, declared that "the Indians are truly men and that they are not only capable of understanding the Catholic faith, but, according to our information, they desire exceedingly to receive it" (quoted in Hanke 1949, 73). However, the church could not stop the conquest of America and the mistreatment of Native American peoples. Even though the Spanish government and the church eventually declared that the Indians were fully human, Vera Cruz was removed as professor and sent to lower Yucatan (Popkin [1974] 1983). The mistreatment of Native Americans did not subside.

As the inhumane conquest of America continued, racial theories remained crucial in justifying the treatment of the local peoples and, a bit later, the enslavement of Africans who often were needed to replace the rapidly dying indigenous Americans as a work force for exploiting the New World. The early Spanish debate was simply a preview of things to come. The two main theories used to explain human differences, pre-Adamite and degenerate, that the Spanish and Portuguese had first proposed in the sixteenth century were later adopted mainly by the English, Anglo-Americans, and the French in the seventeenth and eighteenth centuries. These theories then provided the basis of racist thought in regard to people of color and

Jews for the nineteenth and twentieth centuries. In fact, I will argue that the threads of these two theories survived Darwinian times and the modern synthesis of evolutionary theory. Furthermore, they are still with us today, both in the general public and in Western science.

The Degeneration Theory of Race from Ancient Times to Darwin

Although the pre-Adamite or polygenic theory had a following throughout the period covered here and became the dominant theory in the mid-nineteenth century, the degeneration theory of race was the most accepted version in earlier times. Rather than challenging the biblical account of human origins, a generally unpopular approach, the degeneration theory assumed that all humans were created by God beginning with Adam and Eve. Nonwhites were thought to be inferior and to need the guidance and control of rational, moral men (i.e., white European Christians). Their condition was considered to be caused by some degenerative process that was related to climate or conditions of life, to isolation from Christian civilization, or to some divine action explained in the Bible (Popkin [1974] 1983). This was, in fact, the more liberal point of view, since proponents of this approach believed that these degenerates could be remediated by giving them the benefits of European education and "culture," especially by missionizing them to Christianity.

After the debates between the church and the conquistadores discussed above, one of the earliest well-known proponents of the degeneration theory was John Locke. Locke was the seventeenth-century architect of English colonial policy who drafted the constitution for the Carolinas. He accepted the biblical account of human origins but believed that the equality at creation and the endowment of natural rights to all humans no longer had to be applied because the American Indians were not using their land properly. He also believed that they should lose their liberty because they had unjustly opposed the Europeans. Locke justified the maltreatment and slavery of nonwhites based on what he considered their personal failures (Locke 1690).

In the eighteenth century, many of the early, well-known natural historians attempted to explain just why these peoples were such "failures." These degeneration theorists attempted to explain "that the factors that led some peoples to change from white skinned to dark involved ways of life that were far inferior to those of Europeans" (Popkin [1974] 1983, 133–134). The French nobleman, politician, and political philosopher of the Enlightenment Montesquieu (Charles-Louis de Secondat, Baron de La Brède et de Montesquieu, 1689–1755) was among the first to develop an elaborate climate theory in his *De l'Esprit des Loix* (1748). He believed that climate and geography affected the temperaments and customs of a country's inhabitants and thus accounted for differences among humans and their cultures. However, these differences were not hereditary, and if one moved from one climate to another, one's temperament would change (Bok 2010). Carl Linnaeus (1707–1778), the founder of modern biology and the person who developed the system of zoological classification of species still in use today, also believed in the unity of mankind. Linnaeus, the son of a Lutheran minister, was born in southern Sweden. He studied medicine and later in his life became a professor at the University of Uppsala. Medicine in those days was mainly a matter of herbal remedies, and Linnaeus became a specialist in botany. However, he continued to practice medicine and became the physician to the royal family (Groves 2008). He also revitalized the Uppsala Botanical Garden. In fact, he considered himself God's registrar—his goal was to systematize the naming of all the plants and animals God had created and put them in order. The order so derived, however, was not based on relationship through evolution. It was a creationist concept: all species were created as fixed and separate species whose perfect representations were to be found only in the mind of God (Brace 2005). As anthropologist C. Loring Brace has stated (2005, 28): "The assumption that the world was hierarchically arranged pervaded medieval Christian thought and continued without question in the outlook of the Enlightenment thinkers as well. Linnaeus and his contemporaries simply took that general view and provided a more specific picture of all aspects of the world arranged in a series of steps running from God at the top down through the various entities of the living world to the inorganic. . . . This arrangement was referred to as the *Scala Naturae* or 'Great Chain of Being.' "

Using this concept, Linnaeus published twelve editions of his famous *Systema Naturae* during his lifetime, and in the tenth edition (1758) he established the system of binomial nomenclature in zoology, the starting point for all zoological nomenclature since. (The first edition was published in 1735, before Linnaeus was thirty years old.) He classified all living organisms into named units in descending order of increasing distinctiveness and began the two-name classification of genus and species for the basic name of an organism. Thus, he devised the term *Homo sapiens* for humans and, in fact, considered all humans to be members of the same species. Based on anatomical similarity, he placed humans in the order Primates, along

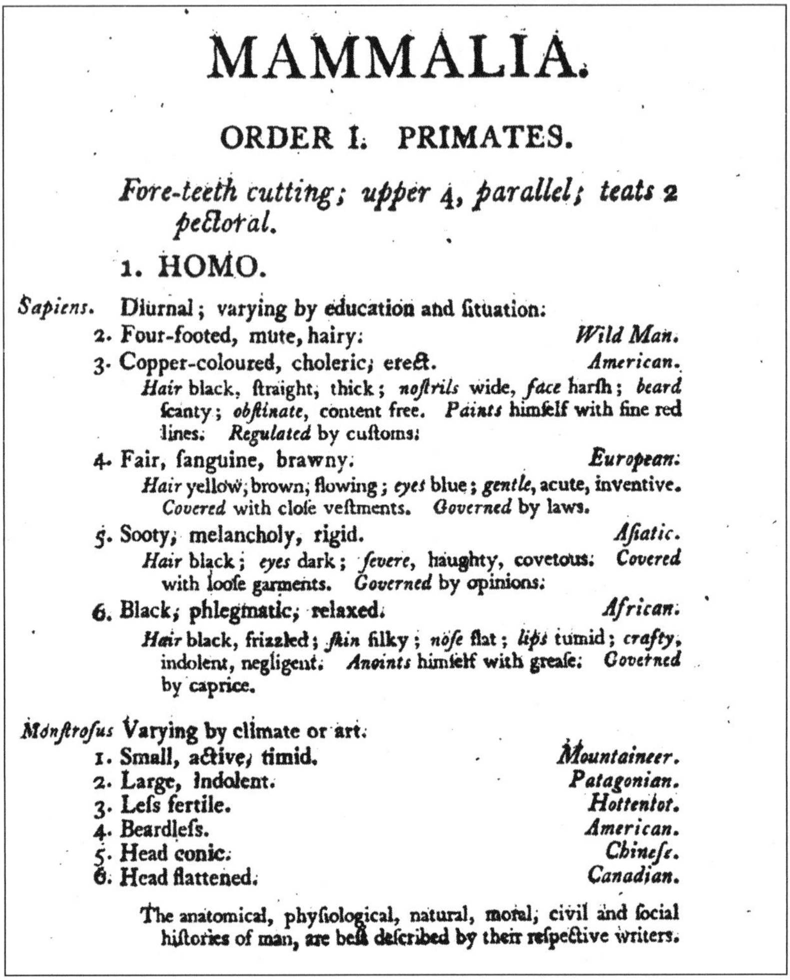

MAMMALIA.

ORDER I. PRIMATES.

Fore-teeth cutting; upper 4, parallel; teats 2 pectoral.

1. HOMO.

Sapiens. Diurnal; varying by education and situation:

2. Four-footed, mute, hairy.   **Wild Man.**

3. Copper-coloured, choleric, erect.   **American.**
   Hair black, ftraight, thick; *noftrils* wide, *face* harfh; *beard* fcanty; *obftinate*, content free. *Paints* himfelf with fine red lines. *Regulated* by cuftoms.

4. Fair, fanguine, brawny.   **European.**
   Hair yellow, brown; flowing; *eyes* blue; *gentle*, acute, inventive. *Covered* with clofe veftments. *Governed* by laws.

5. Sooty, melancholy, rigid.   **Afiatic.**
   Hair black; *eyes* dark; *fevere*, haughty, covetous. *Covered* with loofe garments. *Governed* by opinions.

6. Black, phlegmatic; relaxed.   **African.**
   Hair black, frizzled; *fkin* filky; *nofe* flat; *lips* tumid; *crafty*, indolent, negligent. *Anoints* himfelf with greafe. *Governed* by caprice.

Monftrofus Varying by climate or art.

1. Small, active, timid.   **Mountaineer.**
2. Large, indolent.   **Patagonian.**
3. Lefs fertile.   **Hottentot.**
4. Beardlefs.   **American.**
5. Head conic.   **Chinefe.**
6. Head flattened.   **Canadian.**

The anatomical, phyfiological, natural, moral, civil and focial hiftories of man, are beft defcribed by their refpective writers.

Figure 1.1 The classification of *Homo* as written by Linnaeus in *Systema Naturae* in 1758.

with apes and monkeys (and bats). This made some of his contemporaries quite uneasy. Linnaeus then classified varieties of humans in relationship to their supposed education and climatic situation (see Figure 1.1).

As usual, those who did the classifying, white Europeans, were seen as the superior variety. As did Montesquieu, Linnaeus believed the differences were due to climate and social conditions.

A contemporary of Linnaeus was the French naturalist Georges Louis Leclerc, comte de Buffon (1707–1788). Buffon was perhaps the greatest naturalist of the eighteenth century. Independently wealthy, as were most scholars at that time (a fact that allowed them to take up their scholarly pursuits in the first place), Buffon moved from Burgundy to Paris in 1739, where he became the keeper of the Jardin du Roi (later to become the Jardin des Plantes, in which was housed the Paris Zoo and the Muséum national d'Histoire naturelle). Buffon died at the age of eighty, one year and three months before the beginning of the French Revolution. Since he had worked for King Louis XVI and had been in his favor (he was made a count in 1771), Buffon was not well treated after his death. His coffin was dug up, his remains were scattered, and his monument was smashed. Worst of all, his only son was sent to the guillotine a few years later (Groves 2008). It is quite amazing that Bernard Germain de Lecépède (1756–1825) and Louis-Jean-Marie Daubenton (1716–1800), protégés, coauthors, and close colleagues of Buffon, survived the French Revolution and were instrumental in the appointment of two key members of the next generation of French scholars, Étienne Geoffroy Saint-Hilaire (1772–1844) and Georges Cuvier (1769–1832), who became dominant figures in the study of natural history just before the publication of Charles Darwin's *On the Origin of Species*. This story is nicely told in Groves (2008).

Buffon offered the most complete explanation of human variation of his time in the fourth and fifth volumes of his forty-three-volume *Histoire Naturelle,* written from 1785 to 1787: "Humans are not composed of essentially different species among themselves, but on the contrary there is only one sole species of man which has multiplied and covered all the surfaces of the earth, [and] has been subjected to different changes due to influences of the climate, differences in nutrition, and those of manner of life [lifestyle], by sicknesses, epidemics, and also by the various infinite mixture of individuals more or less similar" (Buffon 1785, 180; my translation).

As one moves away from Central Europe, Buffon explains, these various factors cause increasing degeneration from the ideal, original humans:

> The best climate is found between 40 and 50 degrees; it is here that one finds the most beautiful and most fit humans, it is in this climate that one finds the ideal of the natural color of man, it is here where one finds the model or the origin from which is derived all of the other nuances of color or of beauty. The two extremes are equally far from the true [ideal?] and the beautiful: The countries situated in this zone are Georgia, Circaffie, Ukraine, Turkey, Europe, Hungary, Germany, Italy, Switzerland, France,

and parts of Spain, all these people are also the most beautiful and the best fit of all the earth. (Buffon 1785, 178–179; my translation)

Although the views of Linnaeus and Buffon might seem similar to us today, their approaches were actually quite different, and they were intellectual rivals throughout their lives. Linnaeus was interested mainly in naming the categories of animals, and his task would be complete when all organisms in creation had been classified. The nested hierarchy of life he created in his classification scheme, however, did not imply any particular process. Buffon, on the other hand, was an enemy of all rigid classification and believed that the categories devised by Linnaeus were simply human creations. He stated: "In fact, in nature there are only individuals; genera, orders, and classes exist only in our imaginations" (quoted in Nordenskiöld 1928, 222). Buffon was more interested in process. He seemed to be aware of what we would call ecology and adaptation and noted the relationships among the forms of plants and animals and certain aspects of the environment in the regions in which they lived. "Throughout his writings, there was a continual concern for the processes by which organic form is shaped that was completely missing in the writings of Linnaeus," Brace noted (2005, 31). It is ironic that Buffon formulated an early scientific version of evolution but rejected it. In fact, he rejected Linnaeus's system of higher taxonomic categories because he thought they implied something insidious—that is, microevolution—and this was against his religious beliefs. Jonathan Marks notes, "Even though Linnaeus himself did not espouse such an idea, it was (according to Buffon) simply because he had not 'grasped sufficiently the full scope' of the implications of his system" (Marks 1995). Buffon regarded Linnaeus as a "nomenclateur" rather than what we would now refer to as a "scientist" (Brace 2005).

The German physician and anatomist Johann Friedrich Blumenbach (1752–1840), often thought of as the father of physical anthropology, was a disciple of Linnaeus and idolized him (Gould 1996). However, Blumenbach was also interested in process. Like Linnaeus and Buffon, he was a monogenicist who believed that all humans were created by God. In his dissertation, which he wrote at the age of twenty-three in 1775, a year before the American Revolution, he attempted to classify the varieties of humans and to explain the significance of their physical and mental differences. As did Linnaeus and Buffon, he believed that all humans were the same species. He also insisted that there were no sharp distinctions between groups and that supposed racial characteristics graded continuously from one people to another (Gould 1996; Montagu 1997). He was among the first to refer to race but believed that divisions of human groups were somewhat arbitrary and were used for the convenience of the classifier (Farber 2011).

Following Buffon, in 1775, Johann Friedrich Blumenbach published the first edition of his dissertation, *De generis humani varietate nativa (On the Natural Variety of Mankind)*, in which he stated that he had constructed his human racial classification simply as a matter of convenience. This book became a standard beginning reference point for discussions about human races (Farber 2011). In a greatly expanded third edition, written in 1795, when Blumenbach was serving as a professor of medicine at the University of Göttingen, he wrote: "Although there seems to be so great a difference between widely separate nations, that you might easily take the inhabitants . . . [of different regions] . . . for so many different species of man, yet when the matter is thoroughly considered, you see that all do so run into one another, and that one variety of mankind does so sensibly into the other, that you cannot mark out the limits between them. Very arbitrary indeed both in number and definition have been the varieties of mankind accepted by eminent men" (quoted in Montagu 1997, 62).

Blumenbach went on to specify first four (based on Linnaeus's four geographically noted varieties) and later five varieties of humans associated with major regions of the world. His five varieties—Caucasian, Mongoloid, Ethiopian, American, and Malay—became widely accepted by the educated community, and with some slight variations they are still in use today. In his scheme of the varieties of mankind, Blumenbach developed two major ideas that have endured in the history of racism and, unfortunately, also are still with us today. First, he coined the term *Caucasian* to refer to people of European descent and in doing so defined them as the most beautiful, the closest to representing God's image, and the "original" humans from which other varieties had degenerated. Was this done by any scientific means? Well, no. He developed this on purely aesthetic grounds and, of course, on his own views of aesthetics. "Blumenbach's descriptions are pervaded by his personal sense of relative beauty, presented as though he were discussing an objective and quantifiable property, not subject to doubt or disagreement" (Gould 1996, 411).

Second, even though he had expressed the difficulty of drawing lines between varieties of humans, he accepted the underlying paradigm of the day, as had Linnaeus, Cuvier, and Buffon, that one variety was indeed better and preferable to another in relationship to God's original creation. In fact, unlike Linnaeus and Buffon, his varieties were set up not simply in a geographic system but also in a hierarchical one. Cuvier before him had described three varieties of human species and maintained that the Mongolian race remained stationary with regard to civilization and that the black race had never progressed beyond utter barbarism (Stocking 1968). Blumenbach, with his five varieties of humans, set up a racial geometry with two lines degenerating through intermediary stages from a central Caucasian "ideal" (see Figure 1.2). Stephen Jay Gould (1996, 405) believed that Blumenbach's hierarchical model of human races was a major factor in the creation of the modern racists' paradigm: "The shift from a geographic to a hierarchical ordering of human diversity marks a fateful transition in the history of Western science—for what, short of railroads and nuclear bombs, had more practical impact, in this case almost entirely negative, upon our collective lives and nationalities? Ironically, J. F. Blumenbach is the focus of this shift—for his five-race scheme became canonical, and he changed the geometry of human order from Linnaean cartography to linear ranking by putative worth."

Gould believed this ironic because although many of the monogenicists were opposed to slavery and the mistreatment of the "degenerated" varieties of mankind and believed they could be "regenerated" in one way or another, Blumenbach was among the least racist and one of the most egalitarian of the Enlightenment scholars. In fact, he had a library in his home devoted to the writings of black authors and praised the "faculties of these our black brethren," most likely as a rebuttal to the more common, pervasive Humeian and Kantian mentality (see later in this chapter). He campaigned for the abolition of slavery (a view not popular in his day) and, interestingly, asserted the moral superiority of slaves to their captors (Gould 1996). Nevertheless, in the end, Blumenbach ended up with a system with one single race, Caucasian, at the top. He assumed that race to represent the closest to "original" creation and then envisioned two lines of departure from this ideal toward greater and greater degeneration. As Brace (2005, 46) emphasized: "For the next two centuries, those who have attempted to 'classify' human biological variation have inevitably built on the scheme proposed by Blumenbach."

Inherent in the degeneration theory of race was the concept of change. This was, however, a difficult idea to deal with in the eighteenth and nineteenth centuries because of the basically accepted tenet of the fixity of species as originally created by God. Yet many of the proponents of this theory attempted to explain how degeneration actually had occurred. Buf-

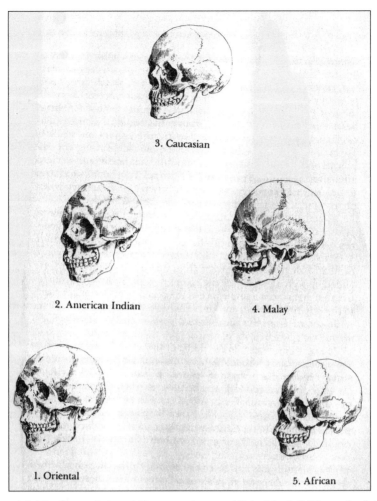

Figure 1.2 The five varieties of *Homo sapiens* established by J. F. Blumenbach in *De generis humani varietate nativa* in the late eighteenth century.

fon noted the relationship between aspects of the environment of particular regions and the forms of plants and animals living there. He accepted the idea that the similarity among differently adapted forms in given regions suggested some kind of "adaptive" relationship, and although he rejected the idea of organic evolution and went on at length to debunk any such theory, he brought it up as a topic of discussion. He was concerned with the processes by which organic forms developed and believed that environmental conditions could cause populations to change within a species but certainly not enough to become another species altogether (Marks

1995; Brace 2005). Similarly, Blumenbach, in writing about human varieties, believed that the farther populations had migrated from their place of origin in the Caucasus the more they were affected by different environments and conditions of life. Factors that could cause these changes were climate, nutrition, and mode of life. Over many generations, differences in these factors led to changes (degenerations) in form from the original (Brace 2005). Cuvier expounded the idea of "catastrophism." He believed that the earth went through a series of invoked catastrophes at the boundaries between geological strata and that new species were created after each of these catastrophes, presumably by divine creation (Marks 1995). Cuvier, however, like Linnaeus, was concerned more with the patterns that had been created by God than with the process by which patterns developed.

The most logical concept of the process of biological change at the time, given the available evidence, was that formulated by Jean-Baptiste Lamarck (1744–1829). Born in the north of France, Lamarck was the youngest of eleven children in an impoverished noble family with a centuries-old tradition of military service. His father and several of his brothers were soldiers. He entered a Jesuit seminary around 1756, but after his father's death in 1761, he bought an old horse and rode off to join the French army. He fought in the Seven Years' War in Germany and at the age of seventeen distinguished himself for bravery under fire and was commissioned as a lieutenant. In 1766, because of an injury, Lamarck was forced to retire to Paris with a meager pension. There, close to poverty, he supported himself as a bank clerk and began to study medicine. Then, possibly influenced by his friend Jean-Jacques Rousseau, he dropped medicine for the study of botany. In 1778, he published a book on the botany of France, and this launched his career in science. Buffon was impressed by Lamarck and engaged him as a tutor for his son (the one who later went to the gallows) and was instrumental in getting him admitted to the Academy of Science. In 1781, Buffon had Lamarck appointed as royal botanist and collector for the Jardin du Roi, and they traveled together collecting plants for the garden in Germany, Holland, and Hungary (Hays 1964).

Lamarck held this position until 1793. In that year Louis XVI and Marie Antoinette were sent to the guillotine and the renamed Jardin des Plantes was reorganized as the Muséum national d'Histoire naturelle. It was to be run by twelve professors of different scientific specialties. Lamarck was appointed as one of those professors, the one in charge of the natural history of "insects, or worms and microscopic animals" (*invertebrates*, a term Lamarck later coined). This was the least prestigious of the professorships and a subject about which Lamarck knew nothing. To his credit, Lamarck took on the enormous challenges of organizing the museum's vast and growing collections and of learning and creating a new field of biology (Clifford 2004). Lamarck published a series of books on invertebrate zoology and paleontology; he also published in the fields of physics, meteorology, and hydrogeology.

Lamarck is most remembered, and often most criticized, however, for his early theories of evolution, which are most clearly stated in his *Philosophie zoologique* (1809). Lamarck expounded the idea that organisms are not passively altered by the environment but that environmental changes cause changes in the needs of organisms that in turn cause changes in their behavior. This altered behavior leads to greater or lesser use of a given structure or organ. Thus use causes increase in size of the structure or organ, and disuse causes it to decrease in size or disappear over several generations. This was Lamarck's "First Law"—that use or disuse causes structures to enlarge or shrink. His "Second Law" was that all such changes were heritable. Lamarck believed in continuous, gradual change of all organisms as they become adapted to their environment (Clifford 2004). One aspect of Lamarck's theory of evolution that differs markedly from modern Darwinian evolutionary theory is that evolution is not driven by chance. He believed instead that in evolution, nature is "attempting" to produce in succession, in every species of animal, a form beginning with the least perfect or simplest to an end product of the most perfect and structurally complex. He posed a specific direction (perfection) to be reached in every lineage, a progressive development in nature.

In this way, Lamarck was attempting to describe a particular process by which change takes place. At the time it was written, this idea was too radical for those who believed in the concept of the fixity of species as created by God. Lamarck was ridiculed for these ideas by his contemporaries and even by his closest colleagues, Buffon and Cuvier. Lamarck struggled with poverty throughout his life. He was married four times and had seven children. He spent his last years totally blind and cared for by two of his devoted daughters. When he died in 1829, he received a poor man's funeral and was buried in a rented grave. His books and the contents of his home were sold at auction. Five years later his body was removed, and no one knows the final location of his remains. His friend Geoffroy Saint-Hilaire gave one of the orations at his funeral.

Two years after Lamarck's death, Cuvier used the forum of a eulogy to discredit Lamarck's scientific beliefs, depicting his theories as able to entertain poets but unable to support the examination of any scientist (Hays 1964). Since Cuvier was so respected in his day, his remarks on Lamarck's views of evolution helped banish them to obscurity at that time. However, by the middle of the nineteenth century, since even the theory of degeneration implied the process of change, many scientists began to develop different views of how changes in biological and social phenomena occur. At that point, Lamarck's theories became acceptable to many of the biologists and social scientists. In fact, at the time of Darwin's *Origin,* Lamarckism had become one of the few scientific theories of degeneration theorists, presenting a coherent explanation of how environment could influence biological and social change. Lamarck was acknowledged as a great zoologist and as a forerunner of the theory of evolution by many of the scientists and evolutionists of Darwin's time, including Darwin, Charles Lyell, Ernst Haeckel, Paul Broca, and the American paleontologist Edward Drinker Cope. In fact, Darwin incorporated Lamarckian explanations in some of his descriptions of the process of change (Rectenwald 2008). Lamarck's view of transformation could be seen as the beginning of modern zoology. Darwin ([1859] 1860, vi) wrote: "Lamarck was the first man whose conclusions on the subject excited much attention. This justly celebrated naturalist first published his views in 1801. . . . He first did the eminent service of arousing attention to the probability of all change in the organic, as well as in the inorganic world, being the result of law, and not of miraculous interposition."

This view of how acquired characteristics could influence change was epitomized in a book written by Richard Louis Dugdale in 1877, with new editions appearing until 1910, entitled *The Jukes: A Study in Crime, Pauperism, Disease, and Heredity.* Dugdale described the history and experiences of the Juke family over many generations. In carefully examining the influence of environment on various members of the same family (actually a compilation of over forty families), Dugdale pointed out the improvement in delinquent behavior that could be brought about by a change in environment. As a believer in the theory of acquired characteristics, Dugdale ([1877] 1891, 55) stated: "Where the environment changes in youth the characteristics of heredity may measurably be altered. Hence the importance of education." To Dugdale, like other degeneration theorists, Lamarckism worked both ways: "Environment tends to produce habits which may become hereditary . . . if it should be sufficiently constant to produce modification of cerebral tissue. . . . From the above considerations the logical induction seems to be, that environment is the ultimate controlling factor in determining careers, placing heredity itself as an organized result of invariable environment" (66). In sum, he argued: "Heredity . . . fixes the organic characteristics of the individual," while "environment . . . affects modifications in that heredity" (11). Many scientists of the time believed that Lamarckian theory meant that environment was a very important factor in producing human behavioral characteristics.

Since Darwin did not provide any single coherent explanation of the processes of biological change in his theory of natural selection, Lamarckism's acquired characteristics could be included as one of many possible processes. In fact, many of Darwin's explanations are quite Lamarckian (his theory of pangenesis, for example). However, with the rediscovery of Mendelian genetics and with the experiments of August Weismann around the turn of the century (see Chapter 2), Lamarckian theory once again was debunked and ridiculed.

The Pre-Adamite/Polygenic Theory from the Sixteenth Century to Darwin

The pre-Adamite or polygenic theory of human variation existed alongside the degenerate theory, and often in direct contrast to it, from the sixteenth century to Darwin's times. Early views that inhabitants of the New World were not descended from the biblical Adam were provided by Paracelsus (1493–1541), who lectured on medicine at the University of Basel and is credited with establishing the importance of chemistry in medicine (Brace 2005). He was a controversial figure who was not reluctant to contradict the traditional view of a single creation of man. In the early sixteenth century, he argued that people in faraway places were from a different source, the same one that had produced nymphs, sirens, griffins, and salamanders, all examples of beings without souls (Popkin [1974] 1983). Later in that century, the philosopher and cosmologist Giordano Bruno claimed that the Indians, Ethiopians, Pygmies, giants, and other strange and far-off beings were not descended from the same progenitor as the rest of the human world (Slotkin 1965; Popkin [1974] 1983, 1976). Bruno was burned at the stake in Rome in 1600, a victim of the inquisition (Yates 1992, Murphy 2012). The polygenic theory that was the most influential in racist ideology and had the most staying power was the pre-Adamite theory of Isaac La Peyrère (1596–1676).

La Peyrère, a French Calvinist from a family of Spanish Jews who had been expelled from Spain at the end of the fifteenth century, had a long and eventful life (see Brace 2005). He was the first to propose this theory, in his work *Prae-Adamitae,* which was written in 1641 and published in Amsterdam in 1655 (Popkin [1974] 1983). It was translated into English in 1656 as *A Theological Systeme upon That Presupposition That Men Were Before Adam.* Popkin (1973, [1974] 1983) gives a detailed synopsis of La Peyrère's theory; I will only point out salient features here. La Peyrère argued that there were millions of people prior to Adam but that they lived in a miserable state. Then God created Adam and began Jewish history in order to save mankind. His book challenged the authenticity of the Bible and was burned and banned. La Peyrère was considered a heretic; his views were condemned and he was imprisoned for six months. When he was released, he was forced to write a formal retraction. Popkin ([1974] 1983, 141–142) states: "Although his work was constantly being refuted from 1655 onward, its polygenetic thesis kept being revived as the best explanation of the new findings in geology, biology, archeology, anthropology, and history that conflicted with the Bible." The racist implications of the pre-Adamite theory began to reemerge during the Enlightenment among theorists who no longer took the Bible literally. By the nineteenth and early twentieth centuries, it had transformed somewhat and developed into a powerful "scientific" defense of racist ideology.

In the eighteenth century, the pre-Adamite or polygenic theory was less popular than the monogenic or degeneration theory because of the latter's closer adherence to the biblical account of human origins (Gould 1996). However, although the polygenic theory was not the most popular view, it was still held as a minority viewpoint (Gould 1996; Smedley 1999; Brace 2005). David Hume (1711–1776), the renowned eighteenth-century Scottish philosopher, economist, and historian, was among the first noted authors to profess the polygenic theory of racism (Popkin [1974] 1983; Smedley 1999). He did so as part of his "inductive" naturalistic philosophy (Craig 1987), or experimental philosophy. "The only solid foundation we can give to this science (science of man) . . . must be laid on experience and observation," he wrote in his *Treatise of Human Nature* (1739–1740, quoted in Beebee 2011, 729). Hume, who is considered among the most important figures in the history of Western philosophy and the Scottish Enlightenment, advocated the separate creation and innate inferiority of nonwhite peoples. In the mid-1700s, Hume wrote: "I am apt to suspect the negroes and in general all the other species of men (for there are four or five different kinds) to be naturally inferior to the whites. There never was a civilized nation of any other complexion than white, nor even any individual eminent either in action or speculation. No ingenious manufacturers among them, no arts, no sciences" (cited in Popkin [1974] 1983, 143). In this statement, Hume was applying his methodology of historical "inductive" reasoning. Human nature was best studied by observations of human historical behavior, and from the European point of view civilization had never existed outside of Europe.

Following in Hume's footsteps but adding and enveloping similar racist ideas into a whole system of philosophical thought, Immanuel Kant essentially created a racist anthropology based on skin color. A contemporary of Linnaeus's Kant (1724–1804) developed his own classification of human races. Kant is acknowledged as one of the most influential philosophers of the Enlightenment. In fact, he is widely thought of as the most important moral theorist of modern times (Guyer [1998] 2004). However, he also can be considered the father of the modern concepts of race and scientific racism (Count 1950; Van de Pitte 1971; Neugebauer 1990; Eze 1995; Mills 1997; Jablonski 2012). Kant, who introduced the term *anthropology* to German science and philosophy, was the founder of what might be considered racist anthropology, which dominated much of anthropology up until World War II. His classification of humans included four races based on color and climate (Kant [1775] 1950). Kant believed that all races of man were created by God but that the characteristics (germs) of each were dependent upon climate. This made his polygenic view more acceptable to biblical interpretations of humanity. Climate determined the natural predispositions or character of each race, and once the process toward each racial disposition had begun, it was irreversible. Existing races and racial characteristics could not be undone by changes in climate or circumstance, "for once a race like the present one has been founded through long sojourn of its original stock . . . it could not be changed into another race by any further influences of the climate. For only the stem-formation can exspeciate into a race; but once the latter has taken root and has stifled the other germs, it resists all further remodeling because the character of the race has now become predominant in the generative power" (Kant [1775] 1950, 24).

Kant's theory of race corresponded to intellectual ability and limitation. He included the typical color-coded races of Europe, Asia, Africa, and Native America, differentiated by their degree of innate talent (Kant [1798] 1974). In Kant's theory, the nature of the white race guarantees its rational and moral order, and they are in the highest position of all creatures, followed by yellow, black, and then red. Nonwhites do not have the capacity to realize reason and rational moral perfectibility through education. To Kant, color is evidence of unchanging and unchangeable moral quality and thus ultimately of free will. White Europeans have the necessary talent to be morally self-educating; Asians have some ability to do so but lack the ability to develop abstract concepts. Innately idle Africans can only be educated as servants (to follow orders) but must be kept in order by severe punishment (and he explains how to properly beat them with split bamboo canes) (Neugebauer 1990; Eze 1995). Native Americans are hopeless and cannot be educated at all (Mills 1997). Furthermore, mixing of races should be avoided because it causes misfortune and damage (Neugebauer 1990).

Although Kant was a champion of the equality of all men and of civil rights, these were only for humans who have the ability to educate themselves and thus have free will—they were only for whites. Full personhood was actually dependent upon one's race. Nonwhites were relegated to a lower rung in the moral ladder (Mills 1997). Eze (1995) summarizes: "The black person, for example, can accordingly be denied full humanity since full and 'true' humanity accrues only to the white European." Kant believed that to be human one must be able to think moral thoughts (reason) and have the ability (free will) to carry them out. Native Americans and blacks did not have these qualities and thus could not be considered fully human. As the philosopher E. C. Eze notes, for Kant, "the ideal skin color is the 'white' (the *white brunette*) and the others are superior or inferior as they approximate whiteness" (Eze 1995, 217). To Kant, nonwhites counted as subpersons who were of considerably less value than whites because they were nonmoral agents (Mills 1997; Hachee 2011). Furthermore, nonmoral agents lacked moral worth and became mere objects to be used as means to the ends of others. They were nothing but irrational animals whom superior moral agents (true humans) could master and rule at will.

As with his contemporaries, Kant's theories were based on travelers' tales and on his own personal opinions. Rephrasing Hume directly, Kant stated: "The Negroes of Africa have by nature no feeling that rises above the trifling. Mr. Hume challenges anyone to cite a single example in which a Negro has shown talents, and asserts that hundreds of thousands of blacks who have been transported elsewhere from their countries, although many of them have been set free, still not one was ever found who presented anything great in art or science or any other praiseworthy quality. . . . So fundamental is the difference between these two races of man, and it appears to be as great in regard to mental capacities as in color" ([1764] 1965, 110–111).

Kant also generalized his depiction of nonpersonhood to Jews. Motives could be good or moral only if they were not motivated by a desire for material benefit, and he saw Judaism as an inherently materialist religion. He equated the Jewish religion with such undesirable traits as superstition, dishonesty, worldliness, and cowardliness (Mack 2003). Of course, Kant's views on nonwhites and on Jews were not original. They supported already existing, long- and widely held stereotypes of Western and Christian thought (Poliakov 1971; Jablonski 2012). The United Jewish Appeal of Toronto (2003) points out that

> going back to at least the 12th century, European culture had developed a rich and ghastly tableau of imaginary Jews. . . . Kant's division of humanity reiterated and reinvigorated the religious and racial hierarchies of the past. . . . He took this earlier religious hostility toward Jews and reformulated it in philosophical language. . . . Kant set the stage for modern secular anti-Semitism . . . [and] provided the framework for future anti-Semites, notably G. W. F. Hegel and the musician Richard Wagner. Since Wagner was a cultural hero for Adolf Hitler, Kant's own anti-Semitism can be seen as having a far-reaching effect.

Kant taught a combination of physical geography and anthropology courses for forty years (1756–1797), introducing a "scientific" concept of race and a particular brand of physical racial anthropology first in Germany, then in Europe and the United States (Kant [1775] 1950, [1798] 1974, 1802; May 1970; Eze 1995; Mills 1997; Elden 2011; Jablonski 2012). Kant had an exalted reputation; there was great respect for his work, and his writings were widely circulated. As the anthropologist Nina Jablonski (2012, 130) stated: "Through his writings and lectures, Kant successfully instilled some of the most trenchant and potent classifications of humanity into the minds of inexperienced and unsophisticated readers and students." He became one of the most influential racists of all times, and his racial philosophy was passed on for centuries. Because Kant is widely thought of as the most important moral theorist of modern times and the father of modern moral theory, his theories on race have, until recently, been essentially ignored in discussions of the history of racism (Eze 1995; Mills 1997; Hachee 2011). However, as Jablonski (2012, 135) states: "In the history of humanity, few intellectual constructs have carried so much weight and produced such a river of human suffering." As philosopher Charles W. Mills summarized (1997, 72, emphasis in original): "*The embarrassing fact for the white West (which doubtless explains its concealment) is that their most important moral theorist of the past three hundred years is also the foundational theorist in the modern period of the division between Herrenvolk and Untermenschen, persons and subpersons, upon which Nazi theory would later draw.* Modern moral theory and modern racial theory have the same father."

The moral contract Kant and his colleagues developed is underlain first by a racial (or color) contract—it only applies to those of the white hue. Furthermore, as Matthew Hachee emphasized (2011): "Not only is it the case that the image of Kant passed on to succeeding generations of philosophers . . . is one excessively sanitized, but it also seems reasonable at this point to infer that this 'selective memory' is simply too extensive to be the result of mere accident or chance. Rather, it appears to be the result of a tradition conveniently blind to its own racism."

By the end of the eighteenth century, as the controversy about slavery and the place of "the Negro" in nature and society heated up, Charles White (1728–1813), an English physician, again focused on the question of whether black Africans were products of the same act of creation as whites in *An Account of the Regular Gradation in Man* (1799). He denounced the views of the degeneration theorists (see Gould 1996). White proposed that black Africans were inferior both physically and intellectually and were an intermediate form between true humans (white Europeans) and apes, with other races intermediate between these extremes. Each race was seen as a separate species, the product of separate creation that was adapted for a particular geographic region. Although others had propounded this view, White's rendition was considered to be the most scientific. As we shall see in Chapter 6, a theory very similar to this one, in which the great apes are invoked to explain the differences among the races, surfaced again in the early 1900s (Urbani and Viloria 2008; Marks 2012).

In the early 1800s in Europe and America there was a revival of Isaac La Peyrère's pre-Adamite theory and continued attempts to reconcile it with the Bible. As described in detail by Popkin ([1974] 1983), Gould (1996), and Brace (2005), the nineteenth-century scientific version of the pre-Adamite theory was developed by the American Dr. Samuel Morton (1799–1851), an accomplished eclectic physician and paleontologist, and his disciples, who became known as the Mortonites. Morton is well known for his technique of measuring the cranial capacity of human skulls by filling them with pepper seeds. His research, which compared different human groups, was published in two volumes, *Crania Americana* (1839) and *Crania Aegyptiaca* (1844).

Morton and his followers, George R. Gliddon, Josiah Nott, and Louis Agassiz, reformulated or resurrected a powerful case for pre-Adamism. Their argument went as follows: (1) the cranial sizes and characteristics of various human racial groups were fixed and remained the same through-out recorded history, at least 3,000 years, as could be seen by skull measurements and Egyptian artwork; (2) the fixed cranial traits included a progressive decrease in cranial capacity from whites to Asians to Native Americans to African blacks; (3) these fixed differences did not fit biblical chronology, and therefore the best explanation was a separate creation of the different types of mankind (Popkin [1974] 1983). Brace (2005) believes that Morton was the founder of what is often called the American School of Anthropology and that he was a true scholar and a careful and innovative scientist and that he has been forgotten because his legacy was carried on in the hands of his followers, who forwarded the cause of slavery and racism. However, Morton, like his followers, strongly professed the inferiority of other races over white Europeans. When he died in 1851, his obituary in the *Charleston Medical Journal* read: "We of the South should consider him as our benefactor, for aiding most materially in giving to the negro his true position as an inferior race. We believe the time is not far distant, when it will be universally admitted that neither can 'the leopard change his spots, nor the Ethiopian his skin.' "

Morton died before *Origin of Species* was published. However, Morton's followers bring us right up to Darwin's then-controversial volume. Josiah Nott (1804–1873) became the main spokesman for the American School of Anthropology. He was a highly respected physician and surgeon from a prominent southern family. His father was a lawyer who graduated from Yale, served one term in Congress, and then became a judge and the president of the South Carolina Court of Appeals. Josiah first practiced medicine in South Carolina and then followed his wife's family to Mobile, Alabama. He married the daughter of a wealthy southern plantation owner. He was brought up on the coast of South Carolina, the region where the largest number of slaves in the world lived, and shared the racist attitudes of the South in the era just before the Civil War. His biographer (Horsman 1987) stated "as a Southern Gentleman, Nott expected to be believed . . . though he had raised his innate prejudices to the level (of what he assumed to be) scientific truth" (87, 296).

Although he claimed to be a scientific realist, Nott's writings on race between 1843 and the outbreak of the Civil War in 1861 and again in 1866 were colored by arrogant racist prejudice. Nott gave two lectures in 1843–1844 that he called his "lectures on Niggerology" and subsequently published in 1844 as a pamphlet entitled *Two Lectures on the Natural History of the Caucasian and Negro Races* (Hammond 1981). In these he claimed there were several species of man that differed in the perfection of their moral and intellectual endowments. Nott claimed he was separating the actual history of mankind from the biblical account by showing that the Bible dealt with the creation and development of the Adamites, the Caucasians, and not with that of the pre-Adamites, the rest of mankind (Popkin [1974] 1983). Morton and Nott corresponded regularly after Morton read Nott's published lectures "with pleasure and instruction." In 1847, Nott wrote to Morton exclaiming, "my niggerology, so far from harming me at home, has made me a greater man than I ever expected to be—I am the big gun of the profession here" (quoted in Erickson 1986, 110). In 1850, Nott's views on polygenics were read at the annual meetings of the American Association for the Advancement of Science (AAAS) in Charleston, South Carolina, again with the backing of Morton. Indeed, he came to be regarded as one of the main articulators of southern views on race in the period leading up to the Civil War and after.

Louis Agassiz (1807–1873) added international scientific credence and respectability to this group of pre-Adamites. Agassiz, son of a Protestant minister in French Switzerland, was a zoologist, paleontologist, and geologist and a disciple of Cuvier. He was most known for his work on fossil fish. Until his death, he was an anti-Darwinist. He believed in "the teleologically egocentric stance of traditional Christianity, which regarded human beings as the object and end of divine creation and assumed that the world and its contents had been put there specifically to be exploited for human use" (Brace 2005, 98). After he experienced some financial difficulties in Europe related to financing his own publications (not unusual at that time), Agassiz came to the United States in 1846. In fact, he came to the United States as a public scientific lecturer in order to rid himself of debt. Soon after arriving in America, Agassiz visited Morton in Philadelphia. Morton was to become an influence on him second only to Cuvier. At the hotel where he was staying, he was served by the first black African he had ever encountered. Having been brought up in lily-white Switzerland and France, he was shocked by his first view of human variation. Shortly thereafter, Agassiz wrote a letter to his mother describing his repulsion at seeing someone so different from himself and the people he was used to seeing in white, upper-class Europe. This emotional reaction was to influence his "scientific" views for the remainder of his life. At this point, he joined Morton and Nott in their views that different types of humans were separate species, not created from Adam, and that mixture between these "species" was leading to biological and intellectual inferiority (Popkin [1974] 1983; Gould 1996; Smedley 1999; Brace 2005).

Agassiz's first American lectures were delivered in Boston in 1846 and were very successful. In fact, as a result of these lectures, he was offered and accepted a professorship of zoology and geology and directorship of the Lawrence Scientific School at Harvard. He remained in the United States for the rest of his life. In 1859, the Harvard Museum of Comparative Zoology was built for Agassiz with the hope that he and this museum would serve as an antidote to the threat of Darwinism. In 1850, at the AAAS meetings in Charleston, Agassiz (who had just been elected as the association's president) heard Nott's paper declaring that people of African descent were innately inferior to Europeans. He rose after the reading of Nott's paper to declare his support of Nott's polygenic position. Soon after that, Nott wrote to Morton: "With Agassiz in the war, the battle is ours. . . . We shall not only have his name, but the timid will come out of their hiding places" (quoted in Brace 2005, 101). In the same year Agassiz wrote an essay (1850b) in the *Christian Examiner and Religious Miscellany* entitled "The Diversity of Origins of the Human Races." Although he insisted that he was dealing with scientific matters and not politics, he wrote: "It seems to us to be mock-philanthropy and mock-philosophy to assume that all races have the same abilities, enjoy the same powers, and show the same natural dispositions, and that in consequence of this equality are entitled to the same position in human society. History speaks for itself" (quoted in Popkin [1974] 1983, 147).

Furthermore, Agassiz asserted that the Bible "never meant to say that all men originated from a single pair, Adam and Eve, nor that the animals had a similar origin from one common centre or from a single pair" (Agassiz 1850a, 185). In the Bible, "there is nowhere any mention of these physical differences characteristic of the colored races of men, such as the Mongolians and negroes. . . . Have we not, on the contrary, the distinct assertion that the Ethiopian cannot change his skin nor the leopard his spots?" (Agassiz 1850b, 135).

Agassiz's basic view was that all humans were created differently, with different talents. People of color had different but inferior talents to those of whites, and these differences should be studied so the best could be gotten out of each race. Just as Hume and Kant had before him, he based his theory on the supposition that Africans had never created a civilization, never developed "regulated societies," had always been slaves and therefore should remain so. Furthermore, Agassiz believed that because of this, it was a waste of time and effort to give Africans the educational and cultural benefits of European civilization (Popkin [1974] 1983; Brace 2005). In fact, he argued along with the pre-Adamites "that God had created blacks and whites as separate species" (quoted in Gould 1977a, 243).

As Brace (2005, 102) stated: "These judgments were not reached by anything remotely like scientific procedure. They were simply the assertions of opinions, and that opinion was largely a reflection of the attitudes held by Agassiz's prominent slave-owning friends of the American South." In addition, these same opinions were being foisted on scientific and popular audiences by the American School of Anthropology, with Morton as the respected American scientist, Agassiz adding Harvard and European scientific distinction, and Nott becoming the principal spokesperson of this polygenics "school." It is interesting that similar unscientific arguments reassert themselves in a very similar manner in the neoracists of today, as we shall see later.

The high point of the American School of Anthropology was the publication of a textbook entitled *Types of Mankind* (Nott and Gliddon 1854). This volume carried the ideas of the Mortonites past the Civil War, past Darwin's *Origin,* and up into the twentieth century. The co-editor, George R. Gliddon (1809–1857), was an English businessman and entrepreneur who had been brought up in Egypt and had provided Morton with Egyptian skulls for his anthropometry collection. He also had befriended Morton and Nott and become a junior associate of the American School of Anthropology and a spokesperson who popularized this group's racist ideas. Sensing the time was right, Gliddon got Nott and others involved in producing a book expressing the ideas and position of the American School. Brace (2005) gives more detail on this interesting character. *Types of Mankind* was first published in 1854. It sold out immediately and went through ten printings by 1871. It was dedicated to Morton and contained a chapter by Agassiz.

The main purpose of the book was to show that the findings of science justified the institution of slavery (Brace 2005). Using Morton's data on fixity of skull size and shape and the "historical approach" of Nott and Agassiz, its theme was that the human races had different origins and in fact were different species, that mixture between the races led to inferior people both biologically and intellectually, that the people of the white race were superior to other races and were the only truly civilized race, and that mixture of whites with other races was causing a deterioration of civilization and a danger to the future. Even with slavery, the book claimed, it was necessary to keep the races apart, and there was no reason to afford people of color an education or other accoutrements of civilization. Agassiz's essay in the book summarized the polygenic views of the group:

> The differences existing between the races of men are of the same kind as the differences observed between different families, genera, and species of monkeys or other animals: and these different species of animals differ in the same degree one from the other as the races of men—nay, the differences between distinct races are often greater than those distinguishing species of animals one from the other. The chimpanzee and gorilla do not differ more one from the other than the Mandingo and the Guinea Negro: they together do not differ more from the orang than the Malay or white man differs from the Negro. (1854, lxxv)

As anthropologist Audrey Smedley (1999, 234) emphasized, *Types of Mankind* "was perhaps the single most important book to set the issue of race into a peculiarly scientific context for the general public. It was the culmination of a trend begun in the latter part of the eighteenth century and was encouraged by the tremendous growth in the reputation of science." For the next few generations, this text was used by students and laypersons as a major source of scientific data on different kinds of human beings. Smedley goes on: "It succeeded in backing with the awesome prestige of science what were actually folk views of the Negro in the nineteenth century, expanded into racial ideology."

Though because of religious tradition Cuvier was a monogenicist, his views were congenial with polygenics, and his protégé Agassiz easily fell into the tradition of polygenism in America. Uninhibited by religious orthodoxy, the American School of Anthropology adopted a Cuvierian static, non-evolutionary, classificatory, comparative anatomy approach to human variation. As the anthropologist George W. Stocking (1968) pointed out, the polygenists assumed environment had no influence in the modification of living forms. They were teleological in their view of biological "types": they based their classifications on skeletal and especially cranial measurements, and they assumed a correlation between cranial and mental differences and racial achievement. By the time Darwin's *Origin* appeared, polygenism was dominant among those who might now be called "physical anthropologists." This form of polygenism was popular in the United States and was epitomized by the American School of Anthropology of Morton, Nott, Glidden, and Agassiz (with influences from Cuvier).

While the Mortonites represented the science of polygenic racism in America, racist theories also were brewing in Europe. The most prominent European biological determinist who was motivated by ideas about race during the time of the American School of Anthropology was the Frenchman Joseph-Arthur, comte de Gobineau (1816–1882), author of a number of novels and nonfiction history books. His most influential work, *Essai sur l'Inégalité des Races Humaines,* was published in four volumes from 1853 to 1855 (Gould 1996) and thus was contemporary with *Types of Mankind.* It was immensely popular in Europe and America in the late nineteenth and well into the twentieth century and in fact outlived the Mortonites' text. With Josiah Nott's assistance, a selectively abridged version of the first two volumes of this book was translated into English in 1856 under the title of *The Moral and Intellectual Diversity of Races* (Biddiss 1970; Brace 2005). Nott (1856) wrote a long appendix to this translation. Ultimately, Gobineau's *Essai* played an important role in Hitler's racial philosophy and horrific politics. Gould (1996, 379) referred to Gobineau as the grandfather of modern academic racism and "the most influential academic racist of the nineteenth century."

Gobineau had a loose relationship to French aristocracy and although he had no proper right to it, he adopted the title of "comte" (Count). He served most of his life as an official in the French diplomatic service, but his aspirations and claim to nobility colored Gobineau's views of the world. He saw the overthrow of the aristocracy during the French Revolution in 1789 as a major symptom of an ongoing deterioration of civilization (Biddiss 1970; Poliakov 1971). In fact, during the eighteenth century, prior to the French Revolution, many writers had set out to explain why certain groups had a divine right to superior status, or nobility. One of these renditions was the "Nordic" myth that may have begun with the writings of an earlier French nobleman, Henri comte de Boulainvilliers (1658–1722) (Poliakov 1971; Smedley 1999). In this myth, the noble classes of Europe were thought to be originally German Franks and Anglo-Saxons, and the Germanic peoples were claimed as most superior. In this argument, the claim of superiority had shifted from being theologically based to being more dependent on biological qualities, although these qualities were seen to be divinely endowed. These writers proposed an inherent biological superiority of those in power. As Smedley (1999, 254) states: "The racial theories of Henri de Boulainvilliers were essentially rooted in the class conflicts of the times, but they carried the invidious notion that each class had distinct and unalterable hereditary qualities derived from separate origins. The weaker classes were naturally inferior to the stronger and owed obedience to them."

Through these writings, there was a popular belief in France that three racial strains inhabited the country: Nordics, Alpines, and Mediterraneans. The light-skinned, tall, blond Nordics were assumed to be the descendants of ancient Germanic tribes, the originators of all civilization, and the only peoples capable of leadership. Gobineau's *Essai* expressed these popular myths vividly and inserted these views into the popular science of the day. His book fed a developing idea that not only were whites superior over all others but also that a certain group among whites was even more superior to other whites. He used the term *Aryan*, coined by a British colonial administrator, to designate the common ancestral language of what is now referred to as the Indo-European language. Around 1819, the term began to gain widespread authority due to the lectures and writings of Friedrich Schlegel, a German poet and scholar. The most influential promoter of the Aryan myth was Jacob Grimm, of Brothers Grimm's fairytales fame, in his *History of the English Language* (1848), which reached a large public audience in the second half of the nineteenth century (Poliakov 1971). Gobineau, however, attributed innate biological and behavioral qualities to Aryan speakers (Biddiss 1970; Brace 2005). He argued that there was a hierarchy of languages that corresponded with a hierarchy of races and that race was a driving force of history. The "Aryan" race was supreme and constituted an aristocratic caste. However, his views were mainly a synthesis of currently popular ideas (Weindling 1989). For example, anti-Semitism existed in Germany long before the Aryan myth, and this just gave the myth a stronger hold (Poliakov 1971). As the historian Léon Poliakov (1971, 233) pointed out, "Gobineau merely systematized in a very personal way ideas which were already deeply rooted in his time. His own contribution consisted mainly in his pessimistic conclusions, which sounded like the death knell of civilization."

To Gobineau, the Aryans were the most noble, intelligent, and vital branch of the white race. Thus, he essentially created a fictitious race of which he imagined himself a member (Hankins 1926). As Marks (1995, 66) stated: "His general theory of the rise and fall of civilization by *recourse* to those different inborn propensities of human groups, his isolation of the single group responsible for *all* civilization, and his identification of cultural decadence and decline with biological admixture, was an original synthesis and made his theory attractive for its simplicity and apparent scholarship."

In his *Essai,* Gobineau proclaimed that the success of civilization was directly dependent upon the purity of "Aryan" blood within it. Those designated Aryans were seen to be the founders of civilization; as more interbreeding occurred, the genius for civilization declined and dissipated. Gobineau believed that the white races, and especially the Aryans, could remain in command only if they could eliminate interbreeding with the morally and intellectually inferior yellow and black peoples (Gobineau 1856).

This notion of racial purity and the dangers of interbreeding became extremely popular in the U.S. South in 1856, being contemporary with the Dred Scott case and the brink of the Civil War. As we shall see, Gobineau had a major influence on the politics of the early twentieth century, both in Europe and America (Biddiss 1970; Gossett 1965; Brace 2005). In 1876, he met German musician Richard Wagner (1813–1883), who was impressed by his work, as was Friedrich Nietzsche (1812–1883) (Engs 2005). In fact, Wagner and Gobineau became close friends and Wagner used Gobineau's theories of racial inequality, anti-Semitism, and Aryan superiority as scientific backing for his own racial theories of culture. Wagner "set the foundations, between 1848 and 1850, of the anti-semitic apocalypse. He evoked the image of the Jew as an agent of corruption, a 'ferment of decomposition' " (Poliakov 1971, 198). In the late 1880s, Wagnerism represented a popular condemnation of liberalism and materialism. A Gobineau Society for aristocrats and other elites was established in 1894, and Gobineau was seen as an inspiration for the regeneration of the German aristocracy (Weindling 1989). Gobineau's brand of racism could be seen as compatible with certain interpretations of the new theories of Darwin, and his writings influenced such writers as Houston Stewart Chamberlain, William Z. Ripley, and Ernst Haeckel (see below), who in turn had a direct influence on Madison Grant and U.S. immigration policies and on Hitler's policies in Europe (Biddiss 1970; Marks 1995; Montagu 1997; Brace 2005).

In fact, these men were the core of the eugenics movement in European and American science. Although I have included Gobineau in this section on polygenism, in reality, he rejected polygenics because he perceived that it conflicted with Catholicism. Although he did no original research, instead echoing Kant's theories and thus producing ideas that were much more compatible with biblical theory, Gobineau claimed that once racial divergence had taken place, the different racial types were permanent and unchangeable (Brace 2005). As Poliakov (1971, 234) stated: "In a word, he was a monogenist in theory and a polygenist in practice."

Thus, with Gobineau and again in the footsteps of Kant, subtle but major changes could take place in this school of racism, moving beyond biblical interpretations and more into "hereditary" ones. Biological determinists could claim biological separation among human races (and other groupings of humans, such as strains and even economic classes) using "blood" or heredity and not necessarily invoking biblical separation. They could use the same "biological" arguments to claim that racial and group distinctions were biologically fixed and unchangeable. Thus, racism could become more acceptable to biblical traditionalists, giving biological deterministic racism a more widespread following and making it compatible with the growth of Darwinism and genetic theory. You might think of Mortonism as the end of an old Bible- (or anti-Bible-) based polygenics and Gobinism (in a Kantian tradition) as a revision of a "blood" or hereditary-based racism that was potentially compatible with Darwinism. The different races of mankind need not have been created separately as explained, or not, by the Bible. They merely needed to be genetically distinct and thus different in their basic biology. In both cases, these biological distinctions were basically fixed, and admixture would lead to inferiority, weakness, and even increased mortality. Little or nothing could be changed by environmental influences.

Just before the turn of the twentieth century, many of Gobineau's (and Kant's) views were introduced to German readers by Houston Stewart Chamberlain (1855–1927) in his *Die Grundlagen des neunzehnten Jahrhunderts (Foundations of the Nineteenth Century)* (first written in German in 1899, translated into English in 1910). Chamberlain was the son-in-law of the German composer Richard Wagner (Montagu 1997; Smedley 1999). The anti-Semitic, racist Wagner had brought Gobineau's views to the German public (Stein 1950). Chamberlain was English, born into a British military family. He had arrived in Germany in his youth, and after meeting Wagner, he increasingly adopted the German culture and language. He became an ardent Wagnerian. Although he wrote a number of books, his most influential was the 1,200-page *Foundations*. In it, following Gobineau, Chamberlain extolled the superiority of the Germanic peoples, related the accomplishments of civilization in the nineteenth century to the Germans, and credited the rise and fall of nations to the amount of "Teutonic" blood in their population. He attacked Virchow's concept of racial equality and criticisms of the concept of Aryan and Jewish races (Weindling 1989). Chamberlain was highly influenced by Immanuel Kant and wrote two volumes on Kant's work (Chamberlain [1905] 1914). Like Kant, Chamberlain was a virulent anti-Semite, claiming that Jews had an inherent morally defective character. His anti-Semitism became a core of Nazi racial philosophy (Oakesmith 1919, Montagu 1997). In the late 1800s and early 1900s, the Pan-German League, a movement of radical German elitists, joined the Gobineau Society in using the popularity of Chamberlain and Gobineau to disseminate Aryan racial theories and popularize nationalist versions of racial anthropology. These organizations and "racial anthropologists . . . rallied to crusade for racial purification. The linking of Aryan theories with the ultranationalist and anti-Semitic right was achieved in the decade prior to 1914" (Weindling 1989, 111–112; see Table 1.1).

A second influential volume, which appeared in 1899, was Haeckel's *The Riddle of the Universe: At the Close of the Nineteenth Century,* translated from German in 1900. Ernst Haeckel (1834–1919) was one of the most respected scientists of his time and an avid Darwinian, Lamarckian, and eugenicist (Shipman [1994] 2002; Spiro 2009). He believed that the living nonwhite races provided links documenting the evolution of humans from apes to the more advanced Europeans (Marks 2010b; 2012). Although not aware of Chamberlain, Haeckel also cited Gobineau in his claims about the superiority of the Aryan race, and like Chamberlain, he was fervently anti-Semitic. He was greatly concerned, as were many of his colleagues of the time, with the dilution of German blood by inferior types who were causing the degeneration of the Aryan race. He called for the halting of immigration of the "filthy" Jews and, claiming that since inferior races are "nearer to the mammals (apes and dogs) than to civilized Europeans, we must, therefore, assign a totally different value to their lives" (quoted in Spiro 2009, 124). As did Galton (see Chapter 2) and Chamberlain before him, Haeckel pointed out that in ancient Sparta the weak, sickly, or those affected with bodily infirmity were killed and only the perfectly healthy were allowed to live and propagate

Table 1.1. Monogenism vs. Polygenism from 1600 to 1900

Monogenism: (degenerate) environmental influence	Polygenism (pre-Adamite): unchangeable, biologically fixed
1600s	
Locke, *An Essay Concerning Human Understanding* (1690)	La Peyrère, *Prae-Adamitae* (1655)
1700s	
Linnaeus, *Systema Naturae* (1758) Buffon, *Histoire naturelle* (1785) Blumenbach, *On the Natural Variety of Mankind* (1795)	Hume, *A Treatise of Human Nature* (1740) Kant, *On the Different Races of Man* (1775)
1800s	

Lamarck, *Philosophie zoologique* (1809)	Mortonites
	    Morton, *Crania Americana* (1839)
	    Nott and Gliddon, *Types of Mankind* (1854)
	    Gobineau, *Essay on the Inequality of Human Races* (1853–1855)

| Darwin, *The Origin of Species* (1859) |	
| Spencer, *Principles of Biology* (1864) |	
Dugdale, *The Jukes* (1877)	Galton, *Hereditary Genius* (1869) Chamberlain, *Foundations* (1899) Weismann vs. Lamarck (1889)
[End of monogenism, or environmental influence]	

their race (Haeckel 1892). Haeckel maintained a huge following among the German public and was seen as a messiah of national and racial regeneration. His book, *The Riddle of the Universe,* sold over 100,000 copies in its first year, was translated into twenty-five languages, and sold 500,000 copies during its revival at the time of World War II (Shipman [1994] 2002).

A third book published in 1899 was William Z. Ripley's *The Races of Europe,* in which the views of Gobineau essentially were translated into English for the U.S. audience. Ripley (1867–1941) taught sociology and anthropology both at the Massachusetts Institute of Technology and (with Franz Boas; see Chapters 5 and 6) at Columbia University. In his book, Ripley divided Europeans into three racial groups, Teutonic, Alpine, and Mediterranean, with each of these having distinct behavioral differences and biological capabilities. Thus, Ripley "introduced Americans to what was generally perceived to be the latest and most sophisticated European thinking on the concept of 'race' " (Brace 2005, 169). Gobineau, through both Chamberlain and Ripley, soon would have a profound influence on Madison Grant (see Chapter 3), and all of these authors would in turn greatly influence Hitler and the Third Reich.

2

The Birth of Eugenics

Up until 1900, Lamarck's theory had been one of the main scientific rebuttals to strict biological determinism. Environment was still seen as a factor that could have an important influence on certain morphological and behavioral traits. This was about to change in most of Western Europe and in the United States.

> The abandonment of the belief in acquired characters was the stimulus for the eugenics movement. . . . By showing that environment could not change behavior based on race or biology, [Darwinism and] the new genetics had given racism a scientific basis it had lacked so long as acquired characters were an accepted principle. (Degler 1991, 24)

Many eugenicists believe that Degler's claim that the abandonment of Lamarckism was a major factor in the development of eugenics is an overemphasis and misleading. For example, in countries such as France and some Latin American countries such as Brazil the eugenics movement was tempered by "neo-Lamarckism." Consequently, in France before 1930, eugenics was often coupled with programs for public health reforms and attention to improving environmental conditions (Paul 1995; Weiss 2010; Science Encyclopedia 2013; Garland Allen, personal communication, 2013). However, the debunking of Lamarck was certainly one of the influential factors in the more radical eugenics movements in Western Europe and the United States, and the countries that were more prone to accepting Weismann and Mendel adopted the harshest eugenics policies (Paul 1995).

Debunking Lamarck

In 1889, embryologist August Weismann put Lamarck's idea of the inheritance of acquired characteristics to rest as an explanation for the evolutionary change in biological characteristics. He showed that no matter what changes occurred in the body or the behavior of an animal, these did not appear in its offspring. From his research, he argued that what he called "germ cells" (often referred to as "germ plasm") could not be affected by environment but were transmitted unaltered from one generation to the next (Paul 1995). Weismann's experiments became very influential among biologists at the turn of the twentieth century with the rediscovery of Mendelian genetics and the prominence of the role of heredity in human life (Degler 1991; Weiss 2010). This was a major blow to monogenism and environmentalists, since they were left with no other explanation of how the differences between the races could have been influenced by environmental factors. As historian Carl Degler stated (1991, 23): "The new genetics . . . effect upon the Lamarckian doctrine of acquired characteristics was nothing less than fatal." The differences between the varieties of humans could be explained only by biological mechanisms, by heredity. At the turn of the twentieth century, although it was no longer necessary to tie these differences directly to Adam and Eve, some of the basic concepts underlying the theories of monogenism and polygenism still persisted; for example, it was thought that the causes of human variation could be influenced by environment or could be strictly biologically determined. To some who might have been considered monogenists in the past, it was the environment that could influence change over generations through the inheritance of acquired social and biological characteristics. To others, reflecting the old polygenist thinking, even though the different races no longer needed to be tied to completely separate origins (though some still believed this to be so), the differences that existed were assumed to be completely hereditary (biological) in nature and could not be changed by environmental influence, at least not in any meaningful time period. As American sociologist Lester Frank Ward expressed it in 1891, "if [Weismann and his followers] are right, education has no value for the future of mankind" (quoted in Paul 1995, 440).

Thus, after the theories of natural selection were combined with Mendelian genetics, it was no longer believed that the degenerative changes that were thought to cause certain races or certain "undesirable" human types to be inferior could be fixed by changing or improving the environment of those races. Within this intellectual atmosphere, it is not hard to understand how the "science" of eugenics began to pervade biological science.

The rediscovery of Mendelian genetics and Darwin's theory of natural selection gave scientific credence to strict biological determinists. In fact, this made the polygenism-like theories acceptable to more people because it did not go against the Bible. The popularity of the elitist Social Darwinism, the fall of Lamarckism, and the combination of Mendelian genetics and Darwin's theory of natural selection created a perfect environment for the eugenics movement—the perfect storm, so to speak.

The connections between old theories of polygenism and the eugenics movement, as one might expect, were very strong. Nott and Gliddon's textbook, *Types of Mankind,* and Gobineau's *Essai sur l'Inequalité des Races Humaines,* although written a half-century earlier, were still popular texts at the beginning of the twentieth century in Europe and the United States, and both Nott and Gobineau had a strong influence on the basic racist aspects of the eugenics movement. As we have seen, at the turn of the century, Chamberlain and Ripley had again popularized Gobineau's ideas for German and American audiences. In the United States, as Brace (2005) has detailed, we also can trace the influence of Agassiz, Nott, and Gobineau directly to Nathaniel Southgate Shaler (1841–1906), professor of paleontology and geography at and subsequently dean of Harvard's Lawrence Scientific School.

Shaler, born in Haiti to a slave-owning family, was brought up in Kentucky. His wife also was from a prominent, wealthy family with ideological and economic ties to the southern slave system. Although he moved to Boston to attend Harvard University in 1859, he maintained a mindset similar to that of Nott's until his death. After a short stint in the Union Army, Shaler returned to Harvard in 1864 and was appointed assistant in paleontology at the Harvard Museum of Paleontology by Agassiz. As a protégé of Agassiz, he became a professor in 1870, shortly before Agassiz's death, and dean in 1891. Over the years, Shaler taught some 7,000 students and was "a popular and flamboyant classroom performer who became 'a figure of legendary stature at Harvard during America's Gilded Age'" (Livingstone 1987, 276).

Shaler's attitude toward human diversity was directly influenced by Gobineau's (and Friedrich Ratzel's) views of Aryan/Nordic superiority and by Nott's pure polygenic racism. In a series of lectures given in 1888–1889 that was published as *Nature and Man in America* in 1891, Shaler expounded Gobineau's theories of Aryan superiority. Adding to this was his Nott- and Agassiz-like American view of African Americans. Harking back to Hume's and Kant's 1700s philosophy and echoing Nott, Shaler (1891, 165–166) believed that "the fixity of race characteristics has enabled the several national varieties of men to go forth from their nurseries carrying the qualities, bred in their earlier conditions through centuries of life in other climes. . . . The Americas, Africa and Australia have shown by their human products that they are unfitted to be the cradle-places of great peoples. . . . These continents have never from their own blood built a race that has risen above barbarism."

Shaler disdained the mixing of various "species" of humans and believed that African American-European hybrids were inferior to their unmixed parents. Echoing Nott, he (1884, 698) stated, in speaking of Haiti, Jamaica, and the United States, that "Every experiment in freeing blacks on this continent has in the end resulted in even worse conditions than slavery brought to them. . . . In the Spanish and Portuguese settlements, the negro blood has to a great extent blended with that of the whites. There the white blood has served for a little leaven, but the mingling of the races has brought with it a fatal degradation of the whole population that puts those people almost out of the sphere of hope. Such are the facts of experience in the effort to bring together the races of Africa and of Europe on American ground. They may be summed up in brief words—uniform hopeless failure."

By the end of the nineteenth century, the views of Gobineau, as echoed by Shaler, had been "picked up with enthusiasm by William Z. Ripley, and others picked it up from him" (Brace 2005, 161). Later, Ripley's classification, which included three European races, was adopted by Madison Grant, who changed the "Teutonic" type into the "Nordic" type, which Grant proposed as the master race and used to help influence American immigration policy in the 1920s (see Chapter 3).

Thus, Shaler played a major role in bringing the theories of Nott, Agassiz, and Gobineau as well as those of seventeenth-, eighteenth-, and other nineteenth-century polygenists into the twentieth century. Furthermore, he shaped the views of a large number of influential Harvard graduates. For example, when three wealthy graduates from the class of 1889, C. Warren, R. D. Ward, and P. F. Hall, formed the Immigration Restriction League in 1894, they named Shaler one of its vice-presidents. Powerful anti-immigration politician Henry Cabot Lodge was a Shaler student, and Theodore Roosevelt learned about human variation from Shaler at Harvard.

More directly relevant to our tracking of strict biological determinism from Nott and Gobineau into the twentieth century is Shaler's influence on Charles Davenport (1866–1944), who earned his PhD from Harvard in 1892, one year after the publication of Shaler's *Nature and Man in America.* Essentially mirroring Shaler's comments over a quarter of a century later in their essay *Race Crossing in Jamaica,* Davenport and Steggerda (1929) were still spewing the same centuries-old views:

> Those who look to the future are naturally concerned with the question: What is to be the consequence of this racial intermingling? Especially we of the white race, proud of its achievement in the past, are eagerly questioning the consequences of mixing our blood with that of other races who have made less advancement in science and the arts. (225)

> one of the results of hybridization between whites and Negroes—the production of an excessive number of ineffective, because disharmoniously put together people. (237)

> It is, however, this burden of ineffectiveness which is the heavy . . . price that is paid for hybridization. A population of hybrids will be a population carrying an excessively large number of intellectually incompetent persons. (238)

As we shall see in the next few chapters, Davenport and his followers in the eugenics movement brought biological determinism, racial prejudice, and the active agenda of eugenics right up to the Nazi regime of Hitler. In fact, it was the "science" of eugenics and strict biological determinism that formed the backbone of Nazism. As anthropologist George Stocking (1968, 45) stated: "From a broader point of view, however, polygenism and monogenism can be regarded as specific expressions of enduring alternative attitudes toward the variety of mankind. . . . One could regard these differences as of degree or of kind, as products of changing environment or immutable heredity, as dynamic or static, as relative or absolute, as inconsequential or hierarchical. Considered in these terms, polygenic thinking did not die with Darwin's *Origin of Species,* nor is it entirely dead today."

It is in this sense that I sometimes use the terms *monogenism* and *polygenism* in later pages of this book, although many historians of eugenics believe that the theories of monogenism and polygenism ended with the publication of Darwin's *Origin* (Garland Allen, personal communication, 2013).

Eugenics Raises Its Ugly Head

Social Darwinism is a belief system, popular in the late Victorian era in England, America, and much of Europe, that espoused that the strongest or fittest should survive and flourish in society, while the weak and unfit should be allowed to die. This elitist theory that proposes that "might makes right" was present in Western society long before Darwin's *Origin*. However, Darwinian ideas were used to reinforce such elitist ideas and ideals. The concept of adaptation was used to reinforce the claim that the rich and powerful were better adapted to the social and economic climate of the time, and the concept of natural selection was used to bolster the idea that it was natural, normal, and proper for the strong to thrive at the expense of the weak. To Social Darwinists, not only was survival of the fittest natural, it was also morally correct. In fact, many argued that it was morally wrong to assist the weak since that would be promoting the survival and reproduction of the less fit. In a general sense, Social Darwinists saw the competitive struggle, or the "survival of the fittest," as taking place among individuals or among different groups within a society and between different societies, or between different racial or ethnic populations.

Though he was not himself a Social Darwinist (see Weinstein 2012), Herbert Spencer (1820–1903) is often thought of as the source of this ideology. Spencer was a social philosopher who coined the phrase "survival of the fittest" in 1864 after reading *The Origin of Species:* "This survival of the fittest, which I have here sought to express in mechanical terms, is that which Mr. Darwin has called 'natural selection,' or the preservation of favoured races in the struggle for life" (Spencer 1864, 444). Even before *Origin*, however, Spencer had used evolutionary concepts in his writings on sociology and ethics, using mainly Lamarckian mechanisms of change rather than Darwin's natural selection. In fact, Darwin also was a Lamarckian of sorts.

One can see the seeds of Social Darwinism in Spencer's writings. He believed that for evolution to produce more perfect individuals, it was necessary for current and future generations to experience the "natural" outcome of their conduct and not be interfered with by government or charity. Only in this way could self-improvement be achieved and be passed on to descendants. Thus, anything that interfered with the "natural" state of things should be resisted, including public institutions that provided education or charity or requiring such things as compulsory vaccination. Assistance or benevolence to the "undeserving poor" would thwart what

Spencer considered necessary for the continued evolution of humans to a higher level.

Both Darwin and Spencer were influenced greatly by Thomas Malthus's *Essay on the Principle of Population* (1798), in which Malthus (1766–1834), a minister and economist, claimed that in nature, plants and animals produce far more offspring than can survive and that even humans are capable of overproducing if left unchecked. For humans, societal improvements resulted in population growth that, Malthus predicted, eventually would be checked by famine, disease, and widespread mortality. It is interesting to note that although Darwin's *Origin* was a very popular and widely read volume immediately after its publication in 1859, Spencer's writings were even more popular. His books sold more than one million copies in his lifetime, the first time that had happened for a philosopher (James 1904). Thus, the writings of Darwin and Spencer were extremely influential in the latter part of the nineteenth century.

The eugenics movement was a direct extension of Social Darwinism, though the two were definitely not the same. The former was proactive and interventionist, whereas the latter was more laissez-faire in its approach to state intervention (Garland Allen, personal communication, 2013). Eugenics, both the term and the movement behind it, was the brainchild of Francis Galton (1822–1911), a cousin of Charles Darwin's; the two shared a common grandfather, Erasmus Darwin. Like Darwin, Galton was born into a wealthy family. The Galtons were highly successful gun manufacturers and bankers. Francis had entered medical school, but upon his father's death, he inherited enough wealth to travel and to dabble in his scientific interests without having to earn a living (Brace 2005). In his lifetime, he made contributions in geography, meteorology, biology, psychology, criminology, and statistics. He also was an explorer and inventor. He was knighted in 1909. Some saw him as a genius and a prolific scientist and believed he had been a child prodigy (a belief Galton shared). Others, however, saw him as an egocentric snob (Brace 2005), a mediocre intellect, a sham and a villain (Graves 2001), and a spiritual fascist (Medawar 1975; Bulmer 2003).

Galton coined the term *eugenics* in 1883 from the Greek words for "well-born" and based the concept on some of Darwin's writings (Galton 1883). Darwin ([1871] 1874, 130) had written in *The Descent of Man:* "We build asylums for the imbecile, the maimed and the sick; we institute poor laws; and our medical men exert their utmost skill to save the life of every one to the last moment. . . . Thus the weak members of civilized societies propagate their kind. No one who has attended to the breeding of domestic animals will doubt that this must be highly injurious to the race of man." However, Darwin believed that any hopes of changing these practices were unlikely and utopian and would not be possible until the laws of inheritance were thoroughly known (Paul 1995). Galton, on the other hand, believed that controlled breeding of humans was not only doable but a highly desirable goal. He spent his professional career, from the 1870s until his death in 1911, investigating and writing about the potential and possibilities of eugenics. Galton did not limit his eugenics ideas to individuals but included racial differences as well. He believed that the average "negro" intellectual standard was two grades below that of extant "Anglo-Saxons," and he saw the intellectual standard of the latter as two grades below that of residents of ancient Greece (Galton 1869). In addition, in his formulation of the concept of eugenics, his goal was to "touch on various topics more or less connected to that of the cultivation of race, or, as we might call it, 'eugenic' questions" and to promote "judicious mating" in order to "give the more suitable races or strains of blood a better chance of prevailing speedily over the less suitable" (Galton 1883, 24–25).

Galton's formulation of eugenics was based on a statistical approach. He and his protégé Karl Pearson (1857–1936) developed a "biometrical" approach to eugenics in which statistical models were used to describe the inheritance of traits. However, with the rediscovery of Mendel's hereditary laws, two separate camps of eugenics advocates emerged, one made up of statisticians and the other of biologists. The statisticians thought that the biologists' mathematical models were exceptionally crude, whereas the biologists believed that the statisticians knew little biology (MacKenzie 1981). Another early split in the eugenics camp divided the positive and negative eugenicists. The former called for the control of human breeding to produce genetically superior people, while the latter lobbied for improving the quality of the human race by eliminating or excluding biologically inferior people from the population through the use of segregation, deportation, castration, marriage prohibition, compulsory sterilization, passive euthanasia, and, ultimately, extermination (Black 2003). Galton was mainly a positive eugenicist, and in the early 1900s, this approach was more popular in the United Kingdom. Negative eugenics was more in vogue in the United States. Among the traits Galton assumed to be inherited were character and personality, general intellectual ability, gregariousness, longevity, strong sexual passion, aversion to meat, craving for drink and gambling, susceptibility to opium, proclivity to pauperism and crimes of violence and fraud, madness, and tuberculosis (Galton 1865, Paul 1995). As historian Paul Weindling (1989, 92) has stated: "Galton's eugenics marked a turning point in the transition from liberal political economy to a biologically based authoritarian collectivism."

In the early 1900s, there were a number of attempts to jump-start radical eugenics movements in Western Europe and the United States. In 1904, Galton endowed a research chair in eugenics at University College London. In Germany in 1905, Drs. Alfred Ploetz and Ernst Rüdin founded the Gesellschaft für Rassenhygiene, the Society of Race Hygiene. In 1907, inspired by Galton, the British Eugenics Education Society (later the Eugenics Society of London) was founded in the United Kingdom. However, in Britain eugenics never received significant state funding before World War I, though it was supported by many prominent figures, including liberal economist William Beveridge; conservative politician Arthur Balfour; authors George Bernard Shaw, H. G. Wells, and Sidney Webb; and future prime minister Winston Churchill (Paul 1995; Okuefuna 2007). Furthermore, in the United Kingdom, eugenics in the early years was focused more on social class than on race (Porter 1999). In Europe generally, national populations were relatively homogeneous and eugenics was focused mainly on class (Paul 1995). The limited popularity of eugenics in Britain is reflected by the fact that only two universities established courses in this field (University College London and Liverpool University). The Galton Institute, affiliated with University College London, was headed by Karl Pearson, who held the Galton Chair at the college, which was endowed by Galton's will (Brace 2005).

The Eugenics Movement in the United States

In the United States, the eugenics movement was much more negative and pernicious even in its early development. In addition, the United States had a large African American population and an influx of immigrants from Asia and southern and eastern Europe (Paul 1995). Both in the United States and Britain, the polygenic-like theory of the eighteenth century became a dominant theory, but instead of the pre-Adamite biblical version, a new scientific version of biological determinism emerged, fortified by Mendelian genetics (Stocking 1968). The unfit were still seen as essentially subhuman, as diseased in some way, but their afflictions were seen as genetically based and not necessarily based on biblical interpretations (Black 2003). After Darwin's *Origin* and the popularity of such views as Social Darwinism, many academics and those with economic, social, and political power began to endorse biologically deterministic theories of human behavior. The rediscovery of Mendelian genetics reinforced these views. For example, many academics of the period believed that much of human behavior was biologically determined. The relatively new academic fields of psychology and criminology epitomized this approach.

In 1881, William James (1842–1910), the Harvard pioneer in psychology, published the first human psychology textbook, *Principles of Psychology*. Drawing from Darwin, James contended that instinct was common to both animals and humans. By the time of his death in 1910, James's emphasis on instinct in accounting for human behavior was accepted as a fundamental insight of the profession. The highly influential psychologist William McDougall (1871–1938) focused on instinct as the center of his explanations of human actions. McDougall, who first taught at Cambridge and Oxford, moved to Harvard in 1920. His influential textbook *Introduction to Social Psychology* was first published in 1909. He argued that such "innate tendencies" as mating, parenting, escape from danger, purposive striving, dominance, and companionship were based on instinct (Degler 1991). From 1900 to 1920, 600 or more books and articles were published in America and England that advanced instinct theory (Cravens 1978). Eugenicists believed that many complex human behaviors had a basis in instinct; that is, they were biologically set (purely genetically determined) behavior patterns triggered by specific environmental stimuli. Many biological and evolutionary explanations of human behavior still provide the intellectual foundations for the fields of sociobiology and evolutionary psychology (Wilson 1975, 1998; Gillette 2011).

A similar pattern can be seen in the field of criminality, or criminal anthropology, as it later was called (Allen 2001a). In 1872, Charles Loring Brace, great-grandfather of C. Loring Brace, the contemporary anthropologist and author of *"Race" Is a Four-Letter Word* (2005), espoused that such "pathological" traits as drunkenness, insanity, prostitution, and criminality could be hereditary. By 1888, the work of Italian criminologist Cesare Lombroso, who claimed that criminals could be identified by physical characteristics (including such features as a sloping forehead, ears of unusual size, asymmetry of the face or cranium, and excessive arm length), had been introduced to American science. In 1891, in his presidential address to the Anthropological Society of Washington, D.C. (which later became the American Anthropological Association), Robert Fletcher explained that the influence of heredity on criminal character was admitted and accounted for 50–75 percent of criminal behavior. Nearly every case of criminal behavior could be explained by ancestral influences: "Good seed generates sound and healthy fruit, and imperfect parentage can only yield defective offspring" (Fletcher 1891, 207).

In 1905, Alfred Binet and his physician colleague Theodore Simon introduced intelligence testing in France in order to help identify school children who might need assistance in their courses. Psychologists, physicians, and educators in the United States had been frustrated in their attempts to find ways to diagnose, classify, and standardize measurements of mental illness and compare "feeble-minded" children with normal ones. After visiting Europe in 1908 and learning about Binet's method, Herbert H. Goddard (1866–1957) was the first to translate the Binet-Simon intelligence tests into English (see Chapter 3). By 1910, he had convinced the American Association for the Study of the Feeble-Minded of the efficacy of the Binet-Simon tests for diagnosing mental deficiency (Zenderland 1998). He subsequently distributed thousands of copies across the United States, and many American social scientists and psychologists began using mental testing. Early in his career, Goddard believed that much "feeble-mindedness" was caused by the environment. Along with his colleague Edward Ransom Johnstone at the Training School for Feeble-Minded Girls and Boys in Vineland, New Jersey, he focused on the education and training of children with mental deficiencies. In fact, Goddard and Johnstone were pioneers in establishing "special education" curricula and programs in classrooms and in public schools.

In 1909, prominent biologist Charles Davenport began corresponding with Goddard and visited him at the Vineland school. Davenport had recommended that Goddard read *Mendelism* (1909), a book by the British biologist R. C. Punnett, and convinced him that he might find relationships between heredity and feeble-mindedness among the children in his school by tracing their family pedigrees. After reading Punnett and Davenport, Goddard increasingly became convinced that Mendelian genetics could answer questions about the causes of inherited conditions, among them mental deficiency. He soon shifted from using statistics and normal, bell-shaped curves to using the concept of "unit characters"; he "began to conceptualize intelligence less as a score along a continuum and more as the presence or absence of a single trait" (Zenderland 1998, 158). Goddard and Davenport quickly developed a strong and long-lasting supportive working relationship (Zenderland 1998).

Thus, from 1909, Goddard believed that hereditary defect was the cause of most law-breaking and that half of all criminals were feeble-minded. It is this that motivated his book on Martin Kallikak (see Chapter 3) in which he stated (53): "No amount of education or good environment can change a feebleminded individual into a normal one, any more than it can change a red-haired stock into a black-haired stock." By World War I, scientists and the public also believed that America and Western Europe was "beset by what came to be called the 'menace of the feebleminded'" (Degler 1991, 37). Intelligence testers believed that between one-third and one-half of all criminals, delinquents, and prostitutes were feeble-minded, or "morons" (Goddard believed that the coining of the latter term was one of his greatest accomplishments [Zenderland 1998]). Soon biologists began to assume that intelligence and feeble-mindedness were single gene traits or "unit characters." For example, in 1922, Vernon Kellogg (18), a prominent Stanford biologist, claimed that the idea that feeble-mindedness "was a unit human trait following the general Mendelian order as regards its mode of inheritance . . . is hardly any longer open to doubt."

These views of the Mendelian inheritance of behavioral characteristics and intelligence were used to discriminate among individuals and ethnic and racial groups; environmental solutions were left out of the picture. The eugenics movement also used these factors as ammunition for its cause. Many worried that the most "valuable individuals and classes" were being outbred by the least valuable (Paul 1995). In 1907, President Woodrow Wilson, who supported the eugenics effort, helped Indiana adopt legislation making sterilization of certain "undesirable" individuals compulsory. More than thirty states adopted such laws.

Beginning with Connecticut in 1896, many states enacted marriage laws prohibiting anyone who was "epileptic, imbecile or feeble-minded" from marrying. Charles B. Davenport, the biologist who influenced Goddard, received his PhD at Harvard in 1892 and then taught at Harvard (1891–1899) and the University of Chicago (1899–1904). He also served as director of the summer school of the Biological Laboratory of the Brooklyn Institute of Arts and Sciences in Cold Spring Harbor, New York, from 1898 to 1923 (Riddle 1947). Davenport became one of the leading proponents of American eugenics. In 1903, after spending part of a sabbatical year (1899–1900) with Francis Galton and Karl Pearson in London (Allen 1986, 2011), Davenport persuaded the Carnegie Institution of Washington (CIW) to fund, with him as director, the Station for Experimental Evolution at Cold Spring Harbor (Allen 1986). The Eugenics Record Office (ERO) opened there in 1910, after Davenport had convinced Mrs. Mary W. Harriman (the widow of E. H. Harriman who had recently inherited her husband's Union Pacific Railroad fortune) of the need for such an office and had gained ample funding from her for its establishment and operation. Davenport was the director and he appointed Harry H. Laughlin (1880–1943) as superintendent. Harriman was among the single largest individual donors to U.S. eugenics causes. She donated seventy-five acres adjacent to the experimental evolution station for the ERO (Engs 2005). After 1916, the ERO was funded by the Carnegie Institution of Washington (Allen 1986, 2011). Michael Barker (2010) has characterized the board of trustees of Carnegie Corporation as dedicated to the umbrella organization that oversaw Andrew Carnegie's various philanthropic activities to preserve Anglo-Saxon prerogatives, customs, and genes. Davenport and Laughlin vigorously promoted eugenics in the United States and internationally. They viewed eugenics as a way "to apply science to the problems of a class-ridden and socially heterogeneous society" (Weiss 2010, 25). The ERO soon became the nerve center of the eugenics movement. It had three main functions: to conduct scientific research on human heredity, to popularize eugenic ideas, and to lobby for eugenics-related legislation (Weiss 2010). The ERO became a meeting place for eugenicists, a repository for eugenics records, a clearinghouse for information and propaganda, a platform from which popular eugenics campaigns were launched, and the home of a number of eugenics publications (Allen 1986). From 1904 to 1929, the CIW invested approximately $3 million in Davenport's eugenics research programs (Patterson 2001; Barker 2010).

Charles Davenport was born on a farm in Stamford, Connecticut, the youngest of eleven children, but he grew up in Brooklyn Heights, New York, where his father worked in real estate. His father was a puritanically religious man and forced Charles to study the Bible in the mornings and evenings. His father also was interested in ancestral study and had traced his family genealogy back to Anglo-Saxon ancestry in 1086. Wanting to go into academics and not theology, young Charles formally asked for his father's permission to study in the sciences. Seven weeks later, his father gave him written permission, stating that "of prime importance is how much money can you make for you and for me" (quoted in Black 2003, 33).

After the rediscovery of Mendelian genetics, Davenport began experimenting on many small marine species. Soon after this, he proposed a research station to study eugenics, with the ultimate intention of focusing on human races. He saw different racial and ethnic groups "as biologically different beings—not just physically, but in terms of their character, nature and quality" (quoted in Black 2003, 35). To Davenport, most of the non-Nordic types swam at the bottom of the hereditary pool, each with its own distinct and undeniable adverse genetic features (Black 2003, 35). Further, he believed that

> permanent improvement of the race can only be brought about by breeding the best. . . . We have in this country the grave problem of the negro, a race whose mental development is, on the average, far below the average of the Caucasian. . . . May we hope at last that the negro mind shall be as teachable, as elastic, as original, and as fruitful as the Caucasians? Or must future generations, indefinitely, start from the same low plane and yield the same meager results? Prevailing opinion says we must face the latter alternative. If this were so, it would be best to export the black race at once. (Davenport to Billings 1903, quoted in Black 2003, 38)

He then told the Carnegie Institution that he proposed "to give the rest of my life unreservedly to this work" (quoted in Black 2003, 38). The Carnegie Institution was impressed by Davenport's proposal and gave him ample funding to pursue these goals for the rest of his life. Even with this lifetime of funding for his eugenics goals, Davenport was usually bitter and disconsolate, seemingly always attempting to prove himself to his father and to God. Davenport became even more bitter and desolate after the death of his son Charles from polio in 1916. After that, he became more insulated and more focused on his work. Davenport was relentless in this work. In fact, from his position at Carnegie Institution's Cold Spring Harbor, he masterminded eugenics research and policy in the United States and throughout the world for thirty years. He retired in 1934 and then worked in a small office at the ERO until his death from pneumonia ten years later.

Harry Laughlin, whom Davenport hired as superintendent of the ERO in 1910, had attended one of Davenport's summer courses at Cold Spring Harbor in 1907 and met Davenport again at an American Breeders Association (ABA) meeting in 1909. Laughlin was brought up in Kirksville, Missouri, a rural town surrounded by agricultural communities. His middle-class parents, who both came from pioneering families, had ten children, five of whom were sons. His father had migrated from Iowa to become pastor of the Kirksville Christian Church but then became chair of the English Department in the First District State Normal School in

Kirksville, the area's main college. Like Davenport's father, Laughlin's father was deeply religious and was caught up in genealogy; he had carefully traced the family's ancestry back to England and Germany. After graduating from the Kirksville normal school, Harry Laughlin began his career teaching in a one-room schoolhouse in nearby but even more rural Livonia, Missouri. He was very disappointed there and hated the town, the school, the local people, and the students. In 1905, he returned to Kirksville, first serving as a high school principal before moving to the Department of Agronomy, Botany and Nature at the normal school from 1907 to 1910 (Black 2003).

It is obvious that Davenport and Laughlin hit it off very well at Davenport's summer course and at the 1909 ABA meeting. Davenport was the idea man and fund-raiser, and Laughlin became the relentless worker who pursued his agenda. Davenport recognized Laughlin's work ethic and gave him living quarters on the grounds of the Cold Spring Harbor station, where Laughlin proceeded to work around the clock. Davenport had certainly chosen the right man, and together their "campaign to create a superior race" was soon launched (Black 2003, 51). Like Davenport, Laughlin spent his life working at Cold Spring Harbor. After working there for twenty-nine years, he was forced to retire on the day the ERO was permanently closed in 1939. He returned to Kirksville and died four years later. Ironically, Laughlin had a "germ plasm" that didn't fit his utopian, eugenic ideals. He was afflicted with a hereditary disease, epilepsy, that he kept as a family secret, even though his seizures occasionally occurred in front of his colleagues and he was forced to quit driving. Epilepsy was one of the diseases that could get you sterilized, according to Laughlin's model sterilization law. In fact, Laughlin died in the grip of epilepsy. He also was not very successful in the survival of the fittest, since he never had any children. Upon his death, Davenport wrote a eulogy in *Eugenical News (EN)*, the official publication of the ERO.

The first meeting of the ABA was held in St. Louis in 1903. It established a Committee on Eugenics in 1906, chaired by David Starr Jordan (1851–1931), a well-known biologist and chancellor of Stanford University (Davenport 1910). Among other luminaries on the committee were Davenport and Alexander Graham Bell, who besides being the inventor of the telephone, was a sheep breeder and eugenics researcher who focused on hereditary deafness (Stansfield 2005). (In fact, many of the early major contributors to the eugenics movement owned and were involved in breeding race horses, including Leland Stanford [the governor of California, who appointed Jordan as head of Stanford University], E. H. Harriman and his wife, and John D. Rockefeller [Stokes 1917]). The committee was to "investigate and report on heredity in the human race" and to make clear "the value of superior blood and the menace to society of inferior blood" (quoted in Degler 1991, 43). By 1910, eugenics was one of the topics most frequently referenced in the *Reader's Guide to Periodical Literature;* good breeding seemed to be everybody's business (Paul 1995). In 1911, Davenport published *Heredity in Relation to Eugenics,* a textbook that was widely used for many years, which helped establish eugenics as a scientific program in America (Marks 1995). The book was dedicated to Mrs. E. H. Harriman, the benefactor who is often considered the founder of the ERO (Engs 2005).

Even with all of these efforts, eugenics was not widely popular in these early years, and many of the efforts of active eugenicists were thwarted. For example, sterilization laws were attempted in Michigan in 1897 and in Pennsylvania in 1905 but were either defeated or vetoed (Degler 1991). Even though some other states had passed sterilization laws, no sterilizations had been performed in Nevada, Iowa, New Jersey, or New York. Thus, aside from the increasing number of eugenics sterilization laws that were being passed in states, the constitutionality of sterilization laws was still in question. As Edwin Black (2003, 69) has noted: "The Cold Spring Harbor stalwarts of the ABA, its ERO and the Carnegie Institution's Experimental Station remained frustrated. . . . Many state officials were clearly reluctant to enforce the laws. . . . The legality of the operations and the question of due process had never been satisfactorily answered." In a 1911 report, the eugenics section of the ABA admitted that the public did not support sterilization, that law officers of states were not anxious to carry out sterilization laws, and that hostility would result from any attempt to carry out these laws in a systematic manner. The report concluded: "We must frankly confess . . . this movement for race betterment is as yet little more than a hobby of a few groups of people. . . . Much more extensive education of the public will be necessary before the practice of sterilization can be carried out to the extent which will make it a factor of importance" (Van Wagenen 1912, quoted in Black 2003, 70).

The First International Eugenics Congress of 1912

The influence of the eugenics movement seemed to change radically after the First International Eugenics Congress held at the University of London in July 1912. This change was somewhat predictable given the efforts surrounding this congress by its advocates. Many of the eugenicists around the world had similar goals and had been communicating for some time. The Eugenics Education Society of London had moved away from Galton's positive eugenics and adopted the more negative approach of the United States, as had other European nations such as Germany and Scandinavia, where theories claiming Nordic superiority, as espoused by Gobineau, were widely accepted. Thus, a major effort was made to bring together an international body of eugenicists and racists and to publicize this meeting widely. The meeting was held at the University of London, hosted by the Eugenics Education Society and the ABA's American Committee on Sterilization, whose preliminary report was considered a highlight of the meeting (Black 2003).

Ultimately, approximately 400 delegates and speakers attended the congress from America, Belgium, England, France, Germany, Italy, Spain, Norway, and Japan. These included some of the most respected, financially powerful, well-known, and influential politicians, academics, university administrators, and scientists of Europe and America. Major Leonard Darwin, Charles Darwin's son, who was president of the Eugenics Society of London from 1891 to 1928, was appointed president of the meeting. Winston Churchill represented the king. Members of the ABA Committee on Eugenics, including David Starr Jordan, Charles Davenport, Alexander Graham Bell, and Bleecker Van Wagenen, the secretary of that committee, were vice-presidents, as was Alfred Proetz, the founder of the German Society for Racial Hygiene, Germany's leading race theorist of the time, and the "founder of eugenics as a science in Germany" (Black 2003, 262). Other vice-presidents of the Congress included Charles W. Eliot, the president of Harvard University, and Gifford Pinchot, a well-known conservationist and future governor of Pennsylvania. Also in attendance from the United Kingdom were Lord Alverstone, the chief justice of England; Lord Balfour; Sidney Webb, a leader of English social democrats; and Sir Thomas Barlow, the president of the Royal College of Physicians. Famous politician Tommy Douglas came from Canada. Other dignitaries attended such as George Bernard Shaw, H. G. Wells, Émile Zola, John Maynard Keynes, industrialist William Keith Kellogg, and the ambassadors of Norway, Greece, and France (Bruinius 2006; Stamm 2009). The meeting was dedicated to Francis Galton, who had recently died. However, because the eugenics leaders had moved away from positive eugenics and statistics, Karl Pearson, Galton's protégé, did not attend. (Pearson's work was statistical, cautious and highly skeptical of the new field of Mendelian genetics as practiced by Davenport [Allen 2011].) It is interesting to note also that in 1912, the current, former, and future U.S. presidents William Howard Taft, Theodore Roosevelt, and Woodrow Wilson were active supporters of the eugenics movement.

With this strong political support, the leaders of the congress attempted to get the U.S. State Department to send a delegation. It also was suggested that the invitation be distributed by the powerful House Appropriations Committee. However, since the meeting was not governmental, it was against the law for the U.S. government to officially participate. Nevertheless, Secretary of State P. C. Knox defied the law and on June 20, 1912, sent invitations to the congress on official State Department letterhead on behalf of the British Embassy, representing the Eugenics Education Society of London, to distinguished American scientists, educators, and politicians across the country. Replies were to be sent back directly to Knox. These invitations were sent to governors of every state, to the presidents of the National Academy of Sciences, the American Academy of Political and Social Sciences, the American Economic Association, the American Philosophical Society, and many other scientific and academic organizations. They also were sent to all the major medical societies, including the American Medical Association (Black 2003).

Since the invitations were sent only a few weeks before the actual congress, it is not likely that the eugenics leaders expected most recipients to actually attend (after all, they could not simply book a flight!). Rather, this was meant as advertisement, publicity, and as a sign of the U.S. government's approval of the congress. "The message was clear. Knox had, for all intents and purposes, turned the State Department into a eugenics post office and invitation bureau. . . . Proper or not, eugenics had overnight been packaged into an officially recognized and prestigious science in the eyes of those who counted" (Black 2003, 72). Knox, the former attorney of the Carnegie Steel Company who supported the eugenics program of the Carnegie Institution, was happy to do this. With this type of backing and with the financial support of such financial giants as Andrew Carnegie, Mary W. Harriman, John D. Rockefeller (the oil industry tycoon), Henry Ford, J. H. Kellogg (the cereal magnate), C. J. Gamble (of Proctor and Gamble), J. P. Morgan (of U.S. Steel), and Mrs. H. B. DuPont (of the chemical company), the eugenics movement gained momentum. The movement now had scientific and political legitimacy; widespread, worldwide recognition; and massive political and industrial financial support.

The scientific and media reviews of the congress were glowing and solidified the legitimacy of the eugenics movement. Positive propaganda about the congress spread throughout the scientific and popular media. The headline of the *Journal of the American Medical Association (JAMA)* read: "The International Eugenics Congress, An Event of Great Importance to the History of Evolution, Has Taken Place." The article emphasized the Social Darwinist approach of eugenics: "The unfit among men were no longer killed by hunger and disease . . . but they [society] must not blind themselves to the danger of interfering with Nature's ways. Cattle breeders bred from the best stocks. . . . Conscious selection must replace blind forces of natural selection" (quoted in Black 2003, 73). The *British Medical Journal* (1912, 253, 255) reported that Lord Balfour "justified the holding of the Congress by contending that its duty was to convince the public that the study of eugenics was one of the greatest and most pressing necessities of our day. . . . The Congress had got to persuade the ordinary man that the task that science had set itself was one of the most difficult and complex it had ever undertaken." The article quoted from the official proceedings' opening remarks by Major Darwin: "They should hope that the twentieth century would be known in the future as the century when the eugenics ideal was accepted as part of the creed of civilization."

The pages of *Nature* (1912, 89) exclaimed: "It is the general feeling of those who attended this Congress (which extended from July 24 to 30) that it has been a complete success. A membership of about 750 is an indication of the widespread interest taken in the subject, though an analysis of motives might reveal that the largeness of the number is partly due to other causes." The *New York Times* (1912) reported: "The first international Eugenics Congress opened here to-day, with 400 delegates, representing twelve countries, in attendance. The congress was formally opened at a banquet and reception, at which speeches were made by ex-Premier Balfour, the Lord Mayor, and Major Darwin. Mr. Balfour said that the study of eugenics was one of the pressing issues of the age. He based his belief in the future progress of mankind on the application of scientific methods to practical life."

In a *London Times* article, one writer wrote that "all of us in this country are immensely indebted [to Charles Davenport], for the work of his office has far outstripped anything of ours." A Scottish physician and eugenics advocate wrote of Laughlin's ERO: "Recent reports of the American Eugenics Record Office . . . have added more to our knowledge of human heredity in the last three years than all former work on that subject put together" (quoted in Black 2003, 213).

Early the next year, Theodore Roosevelt, in a letter dated January 3, 1913, wrote:

> My dear Mr. Davenport: I am greatly interested in the two memoirs you have sent me. They are very instructive, and, from the standpoint of our country, very ominous. You say that these people are not themselves responsible, that it is "society" that is responsible. I agree with you if you mean, as I suppose you do, that society has no business to permit degenerates to reproduce their kind. It is really extraordinary that our people refuse to apply to human beings such elementary knowledge as every successful farmer is obliged to apply to his own stock breeding. Any group of farmers who permitted their best stock not to breed, and let all the increase come from the worst stock, would be treated as fit inmates for an asylum. Yet we fail to understand that such conduct is rational compared to the conduct of a nation which permits unlimited breeding from the worst stocks, physically and morally. . . . Some day we will realize that the prime duty[,] the inescapable duty[,] of the good citizen of the right type is to leave his or her blood behind him in the world; and that we have no business to permit the perpetuation of citizens of the wrong type.

By 1912, Davenport's textbook *Heredity in Relation to Eugenics* (1911) and a few others such as those of Nott and Glidden and Gobineau had become widely used in biology, psychology, and social science departments in American and European universities. Eugenics courses appeared in both the most prestigious and the smallest universities in the United States and in high schools. Black (2003, 75) states that "eugenics rocketed through academia, becoming an institution virtually overnight." By 1914, some forty-four colleges and universities offered eugenics instruction. By 1928, that number had grown to 376, reaching some 20,000 students annually (Cravens 1978). Eugenics became a top-down model of "education reform" for these educators (Stoskepf 1999). A more recent analysis of forty-one high school biology textbooks used through the 1940s revealed that nearly 90 percent of them had sections on eugenics (Selden 1999).

The First International Eugenics Congress had been a remarkable success for the eugenicists. They had been able to publicize, legitimatize, popularize, and give an aura of wide acceptance to their goals. By the 1920s over thirty countries had eugenics movements, though they interpreted eugenics somewhat differently (Weiss 2010). At the time of the conference, the immediate goals of the eugenics movement were: (1) the promotion of selective breeding; (2) the sterilization and castration of the "unfit"; (3) the use of intelligence testing to identify mentally deficient individuals and to identify differences in intelligence between racial and ethnic groups (dubbed racial psychology); and (4) limiting the immigration of various ethnic and racial groups (Degler 1991). In the following years, they vigorously pursued this agenda.

3

The Merging of Polygenics and Eugenics

After the First International Eugenics Congress, with the eugenics movement widely accepted in Europe and America (at least Canada and the United States), the powerful advocates of the movement could get started on their agenda. They could begin fulfilling their goals, and this they did with vigor. The goals of the movement were not mutually exclusive, and the many links among them will become apparent.

Fulfilling the Goals of the Movement

Some of the richest, most powerful, most prestigious academics of the time formed a relatively small but extremely active and integrated group in order to ensure that polygenic policies would permeate public policy. They focused much of their efforts on achieving the goals of eugenics from 1912 to the mid-1940s. They were only slightly slowed down by World War I. For example, during the first international congress, the Permanent International Eugenics Committee was formed, which met for the first time a year after the congress, in 1913. The next meeting of the committee was planned for 1914, and the Second International Congress of Eugenics was scheduled to convene in New York in 1915. Both of these meetings had to be postponed, and the second congress was not held until September 1921, at the American Museum of Natural History in New York, as originally planned (organized and run by Madison Grant, Davenport, and Henry Fairfield Osborn, the president of the museum). However, the eugenics movement, led by the American contingent, was patient and worked persistently and diligently over the next thirty years to fulfill its goals.

Shortly after the first international congress, three major eugenics organizations within the United States began working rigorously to make the goals of the movement successful: the Eugenics Record Office (ERO), the Eugenics Research Association (ERA), and the Eugenics Committee of the United States of America (ECUSA), the latter of which evolved into the American Eugenics Society (AES). The ERO, as we have seen, was established in 1910 by Davenport with Laughlin as its superintendent. It was funded mainly by the Carnegie, Harriman, and Rockefeller families throughout its existence into the 1930s. The ERO sponsored eugenics research, operated a summer school for eugenics field workers, was the repository for data collected by those field workers, and published *EN*, which was edited by Laughlin (Spiro 2009). It had a small board of scientific directors:

Frank L. Babbott
Lewellys F. Barker
Charles B. Davenport
Arthur Estabrook
Irving Fisher
Henry Goddard
Harry H. Laughlin
Mary Harriman Rumsey
William H. Welch
Robert M. Yerkes

In 1912, the year of the First International Eugenics Congress, Henry Herbert Goddard published *The Kallikak Family: A Study in the Heredity of Feeblemindedness*. His intention was to debunk Dugdale's earlier, Lamarckian-oriented volume on the Juke family. Goddard traced the history of the Kallikak family in an attempt to illustrate the biological fixity of certain "hereditary" characteristics. The name Kallikak was fictitious, invented by Goddard by joining the Greek words for "beautiful" *(kalos)* with "bad" *(kakos)* (Paul 1995). The book went through twelve editions and was translated into German (Engs 2005). A new edition was printed in Germany shortly after that nation passed a sweeping sterilization law in 1933, and a thirteenth and final U.S. edition was published by Macmillan in 1939 (Zenderland 1998).

In the book, Goddard wanted to illustrate how feeble-mindedness, because it was completely hereditary, could quickly spread and become a major social problem (see Figure 3.1). Martin Kallikak Jr., he wrote, was

Figure 3.1 Martin Kallikak's good and bad descendants, based on Henry Goddard's 1912 book.
From Garrett and Bonner, *General Psychology*, 1961.

a soldier in the Revolutionary War who had a casual relationship with a presumably mentally defective young woman he met in a tavern. Over time, 480 descendants were produced from this ill-fated union. According to Goddard, of the known descendants, 143 were feeble-minded and only forty-six were "normal." These, Goddard claimed, must have been the result of the genetic contribution of the mother, since Mr. Kallikak later married an upstanding Quaker maiden and the descendants from this marriage were all morally upright and as successful as their father. Goddard completely ruled out environmental influences. He asserted that "the striking fact of the enormous proportion of feeble-minded individuals in the descendants of Martin Kallikak, Jr., and the total absence of such in the descendants of his half brothers and sisters is conclusive on this point. Clearly it is not environment that has made that good family" (quoted in Degler 1991, 39).

Goddard's book was immensely successful and seemed to capture both scientific and popular ideas about the larger meaning of heredity in a time of changing paradigms. As historian Leila Zenderland (1998, 144) has stated, "*The Kallikak Family* seemed to suggest a major scientific breakthrough. This book's publication quickly catapulted Goddard into the most respected ranks, both nationally and internationally, of scientists studying mental deficiency. Of equal importance was its impact on the public. . . . Over the next three decades . . . its story would . . . be told and retold in scientific textbooks, court cases, political speeches, public exhibits, and popular magazines."

In 1913, Davenport and Laughlin established the ERA to supplement the work of the ERO. It held annual summer conferences at Cold Spring Harbor and was affiliated with the AAAS. The ERA was developed to politicize and popularize eugenics—to promote the use of eugenics research in legislative and administrative action and public propaganda for the causes of eugenics, raceology, and Nordic racial supremacy (Black 2003). It had fifty-one charter members (see Black 2003 for a list of these) who constituted the core of the scientific, political, and business leaders of the eugenics community. Throughout its existence, the ERA never exceeded 500 members. In fact, both in America and Britain, the official eugenics societies remained small. However, their members included highly influential individuals (Paul 1995). For example, in 1916, when the journal *Genetics* was established, all of the members of the editorial board endorsed eugenics (Weiss 2010).

In 1916, Arthur Estabrook, a close colleague of Davenport and researcher at the ERO, updated and reanalyzed the field notes from Richard Dugdale's original study of the Jukes family. Similar to Goddard's analysis of the Kallikak family, Estabrook asserted that half the family was feeble-minded and that this was due entirely to their defective "germ plasm" (Estabrook 1916). His message was the exact opposite of the one Dugdale had expressed and "came to symbolize the futility of social change and the need for eugenic segregation and sterilization" (Paul 1995, 49). Estabrook's book was published by the Carnegie Institution and, along with God-dard's *Kallikak* book, became a standard reference for eugenics research and the hereditarian concept of behavior. Estabrook was named to the executive council of the ERA in 1917 and then elected president in 1925. In 1927, he testified at the *Buck v. Bell* Supreme Court case, which upheld Virginia's right to forcibly sterilize individuals against their will (Engs 2005) (see below).

In 1922, Madison Grant and five other leading eugenicists (including Davenport and Osborn) met at the American Museum of Natural History in New York and formed the Eugenics Committee of the United States of America (ECUSA). This began a permanent eugenics organization that was determined to "disseminate the eugenics ideal throughout society and thereby alter the course of history" (Spiro 2009, 180). The members of the ECUSA carefully and patiently proceeded to do this. In 1922, they estab-lished an Advisory Council of individuals who had responded to a letter asking them to serve on the council and thereby protect America "against indiscriminant immigration, criminal degenerates, and race suicide" (Spiro 2009, 181). Ninety-nine individuals, including many of the usual suspects, responded to the call. The Advisory Council was made up of highly edu-cated, wealthy, influential, old Puritan stock Americans. The ECUSA then began to solicit general membership, and by 1924 there were 1,200 mem-bers. Funds also were being raised from major donors such as George Eastman and John D. Rockefeller Jr. using the argument that eugenics would decrease the need for charities (Spiro 2009).

In 1926, the ECUSA was dissolved to form a more permanent organiza-tion, and the AES was officially incorporated in Madison Grant's Man-hattan town house. The organizers had taken their time because they be-lieved that eugenics was not a passing fad but would become the driving force in improving the human race forever. The eugenics movement, in their minds, was "destined to become increasingly significant in molding the history—or more specifically the germ plasm—of future generations of man-kind" (Spiro 2009, 183).

As Spiro (2009, 183–184) summarized the situation, "There were now three major eugenics groups in the United States.... While each group had its specific task to perform, they featured similar boards and overlap-ping membership lists.... The members of the American Eugenics Society were extremely active in the 1920s. They were missionaries, and their goal was a society permeated with the eugenic ideal. They held meetings, wrote reports, made surveys, published pamphlets, provided articles to newspapers, sponsored lectures, and served enthusiastically on the various committees of the AES." Many of these organizations shared members of their advisory committees (Spiro 2009). This relatively small group of racists had a major influence on American policy during the early to mid-twentieth century and negatively affected the lives of an enormous number of people during this time and, as we shall see, even into the present.

Selective Breeding

> Apart from migration, there is only one way to get socially desirable traits into our social life, and that is by reproduction: there is only one way to get them out, by preventing their reproduction by breeding. (Davenport and Laughlin 1915, 4)

In his original conception of eugenics, Galton developed the basic views of selective breeding for this movement. As mentioned in chapter 2, the strategy included both "positive eugenics," in which the fittest members of society would be encouraged to have more children, and "negative eugenics," which discouraged or disallowed the propagation of the unfit. Restriction of marriages between certain types of individuals began even before the eugenics movement and before Darwinism and Mendelian thinking were combined. Throughout history people had been prohibited from marrying for reasons related to kinship, religion, economics, culture, health, age, ethnicity, and "race." In Europe, marriage bans generally focused on close familial relationships, serious health conditions, and age categories, and marriage among persons of different social status was frowned upon. However, in the United States during slavery and after the Civil War, marriage bans were mainly racially driven. By 1913, twenty-nine of the forty states had laws forbidding race mixing. Nineteen prohibited marriage between African Americans and whites; eight added Chinese and Japanese; one included African Americans and "Croatian" Indians; and one (Nevada) did not allow marriage between whites and "Ethiopian, Malay, Mongolian, or American Indian races" (Farber 2011). Some eugenic marriage laws focused on public health. Connecticut, for example, prohibited any epileptic, imbecile, or feeble-minded person from marrying (Lombardo 2008).

With Darwin's theories, Mendelian genetics and Weismann's experiments, eugenicists had modern science to justify such marriage restrictions and push for their expansion. As Davenport (1912, 89) explained, "If we study the pedigrees of [asocial] men carefully (and many of them have been studied for seven generations) we trace back a continuous trail of the defects until the conclusion is forced upon us that the defects of this germ plasm have surely come all the way down from man's ape-like ancestors, through 200 generations or more." Improving environment was no longer seen as a possible cure for problems caused by the poor, the criminal, the uneducated, the physically and mentally unhealthy, and the inferior "races" of the world. From the ERO, trained field workers were sent out throughout the United States to collect anthropometric data and family histories from hospitals, asylums, prisons, charity organizations, schools for the deaf and blind, and institutions for the feeble-minded. The goal was not purely racial; the goal was to identify and put on record the complete family pedigree of any individual or group of individuals Davenport and Laughlin considered to be physically, medically, morally, culturally, or socially inadequate, thus creating a massive underclass of "unfit" individuals and their families. It was estimated that initially this would include approximately 10 percent of the U.S. population (millions of Americans). By the date of its closure, December 31, 1939, the ERO had collected approximately 750,000 records on individuals and families (Black 2003; Lombardo 2008; Spiro 2009). The data collected included hundreds of physical, occupational, and mental characteristics that were listed in a trait book the field workers carried that had been written by Davenport, with the assistance of psychologists E. L. Thorndike and Robert Yerkes (Paul 1995).

Ten groups were targeted as "socially unfit": (1) the feeble-minded; (2) paupers; (3) alcoholics; (4) criminals, including petty criminals and those jailed for nonpayment of fines; (5) epileptics; (6) the insane; (7) the constitutionally weak; (8) those with specified diseases; (9) the deformed; and (10) the deaf, blind, and mute (with no indication of the severity of these disabilities). The remedies proposed to eliminate this inferior "germ plasm" included restrictive marriage laws, compulsory birth control, forced segregation, sterilization, and euthanasia, although it was believed that it was too early to implement the last one (Laughlin 1914a). Goddard, Laughlin, and Davenport all believed that stopping the reproduction of the unfit would greatly reduce their numbers within a few generations (Paul 1995). As was the case during the Spanish Inquisition, the property of the incarcerated could be acquired and sold to help defer the costs of these programs. From 1912 well into the 1930s, eugenicists worked hard to pass marriage legislation. By the mid-1930s, there were laws prohibiting marriage of alcoholics in four states, of epileptics in seventeen states, and of the feeble-minded in forty-one states. Laws also existed in many states that forced newlyweds to post bonds of up to $14,000 (equivalent to over $130,000 today) to cover the cost of producing infants with blindness, deafness, epilepsy, feeble-mindedness, or insanity. The money could be returned once the wife was past reproductive age (Black 2003; Spiro 2009).

Davenport's ERO also collected pedigrees on eminent and racially acceptable families as part of the eugenics strategy to increase the numbers of the superior classes. To do this, polygamy and systematic mating was suggested, as well as financial incentives to increase propagation in acceptable families. Just before the United States entered World War I, in 1917, Davenport and Laughlin's *Eugenical News* praised Germany's eugenic plan to breed a superior race after the war in order to replace fallen German soldiers (Davenport and Laughlin 1917). This plan included setting up special apartment buildings for desirable single Aryan women and giving cash payments to good Aryan men and women as inducement to have babies (Black 2003). As part of the initiative of positive breeding, Mary T. Watts, popular education chair of the committee for propaganda of the AES, had been running better babies contests at state fairs since 1911. In 1920, Davenport convinced her that even prize-winning babies could develop epilepsy by age ten and, thus, that phenotype was not as important as genotype. It was better to award an entire family as a unit for its eugenic fitness. In 1920, the first Fitter Families for Future Firesides Competition was held at the Kansas Free Fair. Each contest was to be judged by a committee consisting of a historian, a pediatrician, a psychiatrist, a psychologist, a dentist, a clinical pathologist, and an ear, nose, and throat specialist (Spiro 2009). These competitions quickly spread and were held in state fairs throughout the country during the 1920s. As Mary Watts explained: " 'While the stock judges are testing the Holsteins, Jerseys, and whitefaces in the stock pavilion, we are judging the Joneses, Smiths, & the Johnsons,' and nearly everyone replies: 'I think it is about time people had a little of the attention that is given to animals' " (quoted in Paul 1995, 11). Davenport and the AES supplied "instructions, rules, scorecards, and equipment, and convinced important politicians to present the winners with a large bronze medal (designed by Madison Grant) depicting the logo of the AES" (Spiro 2009, 186; see Figure 3.2). An image of the medal became the logo of the journal *Eugenics* (Engs 2005).

Although the eugenics effort was not specifically focused on race, eliminating inferior races was always a major goal of the movement. To eugenicists, "unfit" groups included the non-Nordic races and any children produced by a mixture of these races with the superior whites. In 1913,

Figure 3.2 The Fitter Families medal given by the American Eugenics Society for families meeting certain standards of "fitness." The medal is about the size of a silver dollar.
Photo by the author.

Davenport reviewed various state laws on race mixing and stated that "no subject can be so threatening to the social order that it may not be fully discussed to the advantage of society" (Davenport 1913, 31). He went on to stress that intermarriage between African Americans and eastern Europeans, whose genetic stock was inferior, with "Americans" threatened the future of the country (Farber 2011). In 1919, distinguished geneticists Edward M. East and Donald F. Jones, like Davenport, believed that admixture of an inferior stock with a superior one led to a lowering of the superior stock (Farber 2011). Reverting to the philosophy of Hume and Kant, they wrote (East and Jones 1919, 253): "In reality the Negro is inferior to the white. This is not hypothesis or supposition; it is a crude statement of actual fact. The Negro has given to the world no original contribution of high merit. By his own initiative in his original habitat, he has never risen. . . . In competition with the white race, he has failed to approach its standard."

The anti-miscegenation law of Virginia is an excellent example of this racist motivation. As a prelude to the law, in 1916, Alexander Graham Bell, an avid eugenicist and chairman of ERO's board of scientific advisors, suggested that the United States Bureau of the Census begin assisting the

ERO in its efforts to collect data on family lineages by adding father's and mother's names to individual records. The census bureau refused. Soon thereafter, in 1918, Laughlin proposed that the census bureau add surveys of all custodial and charitable facilities and jails. This was agreed to and Laughlin was named special agent of the Bureau of the Census. The bureau had been collecting data on what was referred to as the defective (insane), the dependent (elderly and infirm), and the delinquent (prisoners) since 1880. Laughlin tried to get them to change their terminology to "the socially inadequate," including "adding stratified contingents of the unfit, especially along racial lines" (Black 2003, 159). The census bureau did not agree, and after a number of years of battle between the bureau and the ERO and ERA (which urged the census bureau to create a massive registry of fit and unfit American citizens), the bureau "simply refused to join the movement," one of the few federal organizations to do so (Black 2003, 161).

Laughlin, unable to get the federal census bureau to accept his eugenics classifications, began working with the House of Representatives and with state governments. Virginia was one of the states that was eager to help the eugenics movement achieve its goals. In Virginia, Laughlin had the assistance of Walter A. Plecker (1861–1947), a radical racist and eugenicist. Plecker, an obstetrician, held the office of registrar in Virginia's newly formed Bureau of Vital Statistics from 1914 to 1942 (Lombardo 2008). His main interest was to maintain Virginia's racial purity and prevent racially mixed marriages. He wrote to Laughlin in 1928, "While we are interested in the eugenical records of our citizens, we are attempting to list only the mixed breeds, who are endeavoring to pass into the white race." Thus, while "carrying the banner of eugenics, Plecker's true passion . . . was always about preserving the purity of the white race" (Black 2003, 165). He believed that existing state laws were too permissive; they were too vague about what constituted a "Negro" or "colored" person. Different states barred marriage between whites and persons who were half, one-quarter, or one-eighth black.

Plecker and a few of his white supremacist friends, who called themselves the Anglo-Saxon Club, began to work on new legislation that would ban marriage between whites and any person with even "one drop" of non-Caucasian blood. They had the backing of many well-known eugenicists around the nation, including the members of the ERO, Madison Grant, and Lothrop Stoddard. They also had a powerful ally in the state's leading newspaper, the *Richmond Times-Dispatch*. While the legislation was being debated in 1924, the newspaper published an editorial stating that

"America is heading towards mongrelism. . . . Thousands of men and women who pass for white persons in this state have in their veins Negro blood. . . . It will sound the death knell of the white man. Once a drop of inferior blood gets in his veins, he descends lower and lower in the mongrel scale" (quoted in Black 2003, 167). Virginia's Racial Integrity Act was passed in March 1924, and falsely registering one's race became a felony, punishable by imprisonment of from one to five years. This law survived for over forty years, until 1966 (Lombardo 2008). As Virginia's registrar, Plecker's pursuit of nonwhites continued throughout his life. He worked not only to prohibit marriages between white and nonwhites but also to prohibit nonwhites from attending white schools, from traveling in white railroad coaches, and from being buried in white cemeteries.

Plecker became an immediate hero among racists and eugenicists across the nation. Articles about him and by him were published in eugenics journals and newspapers and in public health literature. In 1925, the *American Journal of Public Health* considered the Virginia law "the most perfect expression of the white ideal, and the most important eugenical effort that has been made in the past 4,000 years" (quoted in Black 2003, 174). Plecker then began campaigning for similar "one drop" laws in other states. Copies of the Racial Integrity Act were distributed to all the governors in the United States with a personal letter from Virginia's governor requesting that similar legislation be introduced in their states. Alabama and Georgia were the first two states to follow, and Wisconsin attempted to pass a similar law. However, due to increasing civil rights activism, other states were slower to follow. Laughlin, hoping to speed up the process, asked Plecker to prepare a chart for *EN* entitled "Amount of Negro Blood Allowed in Various States for Marriage to Whites" (Laughlin 1928). Plecker continued to work on ensuring racial purity even after he retired in 1946 at the age of eighty-four, by publishing racist pamphlets that decried mongrelization and defended the purity of the white race. Plecker "dictated the nature of existence for millions of Americans, the living, the dead and the never born. . . . [He] defined the lives of an entire generation of Virginians—who could live where, who could attend what school and obtain what education, who could marry whom, and even who could rest in peace in what graveyard" (Black 2003, 182). The first state to ban an anti-miscegenation law was California in 1948. However by the 1960s, seventeen states still had anti-miscegenation laws prohibiting interracial marriage. In fact, it was still legal in the United States for states to ban interracial marriage until 1967, when the Supreme Court ruled in the landmark *Loving v. Virginia* case that anti-miscegenation laws were unconstitutional (Farber 2011).

Sterilization of the "Unfit"

Sterilization of those deemed unfit, either physically, mentally, socially, or racially, was one of the major weapons of eugenics in their "war against the weak" (Black 2003). Even before 1900, sterilization was used as a means of controlling crime. The first recommendation for sterilization was made at the Cincinnati Sanitarium in 1887 as a punishment and as a way of controlling criminal proclivities (Degler 1991). However, because eugenicists saw crime as one of those unitary genetic characteristics that defined the unfit, it soon became obvious that sterilization was also justified as a general practice to prevent the propagation of unfit individuals, family lineages, ethnic groups, and races. Furthermore, it was generally believed that the feeble-minded and other unfit individuals would breed with no regard for the consequences, would have larger families, and thus would outnumber normal families (Paul 1995).

However, as there was still public, religious, and, in the South, conservative anti-science sentiment against this invasive, irreversible means of birth control (Larson 1995; Lombardo 2008), it was difficult to get legislation passed. In 1897, a sterilization law that was introduced in the Michigan state legislature was defeated. When the Pennsylvania legislature enacted a similar law, it was vetoed by the governor. A similar law was introduced in Kansas, but it failed to be approved (Lombardo 2008). The first law permitting sterilization was enacted in 1907 in Indiana. In 1909, three more states, Washington, Connecticut, and California, passed laws that ratified eugenics sterilization, and New Jersey and New York passed laws in 1911 and 1912, respectively. The New Jersey law was signed by eugenics enthusiast Woodrow Wilson, who was governor at the time. However, in the early 1900s, Oregon, Illinois, and Wisconsin had rejected sterilization laws, and none of the conservative southern states were in favor of these involuntary sterilization laws. Indeed, the sterilization movement was liberal, progressive, scientific, and essentially elitist (Degler 1991; Larson 1995). Thus, by the time of the First International Eugenics Congress in 1912, the eugenicists of Cold Spring Harbor, the American Breeders Association (ABA), and the ERO were extremely frustrated. Even in the states where sterilization was legal, few Americans had actually been sterilized involuntarily. In 1914, in the wake of the success of the first eugenics congress in publicizing and justifying the eugenics movement and its goals, Laughlin formulated a model sterilization law that was used by many states and, later, in Germany to prepare sterilization legislation. (A similar phenomenon is occurring today, as Kansas secretary of state Kris Kobach drafts model immigration laws; see Chapter 11.) By 1915, although fifteen states had enacted sterilization laws, four of those had been repealed, revoked, or declared unconstitutional (Engs 2005). By 1921, only ten sterilization statutes were still in active use. New or revised laws were introduced in Illinois, Minnesota, New Hampshire, and Ohio, but each of these failed (Lombardo 2008).

Thus, in the early 1920s, many state officials were reluctant to enforce sterilization laws. Earlier, a report of the ABA had noted that these laws had been pushed by a "very small energetic group of enthusiasts, who have had influence in the legislatures . . . [but] public sentiment demanding action was absent. Law officers of the state were not anxious to undertake defense of a law the constitutionality of which was questioned" (Van Wagenen 1912, quoted in Black 2003, 70; see also Larson 1995; Lombardo 2008). In 1914, Laughlin lamented that

> the present experimental sterilization laws have been pioneers—pointing the way—and as such they are to be commended, but as remedies for social deterioration they have not thus far, in a national way, functioned. Indeed less than 1,000 sterilizing operations have been performed under the immediate provisions or even under the shadow of the twelve statutes. . . . Unless all of the states co-operate in the purging of the blood of the American people of its bad strains, and their co-operation is supported by the federal government . . . as indicated in the calculations, we should not expect the program to work out as calculated (1914b, 145–146).

Even by 1922, only 3,200 sterilizations had been performed on inmates of prisons, insane asylums, homes for the epileptic and feeble-minded, and other institutions of social welfare. Almost 80 percent of these were carried out in California (Lombardo 2008). In fact, the "progressive" state of California was a model that eugenicists in other states sought to emulate (Larson 1995), but concern over legislation had slowed the pace of sterilization operations even in California. Laughlin had attempted to resuscitate interest in sterilization laws and policies in 1922 by publishing a 502-page book, *Eugenical Sterilization in the United States*, which included copies of his model sterilization laws. However, the eugenics community was still frustrated by the limited use of sterilization in the United States and soon would test the constitutionality of sterilization operations in the Supreme Court.

Virginia once again served as the staging location for eugenics legislation. Albert Priddy was the first superintendent of the Virginia State Colony for Epileptics and Feebleminded, which opened in 1910 in Lynchburg, Virginia. In 1916, Virginia passed a law allowing colony residents to be sterilized before being discharged, if such operations were safe, effective, and approved by the court. This would ensure that society would be protected from potentially degenerate progeny (Lombardo 2008). Upon passage of this law, Priddy, a staunch supporter of eugenics, began to sterilize women in his colony. The goals and scope of Priddy's program were to prevent female inmates who had trouble with alcohol or were considered to be insane, defective, weak minded, feeble-minded, incorrigible, wayward, backward, illegitimate, homosexual, untruthful, criminal, immoral, promiscuous, or oversexed or who had wanderlust from reproducing. All of these traits were assumed to be hereditary. Priddy and his colony "had 'a duty to society' to prevent such girls from being set free 'to breed defectives' " (Lombardo 2008, 63).

However, Priddy's sterilization program was frustrated and temporarily suspended when he was sued for the harm he had inflicted on a mother and daughter, Willie and Jessie Mallory. He had sterilized these women without a proper court order; skirting the law in this fashion had become a habit with Priddy. When the state court of appeals did not dismiss the lawsuit, Priddy wrote to the court claiming that "there is no use in the State trying to maintain an institution for defectives of the criminal class" if the suit was successful (quoted in Lombardo 2008, 73).

Priddy sought support from his friend Irving Whitehead, a lawyer who was a member of the board of directors of the Virginia colony that had originally appointed Priddy as superintendent. Whitehead also had endorsed Priddy's expansive interpretation that Virginia's law authorized the sterilization of inmates, and he had initiated the board's vote to approve the sterilization of Mallory and her daughter. Whitehead thought the *Mallory v. Priddy* suit was a "splendid case to test the law" (Lombardo 2008, 74). However, the court found that fundamental civil rights were abused when sterilization was performed without permission and that the Mallory surgeries had been performed without consent. Although Priddy lost the case, the jury refused to award damages to the Mallory family. The trial judge, however, warned Priddy to stop operating until the existing law was changed. The lawsuit "frightened all the superintendents in the State and all sterilization was stopped promptly. . . . Without a law that would protect state doctors from being dragged into court, it looked as if a robust plan for eugenical sterilization had no future in Virginia," and this served as a warning for doctors in other states (Lombardo 2008, 76–77).

Priddy turned to Audrey Strode, the colony's advocate in the Virginia General Assembly, in an attempt to remedy his legal vulnerability. Strode was a eugenicist and a boyhood friend of Irving Whitehead. Strode quickly drafted two pieces of Virginia legislation that protected Priddy and other doctors from future lawsuits. However, doctors were still reluctant to perform operations until a new sterilization law was in effect. In 1923, after a new governor, E. Lee Trinkle, had been elected, public sentiment about sterilization in Virginia changed. Trinkle was a longtime friend of Audrey Strode and a strong supporter of eugenic sterilizations. In 1924, he told Strode to draft a new sterilization bill. Strode drew upon the language and arguments from Laughlin's recently published *Eugenical Sterilization in the United States* as he drafted the new law. In many places, he used the language of Laughlin's model sterilization law verbatim (Lombardo 2008). He was careful in laying the political groundwork of the law, and the bill quickly and unanimously passed the Senate on June 17, 1924.

Priddy was anxious to begin sterilization surgery in his colony once again, but Strode advised him not to perform any more sterilizations until the constitutionality of the Sterilization Law could be tested in the Supreme Court of Appeals of Virginia and possibly in the U.S. Supreme Court. Strode was retained to litigate the test case. From Priddy's list of sixteen women candidates for sterilization, the eugenics advocates chose one to be the test case for the law, eighteen-year-old Carrie Buck. By the time she was chosen for the test case, Carrie had been in the colony for three months. Under Priddy's direction, the colony's board voted to sterilize her. In order to complete the sterilization procedure, the board was required to appoint an attorney to appeal the sterilization order and to defend Carrie in the litigation. Irving Whitehead was chosen. Thus, Carrie was to be defended in all of her legal proceedings by a close confidant and friend of Priddy, a boyhood friend of Audrey Strode, a former director of the colony, and an avid sterilization advocate (Lombardo 2008).

On January 23, 1924, after a brief hearing (inquisition might be a better word), Virginia's justice of the peace, C. D. Shackleford, condemned Carrie Buck as feeble-minded. Miss Buck was actually not a bad student and came from a family of average or above-average students. However, at the time of her inquisition, she was considered "poor white trash" from the back streets of Charlottesville, Virginia. Her mother, Emma, who had given birth to Carrie in 1906, had become poverty stricken after her husband died in 1909. In 1920, Shackleford sent Emma to the Colony for Epileptics and Feebleminded with an initial diagnosis of "Mental Deficiency, Familial: Moron" (Lombardo 2008). She spent the remainder of her life there. John Dobbs and his wife, Alice, who had a child who was close to Carrie's age, took the girl into their home to help with chores, although they did not adopt her. She performed adequately in school and in her housework. However, in 1923, at the age of seventeen, Carrie was discovered to be pregnant. She insisted that she had been raped by Dobbs's nephew. However, the Dobbses were shamed and wanted Carrie out of the house. John was temporarily deputized as a "special constable" in Charlottesville and filed commitment papers with Justice Shackleford, claiming that the girl, like her mother, was feeble-minded, epileptic, or both. Carrie was quickly declared feeble-minded and sent to the Virginia colony, as her mother had been earlier.

On March 28, Carrie gave birth to a daughter, Vivian, but since she had been declared mentally incompetent, she could not keep the baby. Ironically, the Dobbs family took the girl, with the agreement that she would be returned to the colony if she was deemed to be feeble-minded (Black 2003; Lombardo 2008). On September 10, 1924, under the new Virginia law, Carrie was condemned to sterilization. By this point, Robert Shelton, the attorney for the Colony for Epileptics and Feebleminded, had been appointed as Carrie's legal guardian. It was his responsibility to initiate an appeal on her behalf "in order that we may test the constitutionality through our state courts, even to the Supreme Court of the United States" (quoted in Black 2003, 114). The case was a setup by the eugenics advocates from the start. Audrey Strode was the prosecuting attorney and Irving Whitehead was appointed by Shelton to represent Carrie. Laughlin provided a long written deposition that relied on IQ test results for Carrie, her mother, and her seven-month-old baby to illustrate that three generations of the Buck family were feeble-minded. In it he stated that "the family history record and the individual case histories . . . demonstrate the hereditary nature of the feeblemindedness and moral delinquency described in Carrie Buck. She is therefore a potential parent of socially inadequate or defective offspring" (Lombardo 2008, 135). After less than five hours, the court decided in favor of Priddy and upheld the new Virginia sterilization law. Whitehead had presented an extremely weak defense. As

Lombardo (2008, 148) states: "He did not fail in advocacy of Carrie Buck simply because he was incompetent; Whitehead failed because he intended to fail." Whitehead reported the decision to a joint meeting of the State Board of Hospitals and the colony board but reminded them that the matter would eventually be put before the U.S. Supreme Court for a final decision.

After Virginia appeals courts rejected weak appeals, the case was prepared for the U.S. Supreme Court. Whitehead presented the status of the test case and the conclusive argument that it should be heard in the U.S. Supreme Court, stating that it was in "admirable shape to go to the court of last resort, and that we could not hope to have a more favorable situation than this one" (quoted in Black 2003, 117). By this time, Priddy had died of Hodgkin's disease, and Dr. John Bell, Priddy's successor as colony superintendent, was named as the defendant in the lawsuit, which then became *Buck v. Bell*. Thus, the fate of Carrie Buck and indeed, that of a major policy of the eugenics movement and of an unfathomable number of people, rested with nine men. Furthermore, the defense of the case was in the hands of Whitehead, who had "violated every norm of legal ethics . . . [acting] as if his real client was not Carrie Buck but his now-deceased friend Albert Priddy, whose sterilization program he had supported over a decade" (Lombardo 2008, 155).

The chief justice of the Supreme Court in 1927 was Oliver Wendell Holmes Jr. (1841–1935). Holmes, who had been appointed chief justice by Teddy Roosevelt in 1902, was considered "the most celebrated judge in America" by 1931 (Lombardo 2008, 163). However, he advocated Social Darwinism, eugenics, and the ideas of Malthus. In 1922, Holmes stated: "All society rests on the death of men. If you don't kill 'em one way you kill 'em another—or prevent them from being born. . . . Is not the present time an illustration of Malthus" (quoted in Black 2003, 120)? Although most people in the United States had turned against radical proposals of eugenicists such as euthanasia, Holmes still held to the radical views of Laughlin and some of his colleagues. He had no compulsion about state-controlled euthanasia of weak and inadequate "undesirables" (Lombardo 2008). On May 2, 1927, shortly after the Virginia "one drop of blood" marriage legislation was passed, the Supreme Court, with only one dissenting vote, approved the Act of Virginia that stated that "the health of a patient and the welfare of society may be promoted in certain cases by the sterilization of mental defectives." Holmes wrote the opinion of the majority (*Buck v. Bell*, 274 U.S. 200, 1927), stating that "it is better for all the world, if instead of waiting to execute degenerate offspring for crime, or let[ting] them starve for their imbecility, society can prevent those who are manifestly unfit from continuing their kind. The principle that sustains compulsory vaccination is broad enough to cover cutting the Fallopian tubes. Three generations of imbeciles are enough."

Edwin Black (2003, 122) summarized the situation: "Eugenical sterilization was now the law of the land. The floodgates opened wide." By 1930, thirty states had sterilization laws and approximately 36,000 people had been sterilized, 30,000 of these after 1927. Grounds for sterilization included being judged feeble-minded, insane, medically unacceptable, morally degenerate, and criminal. Some were classed as "other" and others were sterilized for being poor. Just as during the Spanish Inquisition, the trials of these "unfit" individuals were often termed inquisitions, characteristics of guilt were ambiguously defined, and evidence was vague, scanty, and mainly based on hearsay. Property was confiscated and families were ruined.

The first sterilization law in Europe was passed in Denmark in 1929. Germany passed its first compulsory sterilization laws, precursors to its practices of euthanasia, in 1933. Other sterilization laws were passed in the Canadian province of British Columbia (1933), Norway (1934), Sweden (1934), Finland (1935), Estonia (1936), and Iceland (1938), and Denmark made its voluntary program compulsory in 1934. These laws corresponded with deteriorating economic conditions worldwide (Paul 1995). The German law was based, essentially word for word, on Laughlin's model sterilization law (Brace 2005). By the end of World War II in 1945, Germany had sterilized approximately two million people without their consent (Chase 1977). At the same time that the United States was using applied genetics as the basis of its sterilization policy and racial eugenics to support its immigration policies (see below), Hitler and the Nazis began using racially driven eugenics policies first to deport and then to exterminate certain ethnic and racial groups. As we shall see, the racial classifications of Hitler's Germany were those invented by the racists of the nineteenth century (Nott, Gobineau, Chamberlain, Ripley) and resurrected in the early twentieth century by the leaders of the eugenics movement (Shaler, Davenport, Brigham, Grant, Stoddard). Although Germany was the first to use euthanasia for its racist eugenics, the idea was first proposed in 1911, as point 8 in the ABA's plan for eliminating the defective germ plasm in the American population (Black 2003, 60). By the end of World War II, the Nazis had exterminated over six million people.

Intelligence Testing and Madison Grant's *Passing of the Great Race*

Although Gobineau was convinced of the correctness of his racist worldview, in the final chapter of his *Essai*, he called for scientific measurement of racial differences. The fact that Humeian/Kantian claims of the superiority of Western and especially Aryan civilization were often countered with tales of the accomplishments of individuals of "inferior races" worried him. He wrote (quoted in Gould 1996, 381–382):

> In the preceding pages, I have endeavored to show that . . . the various branches of the human family are distinguished by permanent and ineradicable differences both mentally and physically. They are unequal in intellectual capacity, in personal beauty, and in physical strength. . . . The discussion has not rested upon the moral and intellectual worth of isolated individuals.
>
> I am prepared to admit . . . nay, I go farther than my opponents, and am not in the least disposed to doubt that, among the chiefs of the rude negroes of Africa, there could be found a considerable number of active and vigorous minds, greatly surpassing in fertility of ideas and mental resources the average of our peasantry and even of some of our middle classes.
>
> Once and for all, such arguments [about individuals] seem to me unworthy of real science. . . . Let us leave such puerilities, and compare, not the individuals, but the masses. . . . The difficult and delicate task cannot be accomplished until the relative position of the whole mass of each race shall have been nicely, and so to say, mathematically defined.

One of Sir Francis Galton's major goals was to fulfill Gobineau's wish and get real measurements of the masses—scientific, mathematical proof of the inequality of the races and of the superiority of the Anglo-Saxons (Gobineau 1853–1855; Gould 1996). He believed that the mental abilities of both individuals and races could be ranked according to a single objective measurement called "intelligence" (Brace 2005; Spiro 2009), even going so far as to rank the intelligence of some breeds of dogs higher than that of some Englishmen and most Africans (Galton 1869). To achieve his goal of quantifying human differences, Galton established an anthropometric laboratory in 1884 and collected detailed measurements of Londoners. He recorded the physical characteristics such as height, weight, arm span, and lung power of some 9,000 individuals and families (Shipman [1994] 2002; Black 2003). He was among the first to attempt to invent a number of tests designed to identify levels of intelligence. However, in this he was not successful. His tests actually measured only physiological reaction time to certain stimuli and did not test mental activity, ability, or reasoning (Gould 1996; Smedley 1999).

In fact, it was Alfred Binet (1857–1911), a French psychologist at the Sorbonne, who first developed a test to measure reasoning ability. In 1904, Binet was commissioned by the French minister of public education to develop a means of identifying children who were performing badly in school and might need some form of special education. Along with his student Theodore Simon (Binet and Simon 1905, 1908), Binet developed a series of tests that combined a number of tasks related to different abilities in order to arrive at a single score that indicated the child's general potential. Unlike Galton, however, Binet did not believe that he was measuring a single, inherited, and unchangeable entity labeled intelligence (Zenderland 1998). In fact, he warned against such a simplified view. The single score that he had developed was only a rough, empirical guide to be used for a practical purpose, that of improving the education of those in need. Binet believed that intelligence was too complex to be measured by a single score. He insisted that intelligence could not be measured on a scale like height and he worried that if the practical score he had developed were to be reified as a unitary entity, it could be "used as an indelible label, rather than as a guide for identifying children who needed help" (Gould 1996, 181). As Brace (2005) has commented, unfortunately, the cautious and nuanced interpretation of Binet's approach to measuring mental capabilities did not survive his death in 1911.

It was American psychologists who ignored Binet's warnings and reverted to Galton's belief that intelligence is inherited in a simple fashion and could be measured by a single score. Unfortunately, the IQ test has been used by many in such a misguided fashion ever since (see Fish 2002; Cravens 2009). Just as Gobineau and Galton did, the avid American eugenicists Davenport and Goddard craved more scientific means of measuring those individuals and races that they deemed "unfit." When Goddard, the author of the book on the Kallikak family, began using similar intelligence tests in the United States, he ignored Binet's warnings, using the test scores as if they identified a simple unitary innate intelligence. He also used his findings not to improve the education of those scoring low on his scales but rather to label them, assuming that intelligence was an unchangeable, biologically fixed characteristic. By 1914, he believed that he had determined the cause of the mental condition of every person in his Vineland institution and had stated that it was necessary to prevent the biological and sociological consequences of poor heredity (Goddard [1914] 1973). Finally the eugenics movement believed it had a simple measurement that made it possible to identify the mentally and morally "unfit" individuals and races they so intensely wanted to isolate in colonies, castrate, sterilize (and even euthanize), and prevent from immigrating into the United States. As Zenderland (1998, 262) notes, "American intelligence testers had turned their attention from diagnosing children to classifying adults. They had also moved far beyond the population of "defectives, dependents, and delinquents," for they now considered themselves capable of gauging the minds of everyone. Psychological science, testers insisted, had much to say about the intelligence of the entire American nation, and about racial, ethnic, and class differences found within it."

Goddard, as exemplified in his book on the Kallikaks, was worried about the "menace of the feebleminded," individuals with the mental ages of eight- to twelve-year-olds. He believed that these individuals were "incapable of adapting themselves to their environment and living up to the conventions of society or acting sensibly" (Goddard [1914] 1973, 571) and that they lacked good judgment and could not make sound moral decisions. For this reason, he considered most criminals, alcoholics, prostitutes, poor people, and even common laborers to be feeble-minded. In this light, Goddard considered that the intelligence score of the average American was equal to that of a twelve-year-old child and that 45 percent of Americans were feeble-minded or of the "moron" class (Goddard 1919, 1920). Foreign-born immigrants were considered to be much worse than native-born people. With these considerations already in mind, and with the very successful First International Eugenics Congress, Goddard set out to measure individuals in institutions with his translated Binet tests and to prove that those immigrating into the United States were indeed feeble-minded.

In 1910, the United States Public Health Service invited Goddard and Johnstone to observe the procedures it used for recognizing and detaining "mentally defective" immigrants entering the United States at Ellis Island, New York (Zenderland 1998). In 1912, Goddard did some preliminary intelligence testing at Ellis Island, but he ran out of funding. In a letter to Davenport on July 25, 1912, the opening day of the first eugenics congress, Goddard asked Davenport for financial assistance for work that would prove the reliability of using intelligence testing in the field so it would be possible to connect feeble-mindedness to heredity (Black 2003). In 1913, he received funding and an invitation from the United States Public Health Service to continue administering intelligence tests at Ellis Island. Given the contents of the tests and the logic behind them (see Gould 1996; Zenderland 1998; and Black 2003), Goddard of course found his predictions to be correct. Eighty-seven percent of Russians, 80 percent of Hungarians, and 79 percent of Italians tested as feeble-minded (Goddard 1913). Russians and Poles scored among the lowest, and these designations were at least in part euphemisms for Jews, of which Goddard's test showed that 83 percent were feeble-minded (Zenderland 1998; Brace 2005). Deportation of these immigrants increased significantly between 1912 and 1914 (Zenderland 1998). As Brace has stated (2005, 213), to the eugenicists "the results of his survey were taken to indicate a spectrum of inherent intellectual worth in the 'races' of Europe." Goddard's research results provided material for the next influential tome that was written to justify and support the blatant racism of the eugenics movement: Madison Grant's *The Passing of the Great Race* (1916).

Madison Grant (1865–1937), as were many of the elitist eugenicists of his day, was born into a wealthy family that was very impressed with its own heritage. One might say they considered themselves among the American aristocracy. Much as Gobineau had deplored the French Revolution for dethroning the aristocrats of France, Grant thought that democracy and the idea of equal rights would lead to the end of civilization. In democracy, the "genius of the small minority is dissipated" and power is shifted from the Nordic "race" to racial inferiors (Grant 1916; Paul 1995, 104). Grant was the eldest son of a very rich and distinguished family whose forbearers were Puritans who had helped settle New England and New York. He was a Yale graduate with a law degree from Columbia, though he never needed to earn an income and never practiced law. Being a big game hunter and a friend of Teddy Roosevelt, he was very involved in the early conservation movement in the United States (Paul 1995). Although he was not trained in biology, Grant's main vocation, besides writing on hunting, conservation, eugenics, and racism, was founding and chairing the New York Zoological Society and helping to create and maintain the Bronx Zoo. Even though Grant had no scientific training, he and his friend, colleague, and (to some extent) mentor, eugenicist Henry Fairfield Osborn (1857–1935), who was president of the American Museum of Natural History and a professor at Columbia University, were among the most prominent and influential natural historians of the time. Cravens (2009, 152–153) has described Grant as "a tall, handsome man and a natty dresser with a handlebar mustache, piercing eyes, and a straight 'Nordic' nose . . . a symbol for an entire era, stretching from the 1870s to the 1920s and even into the 1930s, of open, brutal racism, segregation, and disparagement of persons of color as well as those from alien lands."

In 1908, William Z. Ripley, author of *The Races of Europe* (see Chapter 1), gave a lecture entitled "The Migration of Races" to an exclusive club of which Grant was a prominent member, the Half-Moon Club (Spiro 2009). Ripley talked about how the dangers of the influx of hordes of Jews and other inferior types into America posed an evolutionary danger to native Anglo-Saxons via intermarriage and a reversion to more primitive types of humanity through hybridization (Spiro 2009). This must have struck a deep chord with Grant the conservationist, who was very worried about the importation of nonnative mammalian species into the United States and how this would lead to the mongrelization and displacement of native species. Ripley asked: "Is it any wonder, that serious students contemplate the racial future of Anglo-Saxon American with some concern? They have witnessed the passing of the American Indian and the buffalo. And now they query as to how long the Anglo-Saxon may be able to survive" (quoted in Spiro 2009, 96). Perhaps this was the stimulus that set Grant off on a new course of interest in human eugenics and race and that stimulated his writing of *The Passing of the Great Race*.

What was the context of Grant's worries? In 1907, the year before Ripley's lecture, 1.28 million immigrants entered the United States. From 1900 to 1908, over 6 million immigrants had come to the United States. Even in the early 1880s, well over 500,000 immigrants entered the United States annually; by 1894, at least 1.4 million of New York's 1.8 million residents had at least one foreign-born parent. New York had more Italians than Rome and twice as many Irish as Dublin, and it was soon to become the largest Jewish community in the world (Spiro 2009). Grant had long been involved in attempting to introduce civic reform to help reduce the "detrimental" effects he believed the immigrants from Eastern Europe were having on New York City. Spiro (2009, 11–12) describes his state of mind at that time:

Grant felt increasingly beleaguered by the waves of swarthy immigrants engulfing his city. They were filling up the almshouses, cluttering the streets, and turning Manhattan into a dirty, lawless, turbulent cacophony of foreign barbarians. . . . Grant was disgusted by what he saw as he braved the sidewalks of his native city. He was repulsed by the bizarre customs, unintelligible languages, and peculiar religious habits of the foreigners. . . .

Grant knew full well that classical Rome had fallen when she opened her gates to inferior races who "understood little and cared less for the institutions of the ancient Republic," and he feared for his country.

Thus, we might suspect that it was Ripley who stimulated Grant to become more involved in human eugenics and racism and begin writing his infamous tome. Grant published *The Passing of the Great Race* in 1916. His thoughts in that volume were greatly influenced by Gobineau, Chamberlain, and Ripley; by Galton on eugenics; and by Davenport's take on Mendelian genetics, although he rarely used citations (Brace 2005). In fact, since the books of the first three were written before the rediscovery of Mendel, Grant was the first of the polygenicists to combine the folk racism of the time from Western Europe and the United States directly to Mendelian genetics.

Grant became more and more active in the eugenics movement. In 1918, he and Osborn organized the Galton Society for the Study of the Origin and Evolution of Man, to which Davenport was elected chair (Chase 1977). Osborn was what might be called a neopolygenicist in the sense described by Stocking (1968; see Chapter 2). For example, he wrote: "If an unbiased zoologist were to descend upon the Earth from Mars and study the races of man with the same impartiality as the races of fishes, birds, and mammals, he would undoubtedly divide the existing races of man into several genera and into a very large number of species and subspecies" (Osborn 1926, 3).

Passing reintroduced racism as a major focus of the eugenics movement. It was stimulated by the goals of the First International Eugenics Congress and became a major tool for eugenicists, in relation to both the use of IQ tests and immigration policy (see later in this chapter). The book received accolades from the political and scientific establishments and became immensely popular. In fact, Hitler considered it his "bible." Grant joined many eugenics committees. He also was appointed by his friend Davenport to chair the Second International Congress of Eugenics, which was finally held in 1923. It was hosted by Grant's friend Henry Osborn at the American Museum of Natural History.

Passing was introduced with a preface written by Osborn in which he spelled out the two friends' elitist philosophy: "In the new world we are working and fighting for, the world of liberty, justice and humanity, we shall save democracy only when democracy discovers its own aristocracy as in the days when our Republic was founded" (Osborn 1918, xiii). In the book, Grant brought eugenics, which had begun to focus on individuals and family lineages, back to a stronger focus on race.

Grant's use of taxonomic terms was careless. He divided mankind into three primary subgroups (sometimes referred to as subgenera of the genus *Homo*): the Caucasians, the Mongoloids, and the Negroids; these then were subdivided into subspecies or races (Grant 1916, 19). According to Grant, the Caucasians, for example, consisted of three subspecies: the Nordics, the Alpines, and the Mediterraneans. However, it seems that he did consider all humans as one species and in fact often used the terms *races, subspecies,* and *subgenera* interchangeably. This classification, ultimately, was based on Ripley's types as outlined in *The Races of Europe* (1899). After *Passing* was published, Grant's classification of human variation was immediately adopted worldwide, with three primary subgroups—Mongoloids, Negroids, and Caucasoids, and Caucasoids divided into three "tertiary" subgroups—Nordics, Mediterraneans, and Alpines (Spiro 2009).

Grant combined the typical eighteenth-century, Humeian methodology of "historical inductive reasoning" (see Chapter 1) with a simple-minded Mendelian eugenicist approach, as described in Davenport's *Heredity in Relation to Eugenics* (1911), in which complex characteristics and behaviors were seen as unitary genetic traits. He assembled a "full-blown eugenics platform, incorporating the Nordicism of Gobineau and the breeding program of Davenport" (Marks 1995, 84). In this conglomeration of "modern" science, he regurgitated the racist folktales of Gobineau, Chamberlain, and Ripley, who in turn had echoed the Mortonites' earlier American version of racism. As Brace (2005, 175) states: "The credibility of that portion of the replacement which dealt with 'race' was enhanced by its assumed European lineage, but its real strength derived from the fact that it was essentially an American turkey come home to roost."

Grant's thesis was that each of the "races" or "subspecies" shared immutable hereditary traits: "The great lesson of the science of race is the immutability of somatological or bodily characters, with which is closely associated the immutability of physical dispositions and impulses" (Grant 1918, xix). For example, Nordics were natural rulers, the apex of the development of the white race; Alpines were always and everywhere the race of peasants; and Mediterraneans were inferior to both Nordics and Alpines in stamina but were superior in the arts. Jews, a subset of Mediterraneans, were small in stature and had a peculiar mentality and ruthless self-interest. According to Grant, their swarming presence in New York

City was crowding out native-born Americans and the more desirable original Nordic inhabitants.

Grant argued that even though two of the Caucasian races were inferior to Nordics, they were all superior to the other subspecies of humans, the Negroes, American Indians, and Mongols (Grant 1918) and that crossing among the subspecies or races always led to degradation. Following the thought of Gobineau, Chamberlain, Ripley, Shaler, and Davenport, Grant wrote that civilization was perpetuated through purity of blood and that civilizations had fallen because of racial mixing and the dilution of superior Nordic blood: "The result of mixing of two races, in the long run, gives us a race reverting to the more ancient generalized and lower type. The cross between a white man and an Indian is an Indian; the cross between a white man and a Negro is a Negro: the cross between a white man and a Hindu is a Hindu; and the cross between any of the three European races and a Jew is a Jew" (Grant 1918, 18).

Grant approved of slavery and echoed the segregation policies of the American South and the earlier views of Nott and Shaler: "As long as the dominant imposes its will on the servient race and as long as they remain in the same relation to the whites as in the past, the Negroes will be a valuable element in the community but once raised to social equality their influence will be destructive to themselves and to the whites. If the purity of the two races is to be maintained they cannot continue to live side by side and this is a problem from which there can be no escape" (Grant 1918, 87–88).

In his approach, Grant presented his views of racial purity and the reversion to primitive types with miscegenation in Mendelian terms, but, as Stocking (1968, 68) noted, his polygenics goes back to pre-Mendelian, pre-Darwinian concepts of human diversity: "Ultimately, they had their basis in the polygenist attempt to apply prerevolutionary concepts of species as absolute, supraindividual, essentially distinct and hierarchical entities to the study of mankind."

Grant's views were simple: recent immigrants into the United States from southern and eastern Europe, especially Poles, Italians, and Jews, were decidedly inferior, physically, mentally, and morally to those who had entered the country in earlier times. Those earlier immigrants, the Nordics, were now on the verge of being outnumbered by the massive influx of these inferior types (Degler 1991). In addition, it was necessary to segregate African Americans and Native Americans from the white population to prevent mongrelization and degradation. Unlike Gobineau, however, Grant did not have an entirely pessimistic view of the future. He offered instead the rational, efficient, and "scientific" remedy of eugenics and anticipated the rise of fascism, the fall of democracy, and the return of power to the patricians. In addition, he foresaw a program of birth control that could reduce the number of offspring of the undesirable classes. Anti-miscegenation laws and segregation of the races would reduce the possibility of mongrelization and protect the purity of the Nordic race, and sterilization on a massive scale would deprive inferior types of the capacity to procreate (Spiro 2009). When one reads Grant's (1918, 47) assessment of eugenics, it is no wonder that Hitler saw *Passing of the Great Race* as his bible: "This is a practical, merciful, and inevitable solution to the whole problem, and can be applied to an ever widening circle of social discards, beginning always with the criminal, the diseased, and the insane, and extending gradually to types which may be called weaklings rather than defectives, and perhaps ultimately to worthless racial types."

Spiro (2009) sees Grant as combining seven disciplines (wildlife management, anthropology, paleontology, the study of race suicide, Aryanism, eugenics, and genetics) into a new amalgam called scientific racism. Scientific racism involves three basic axioms: (1) The human species is divided into distinct, hierarchical subspecies and/or races, with the Nordic race at the top of the hierarchy; (2) intellectual, moral, temperamental, and cultural traits of each race are immutable and correlated to, and inherited with, immutable physical traits, and the genes for these traits are unaffected by the environment; (3) the mixture of races always results in reversion to the primitive, inferior type, and thus eugenic measures must be taken to prevent the degeneration of the superior race. In the minds of its proponents, this scientific racism was new and differed from the past, popular racism in that it purported to employ physical anthropology, Darwinian evolution, and Mendelian genetics to explain why non-Nordic races were biologically inferior. In actuality, this scientific racism, which is still popular in some circles today, used bad anthropology, a misunderstanding of Darwinian evolution, and a simple-minded and poorly understood version of genetics. This led to the same false conclusions that drove polygenics during the Spanish Inquisition: that certain groups of people were inferior, that the characteristics (physical, mental, and behavioral) that led to this inferiority were biologically fixed and immutable, and that no alteration of the environment could change these fixed characteristics. Spiro summarizes the conclusions scientific racists drew from their beliefs: "There was no action you could take that would modify the facts of your heredity. A popular racist might admit (and hope) that a Jew could become a Christian, but a scientific racist would point out that a Jew could never become a Nordic" (Spiro 2009, 140). Grant had embraced Darwinism and added Mendelian genetics, at least the version that Davenport and his colleagues used, to the old polygenic racism, creating a new scientific racism.

Grant's book was reprinted and/or revised at least seven times before World War II (Shipman [1994] 2002). *Passing* became the core of a resurgence of American racism and one of the essential books of Hitlerism. Although it had all been said before, at least since Darwinism had been combined with the rediscovered Mendelian genetics, no one had brought it all together "in one place and presented the whole with such esprit, audacity, and clarity" (Spiro 2009, 157). As a result, Spiro states, "what had been the province of a few obscure academics was now made accessible to the general reader. . . . While *Passing of the Great Race* was never a best seller, its ideas began percolating throughout U.S. society soon after the war, and became part of the common intellectual currency of the 1920s" (Spiro 2009, 157–158, 167). Up until the time of World War II, reference to Grant's tome appeared in both popular writings and scholarly works by prominent scientists. References to it showed up in pamphlets of the Ku Klux Klan and in women's magazines. Margaret Sanger (1879–1966), who led the early birth control movement in the United States, was an ardent eugenicist. She advocated mass sterilization, incarceration of the unfit, and draconian immigration restrictions, and she included *Passing of the Great Race* on the eugenics reading list of her journal *Birth Control Review* (Black 2003; Spiro 2009). Although *Passing* itself might not have been read by a massive number of people, the works of Grant's disciples were read by millions.

Measuring Intelligence

One of the imagined immutable, unit characteristics that was a centerpiece of scientific racism was intelligence. Just as Galton and Gobineau believed that it was necessary to get measurements of the physical and mental characteristics that separated the races, so the modern eugenicists believed that measuring the differences in the unit characteristics of intelligence would at last give them quantitative scientific validation of their racial hierarchies. After Goddard had introduced the Binet tests into the United States and had attempted to determine the intelligence of immigrants coming through

Ellis Island, he sent his version of the tests to a colleague, Stanford professor of psychology Lewis Terman (1877–1956). In the early twentieth century, Terman was an active eugenicist and an avid supporter of Grant's views (Degler 1991; Zenderland 1998). While he was serving as president of the American Psychological Association in 1923, Terman said, "The ordinary social and political issues which engross mankind are of trivial importance in comparison with the issues which relate to eugenics" (quoted in Spiro 2009, 137). Echoing Grant, he also believed that "the immigrants who have recently come to us in such large numbers from Southern and Southeastern Europe are distinctly inferior mentally to the Nordic and Alpine strains" (Terman 1922, 660). Like Galton, Terman's intellectual hero (Spiro 2009), he believed that measuring intelligence was extremely important for the science of eugenics.

Terman and his fellow psychologists and testers, such as Henry Goddard and Robert Yerkes (see below), wanted a simple, more identifiable number to represent the unitary character of intelligence. Both Terman and Goddard had received their doctorates at Clark University under the tutelage of G. Stanley Hall (1844–1924). Hall was the first to have received a PhD in psychology in the United States (under William James at Harvard in 1879). He was one of the most influential educational psychologists of his time and was among the foremost exponents of genetically determined behavioral traits (Chase 1977; Brace 2005; Zenderland 1998). In 1916, Terman developed the Stanford-Binet intelligence test, the direct ancestor of intelligence tests used today (Brace 2005). Using this test, he divided "mental age" as determined by his test by "chronological age" and then multiplied by 100 (those below a score of 70 were graded as feeble-minded, or as morons, idiots, and imbeciles). This became the American version of the *intelligence quotient,* which Terman called IQ. Terman published his initial work on testing in *The Measurement of Intelligence: An Explanation of and a Complete Guide for the Use of the Stanford Revision and Extension of the Binet-Simon Intelligence Scale* (1916).

As did most eugenicists and intelligence testers of the day, Terman believed that intelligence was a "unit character" inherited by a single-gene trait in simple Mendelian fashion and that this characteristic, just like skull size and shape, was fixed and immutable. Furthermore, following Goddard's claims in his book on the Kallikak family, Terman eliminated environment from the equation: "The common opinion that the child from a cultured home does better on tests solely by reason of his superior home advantages is an entirely gratuitous assumption. . . . The children of successful and cultured parents test higher than children from wretched and ignorant homes for the simple reason that their heredity is better" (Terman 1916, 115).

Terman's IQ test scores gave eugenicists a single number that could be used to identify an individual's intelligence and compare the scores of different racial groups. They could use this method to test adults. They also finally had a quantitative, scientific means to put races into the hierarchical order that would mirror century-old beliefs about the different mental capabilities of races. They could essentially test age-old beliefs about differential racial mental abilities, as recently outlined in Madison Grant's *Passing of the Great Race*. This is precisely what a group of highly respected psychologists who would soon be referred to as racial psychologists began to do.

In 1917, days after the United States entered the war, Robert M. Yerkes (1876–1956), the current president of the American Psychological Association, suggested that the U.S. Army test the intelligence of its draftees. Yerkes was an eminent Yale psychologist, a Harvard graduate, and a protégé of Charles Davenport. His main interest was the evolution of intelligence. He also founded what was to become the Yerkes National Primate Research Center and could be considered a founder of American primatology (Sussman 2011).

After some debate, the idea of testing U.S. Army recruits was accepted and Yerkes was commissioned as a major (soon to be colonel) of the U.S. Army Sanitary Corp, which administered the tests (Zenderland 1998). Yerkes assembled fellow hereditarians and psychometricians Terman, Goddard, and Bingham, among others, to write and organize the tests. As were Terman and Goddard, Yerkes was an active eugenicist, and all three were members of Davenport's militant eugenic research body, the ERA. Going beyond other eugenics groups, the ERA was determined to turn the results of eugenics "research" into legislative and administrative action "and public propaganda for the causes of eugenics, raceology, and Nordic race supremacy" (Black 2003, 90).

The army tests were ideal for this purpose. The team of army testers developed three types of tests, the Army Alpha test for literate recruits, the Army Beta test for the illiterate or those who did not speak English, and individual Binet-type tests for those who failed the beta test. This was the first use of mass testing and, for Yerkes, who equated scientific rigor with numbers and quantification, it helped establish psychology as a hard science that was equivalent to physics (Gould 1996; Zenderland 1998).

Approximately 1.75 million recruits were tested, and the most lasting results, given the assumptions of the eugenicists and the cultural bias of the tests, were predictable.

The average mental age of white American adults as indicated by these tests was just above thirteen, a score just above the mental age limit that intelligence testers had been using for years to define feeble-mindedness. In fact, according to the logic of the test, 47 percent of the whites who were tested would be considered morons, as would 89 percent of African Americans. The scores of European immigrants could be graded by their country of origin. The average mental age of immigrants from many countries was in the moron range, and darker peoples from southern Europe and Slavs from Eastern Europe were less intelligent than the more fair-skinned people from Western and northern Europe. The average mental age of Russian immigrants was 11.34, for Italians it was 11.01, and for Poles it was 10.74. Blacks were at the bottom of the scale with an average mental age of 10.41 (Gould 1996; Zenderland 1998).

The tests included such questions as these (Paul 1995, 66; Yerkes 1921):

Five hundred is played with: rackets, pins, cards, dice.
Becky Sharp appears in: Vanity Fair, Romola, The Christmas Carol, Henry IV.
The Pierce Arrow car is made in: Buffalo, Detroit, Toledo, Flint.
Marguerite Clark is known as a: suffragist, singer, movie actress, writer.
The number of a Kaffir's legs is: 2, 4, 6, 8.
Christy Mathewson is famous as a: writer, artist, baseball player, comedian.
The pitcher has an important place in: tennis, football, baseball, handball.
The Wyandotte is a kind of: horse, fowl, cattle, granite.
The Knight engine is used in the: Packard, Stearns, Lozier, Pierce Arrow.
Isaac Pitman was most famous in: physics, shorthand, railroading, electricity.
"There is a reason" is an ad for a: drink, revolver, flour, cleaner.
Ensilage is a term used in: fishing, athletics, farming, hunting.

In the open-ended questions of an "Information Test," soldiers might be asked the color of chlorine gas (green), the author of *Robinson Crusoe*

(Defoe), the occupation of Jess Willard (boxer), the origin of silk (worm), or a first-class batting average (.300) (Zenderland 1998). These were among the questions for the literate recruits. The beta tests consisted of pictures in which the recruits were asked to correct the unfinished pictures (see Figure 3.3). For both alpha and beta tests, the conditions in which the tests were taken, the timing of the tests, and the atmosphere were extremely stressful and often completely outside the experience of the individuals taking the tests. In many cases, the recruits were from very rural areas or were recent immigrants and were not at all familiar with the items depicted on the tests. The tests could (should) have been interpreted as culturally biased, but since they had been written by America's leading eugenics psychologists, and the concept of culture was not widely accepted (see Chapter 5), all the publicity written about the results used hereditarian interpretations.

The tests reconfirmed the eugenicists' assumptions about the inferior intelligence of African Americans, the lower economic classes, and certain immigrants: psychological testing could now easily be transformed into explicit political philosophy, confirmed by simple hereditarian explanations. Gould observes that "[the tests] had been constructed to measure innate intelligence, and they did so by definition" (1996, 228). As Yerkes (quoted in Chase 1977, 249) described them, "Examinations Alpha and Beta are so constructed and administered as to minimize the handicap of men who because of foreign birth or lack of education are little skilled to use English. These group examinations were originally intended, and now are definitely known, to measure native intellectual ability. They are to some extent influenced by educational acquirements, but in the main the soldier's inborn intelligence and not the accidents of environment determines his mental aging or grade in the army."

In his massive, nearly 900-page official publication on the results of these tests, Yerkes (1921, 742), evaluated his findings on African American recruits: "All officers without exception agree that the negro lacks initiative, displays little or no leadership, and cannot accept responsibility. . . . All officers further agree that the negro is a cheerful, willing soldier, naturally subservient. These qualities make for immediate obedience, although not necessarily for good discipline, since petty thieving and venereal disease are commoner than with white troops."

Because Yerkes's official report was very dense, quantitative, and pedantic and was not widely distributed or widely read, one of the officers who assisted Yerkes with the testing project, Carl Brigham (1890–1943), wrote

5 • 5 Part six of examination Beta for testing innate intelligence.

Figure 3.3 The beta intelligence test, given to army personnel who could not speak English. They were to fill in missing portions of the figure. Often they would put a cross on the house rather than adding the top of a chimney or not realize that balls or nets were used in American games of bowling or tennis. The test was developed by Robert M. Yerkes in 1917.

a book entitled *A Study of American Intelligence* (1923) that summarized the results of the army tests and presented them to the public in layperson's terms (Brace 2005). Yerkes (in Brigham1923, vii) wrote a glowing introduction to Brigham's book, stating: "The author presents not theories or opinions but facts. It behooves us to consider their reliability and their meaning, for no one of us as a citizen can afford to ignore the menace of race deterioration in the evident relations of immigration to national progress and welfare." This quote well represents the propaganda, the message, and the practical purposes to which the eugenicists proceeded to use the U.S. Army test results.

Brigham acknowledged that his treatment of race in the book relied on the writings of prominent figures such as William Ripley and Madison Grant. In fact, the eugenicists saw the army IQ testing project as a test of Grant's theories on race as put forth in *Passing of the Great Race*. Brigham (1923, 182) claimed: "In a very definite way, the results which we obtain by interpreting the army data by means of the race hypothesis support Mr. Madison Grant's thesis of the superiority of the Nordic type." Brigham wrote that the results of these tests clearly showed that the "races" as defined by Grant could indeed be placed in order of intelligence, with Nordics on top and "Negroes" on the bottom. Among the foreign born, southern and eastern Europeans were proven to be innately inferior in intelligence to Western European and white Americans. Brigham reported that test results showed that more recent immigrants were less intelligent than earlier ones. This could have been interpreted as related to the fact that earlier immigrants had become more familiar with the English language and American culture, but instead, it was linked to the idea that more recent immigrants were drawn more and more from less competent and "degenerate" segments of the European population (Brigham 1923; Brace 2005).

Thus, Grant's worries about the deteriorating effects on the Nordic race and on civilization by the recent immigration of southern and eastern European immigrants into the United States and the earlier immigration of African slaves were supported by the scholarly analysis of the army tests. Brigham (1923, xxi) reported: "The most sinister development in the history of this continent was the importation of the negro." Furthermore, he claimed that "American intelligence is declining and will proceed with an accelerating rate as the racial admixture becomes more and more extensive. . . . The presence of the negro . . . will make that decline in intelligence even more precipitous. . . . These are the plain, if somewhat ugly facts that our study shows" (quoted in Degler 1991, 52).

Brigham's book, along with two by Lothrop Stoddard, *The Rising Tide of Color: Against White World-Supremacy* (1920), which included an introduction by Madison Grant, and *The Revolt against Civilization* (1923), helped popularize the results of the army I.Q. tests (Zenderland 1998; Paul 1995). In his popular novel *The Great Gatsby,* F. Scott Fitzgerald (1925, 17), using literary license, refers to Stoddard's book. The character Tom Buchanan says to his wife and a dinner guest: "Civilization is going to pieces. . . . Have you read 'The Rise of the Coloured Empires' by this man Goddard? . . . If we don't look out the white race will be—will be utterly submerged. It's all scientific stuff. . . . It's up to us who are the dominant race to watch out or these other races will have control of things."

Finally the eugenicists had the "scientific" ammunition they needed to step up their attacks on the "unfit" individuals and races. Their acquisition of this ammunition, however, was no accident. Brigham's book was financially supported and carefully orchestrated by the eugenics movement. Charles W. Gould, a wealthy disciple of Madison Grant and a member of the AES's Committee on Selective Immigration, believed that data from the army tests could be used to justify discrimination and legislate immigration policies against non-Nordic "races." To further that goal, he invited Grant, Yerkes, and Brigham to a dinner at which it was agreed that C. W. Gould would underwrite a book to be written by Brigham that would analyze the tests along racial lines (Spiro 2009). Before this, the cephalic index had been the main measurement race scientists had used as a proxy for mental abilities. Spiro notes that "now that it was possible to quantify intelligence itself, mental testing replaced calipers as the instrument of choice among scientific racists" (Spiro 2009, 217). Brigham's book "provided the most important single scientific buttress for the racism of the 1920s" (Stocking 1968, 300–301), and, as Stephen J. Gould (1996, 254) remarks, it "became a primary vehicle for translating the army results on group differences into social action." Finally, the racist theories that began in the sixteenth century, were dragged into the eighteenth century by the Mortonites, evolved during the nineteenth century through the works of Spencer, Galton, Gobineau, Chamberlain and Ripley, and ultimately were translated into the twentieth century by Grant, had a quantitative, "scientific" basis.

Virginia's Racial Integrity Act was passed in March 1924, shortly after Brigham's book was published, and the *Buck v. Bell* case testing sterilization laws began in November 1924, with Harry Laughlin giving a deposition relating to the mental age of Carrie Buck that used the Stanford

Revision of the Binet-Simon IQ test (see earlier in this chapter). In reaction to increased immigration from Catholic countries and by Jews, and fueled by this "scientific racism," American racism was bolstered, reinforced and heightened. Anti-Jewish, anti-Catholic, and anti-black sentiment reached new heights. Membership in the Ku Klux Klan grew to over 6 million members during the mid-1920s (Smedley 1999). The separation, segregation, and sterilization of "unfit" individuals and races intensified after the polygenic eugenicists had their "scientific" proof. The eugenicists now could begin work on their fourth goal, restricting immigration.

It is interesting to note, however, that as early as 1920, before the *Buck v. Bell* case, Goddard had abandoned the position on which the trial was decided and arrived at the decision that IQ tests could not be used to measure feeble-mindedness (Zenderland 1998). By 1927, he had recanted almost everything he had said about the treatment of the feeble-minded, stating that their child-bearing posed no danger to society and questioning the practices of segregating the so-called feebleminded into colonies and sterilizing them. He admitted that most of his influential conclusions about feeble-mindedness had been in error, and he once again noted the importance of education (Goddard 1927; Zenderland 1998; Lombardo 2008). By the early 1930s, Carl Brigham and Lewis Terman had also begun questioning the validity of IQ testing. In 1930, Brigham denounced his views on the intellectual superiority of the "Nordic Race" and disowned the findings of his 1923 book on the U.S. army IQ tests and American intelligence (Brigham 1930; Barkan 1992; Gould 1996). Terman found Nazi racial policies to be "beneath contempt" and became a strong supporter of Franklin D. Roosevelt's liberal New Deal politics, the welfare state, and the civil rights movement (Zenderland 1998).

Restriction of Immigration

The immigration of foreigners into the United States had begun to be seen as a problem as early as the 1870s, when there was a movement to limit immigration of Chinese and Japanese because of the economic competition they posed. Chinese immigrants had first arrived during the gold rush and then were imported as laborers, mainly to work on building railroads and in other low-wage jobs. By the 1870s, as the post–Civil War economy declined, anti-Chinese sentiment had become politicized, and Chinese were blamed for depressed wage levels, especially in California. In 1882, the Chinese Exclusion Act was passed. As more Japanese immigrants began to enter the United States, opposition to them increased as well, and an executive order was passed in 1907 to restrict their immigration. In 1913 and again in 1920, the state of California passed Alien Land Laws designed to prevent Japanese farmers from owning agricultural land. In 1917, an Asiatic Barred Zone was created, barring immigration from much of the rest of Asia (Smedley 1999; Paul 1995).

Eugenicists believed that most immigrants who entered the United States after 1890 were genetically undesirable (Black 2003). In the three decades after that year, over twenty million immigrants had entered the United States. These new immigrants were mainly from Europe, where economic and ethnic problems were causing upheavals. More than eight million of these had arrived from 1900 to 1909; more than a million arrived each year in 1910, 1913, and 1914 (Black 2003).

This was a new group of immigrants, mainly from southern and eastern Europe. They were different from the old-line American families that could trace their ancestry to colonial times and from other earlier immigrants, who mainly came from Western Europe, and many of the old guard were threatened by the newcomers, who were generally poor, Catholic and Jewish; often were illiterate and unskilled; and tended to congregate in large cities. The demography of the United States was also changing in other ways. At the end of the Civil War in 1865, the United States was basically an agricultural country. By 1880, about a quarter of Americans lived in cities, and by 1900 that proportion had increased to 40 percent (Paul 1995). After World War I, as immigration continued to boom, for the first time in American history, a majority of the population lived in urban areas. War industries began to lay off workers at the same time that soldiers were returning from the war. African Americans who had served in the war were tired of racism and wanted civil rights and expected employment. In 1919, there were massive labor strikes; up to 22 percent of the workforce joined labor actions at some point during that year. Thus, after World War I, the United States was in economic, ethnic, and demographic turmoil (Black 2003).

The eugenicists saw an opportunity to push their goal of setting immigration policy. As early as 1912, Robert DeCourcy Ward, a chief immigration strategist of the eugenics movement, developed a plan to screen immigrant candidates while they were still in their home country. Davenport wrote to a colleague: "I thoroughly approve of the plan which Ward urges of inspection of immigrants on the other side" (quoted in Black 2003, 187). Harry Laughlin proceeded to reformulate model immigration laws in eugenic terminology (Degler 1991; Black 2003, 187). Madison Grant was very much involved in the eugenics immigration plans, as president of the ERA and vice-president of the Immigration Restriction League. He was also a close friend and confidant of Representative Albert Johnson (1869–1957), chair of the House Committee on Immigration and Naturalization. Eugenicists recognized Grant's influence in Congress regarding immigration policy and used it well. Throughout this time period, Davenport sent Grant a flow of materials pertaining to eugenics research on immigration so that he could pass this information on to Johnson.

In 1923, Madison Grant advised Representative Johnson that more permanent and stricter immigration laws were needed. Johnson was a fanatic racist and eugenicist, and during his tenure in Congress he shaped American immigration policy in ways that lasted well into the future (Black 2003). Grant explained that "you have the country behind you and a most popular cause. It is growing in strength every day and it is only a question of time when even greater restrictive measures will be put through" (quoted in Spiro 2009, 224). The two men devised a plan that would lower the percentage of immigrants from various countries to a quota equivalent to the percent of immigrants from each country who lived in the United States at the time of the 1890 census. At that time, there had been far fewer immigrants from eastern and southern Europe. This would significantly reduce (essentially eliminate) immigration from those countries.

Historians Franz Samelson (1979) and Carl Degler (1991) believed that the army intelligence tests and Brigham's book did not have a great deal of influence on the 1923–1924 immigration debate. The tests were rarely mentioned in the 600 pages of congressional debate, and none of the actual testers (Brigham, Goddard, Terman, Yerkes) testified in the hearings (Degler 1991). However, during the hearings on the Johnson bill before the Senate Immigration Committee, Senator Francis Kinnicutt referred extensively to the just-published *Study of American Intelligence* (Brigham 1923), calling it the "most important book that has ever been written on the subject." The chair of the committee responded: "I think every member of the committee ought to read that book." Kinnicutt then gave each member of the committee a copy of Brigham's book (Spiro 2009, 220).

As intended, Brigham's interpretation of the army intelligence tests carried great weight with congress and helped to crystallize the sentiment in favor of more extreme restrictions. . . . By 1923, thanks to the Second

International Eugenics Congress, Harry H. Laughlin's *Analysis of America's Modern Melting Pot,* and Carl C. Brigham's *Study of American Intelligence,* all congressmen were aware that science had proved that southern and eastern Europeans were biologically and intellectually inferior to the Nordics, and that the genetic health of the nation would be jeopardized if any more New Immigrants were permitted to enter the country.

On December 5, 1923, Johnson introduced the bill he and Grant had drafted to Congress. On the next day, President Coolidge, in his first address to Congress, said that "America must be kept American. For this purpose, it is necessary to continue a policy of restricted immigration" (quoted in Garis 1927, 170). Although Madison Grant was bedridden at the time, he was extremely active behind the scenes. Grant worked closely with Johnson, sending "a parade of eugenicists" to Capitol Hill to testify that the 1890 formula was the only eugenically sound plan for conserving the Nordic race in the United States. Grant also sent a lengthy letter to the House Immigration Committee defending the 1890 formula as scientific. He argued in the letter that this was the best and most just way of preventing "native" Americans from mixing with lower types. Johnson read the letter to the committee, and it was officially entered into the record (Spiro 2009). One of the eugenicists working with Grant and Johnson was Harry Laughlin, who had been working with Johnson on immigration policy since 1920 and was repeatedly invited to testify before the committee. In fact, he presented over 200 pages of testimony, including data based on Carl Brigham's book, showing that the Nordic race was at the top and Jews were at the bottom of the intelligence scale (Spiro 2009). In 1922, Laughlin had become the designated eugenic authority for Johnson's committee, and Johnson created a new title specifically for him: "Expert Eugenics Agent." With this title, Laughlin was empowered to conduct racial and immigration studies and to present them to Congress as reliable scientific data.

Coincidentally, in 1923, Senator Johnson was becoming increasingly active in the leadership of the eugenics movement. He joined the elite, private, nongovernmental ad hoc Committee on Selective Immigration, of which Madison Grant was chair, R. D. Ward (a Harvard graduate climatologist and eugenicist) was vice-chair, and Laughlin was secretary. In that same year, Johnson was elected president of Davenport's ERA. Laughlin greatly influenced the immigration law passed by this committee. He gave his final testimony, entitled "Europe as an Emigrant-Exporting Continent and the United States as an Immigrant-Receiving Nation" (Laughlin 1924), before Johnson's committee on March 8, 1924. In it he warned that "immigration into the United States, in the interests of national welfare, is primarily a biological problem. . . . The character of a nation is based primarily upon the inborn racial traits of the inhabitants" (1339, 1297). Laughlin also influenced the passage of the law by presenting skewed data in support of his assertion that the percentage of recent immigrant populations in prisons and mental institutions was far greater than their percentage in the general population would warrant (see Gilman 1924 and Allen 2011 for detailed critiques of Laughlin's prison data and conclusions). As Spiro (2009, 229) has stated:

> Every day, congressmen received more and more material from the ADS, the IRI, the ERA, the ERO, and the ECUSA, all of it clamoring for restriction. . . . What may not have been clear to the congressmen, however . . . was that the same elite coterie was running each organization. . . . What was also not obvious (if they did not look carefully at the statements, articles, editorials, speeches, and letters) was that they were all drawing on a common well of influence. For the center of all these people and organizations, like a spider perched in the middle of its increasingly intricate web, was one figure, Madison Grant. . . . Crippled Grant lay in his bed in Manhattan and used the telephone, the telegraph, and the U.S. mail to masterfully coordinate the activities of the interlocking directorate of scientific racism.

And Spiro (2009, 227) further points out: "Like his colleagues, Grant bolstered his argument for Nordic superiority by referring to the army intelligence tests and the testimony of Harry H. Laughlin, and thus the circle was complete: Grant cited Laughlin, who had based his analysis on Brigham's statistics, which were in turn based on Grant's calculations of the racial composition of the European population. What seemed . . . to be a plethora of independent studies by reputable scientists was actually a series of self-referential claims that, like the worm Ouroboros, constantly fed upon itself."

Thus, after massive rhetoric and propaganda was dumped on the politicians, a combined House and Senate bill, the Johnson-Reed Act, passed on

May 15, 1924, by a vote of 308 to 69 in the former and 69 to 9 in the latter. The bill was approved as the Johnson-Lodge Act and was signed into law by President Calvin Coolidge on June 30, 1924. This law remained the major immigration policy in the United States until the Immigration and Nationality Act of 1952.

The national origins formula of 1924, according to the Johnson-Lodge Act, established the "national origins quota system." The act set the annual quota of any nationality at 2 percent of the number of foreign-born persons of that nationality who lived in the continental United States in 1890. This essentially stopped immigration from eastern and southern European countries. The Johnson-Lodge Act also halted other "undesirable" immigration using quotas. It barred immigration from the Asia-Pacific Triangle. Thus, Chinese and Japanese could no longer immigrate into the United States. However, this also included peoples from the Philippines (then under U.S. control), Laos, Siam (Thailand), Cambodia, Singapore (then a British colony), Korea, Vietnam, Indonesia, Burma (Myanmar), India, Ceylon (Sri Lanka), and Malaysia. According to the Naturalization Act of 1790, these nonwhite immigrants were not eligible for naturalization, and the Johnson-Lodge Act forbade further immigration of any persons who were not eligible to be naturalized (Guisepi 2007). The Japanese government issued heated protests, and it declared May 26, 1924, days after the Johnson-Lodge Act was signed into law, a day of national humiliation in Japan, adding another item to a growing list of grievances against the United States. A number of scholars have long regarded this law and the inability of politicians to modify the Japanese exclusion clause in an effort to stabilize relations between the two countries as major contributing factors in the final collapse of U.S.-Japan relations in 1941 and the ultimate entry of Japan into World War II against the United States (see Hirobe 2001).

It is difficult to estimate how many eastern and southern Europeans this legislation kept from being able to leave Germany during Hitler's reign. Although two million Jewish immigrants had arrived in the United States from Eastern Europe between 1881 and 1924 (Hardwick 2002), this, of course, was before the Nazis took over Germany. The immigration policies of the United States were not lifted during the Holocaust, news of which began to reach the country in 1941 and 1942. It has been estimated that at least 190,000–200,000 Jews could have been saved during World War II alone (Weiner 1995), not including those who were unable to leave Europe between 1924 and the beginning of the war. Chase (1977) estimates that 6,065,704 Italians, Jews, Poles, Hungarians, Balts, Spaniards, Greeks, and other European people were trapped in Europe when the war began in 1939. He states (Chase 1977, 301): "How many of these 6,065,704 immigrants excluded by the racial quotas of 1924 would have emigrated to America between 1924 and 1939 will, of course, never be known. One thing is certain: most of the Jews, Poles, Russians, and other people marked by the Nazi race biologists as dysgenic who were trapped in territories controlled by Germany between 1933 and 1945 were rounded up and thrown into Auschwitz and other Nazi extermination camps."

The plight of the passengers of the German transatlantic ship MS *St. Louis* is a good example of the fallout of U.S. immigration policies. On Saturday, May 13, 1939, 937 passengers boarded the MS *St. Louis* in an attempt to escape persecution and start a new life, first in Cuba and ultimately in the United States. Almost all were Jews fleeing the Third Reich. Most were German citizens, some were from Eastern Europe, and a few were officially stateless. At great expense and after giving up their homes, possessions, and past lives, they were subjected to a long and very disappointing journey. Upon reaching Cuba on May 27, all but twenty-eight passengers were refused entry (one passenger had died during the voyage). They then attempted to seek asylum in Florida. However, because of the quotas set by the 1924 immigration laws, the U.S. government refused to allow the passengers to disembark. The MS *St. Louis* was forced to sail back to Europe on June 6, 1939. The passengers were not forced to return to Germany. Instead, Jewish organizations (particularly the American Jewish Joint Distribution Committee) were able to secure entry visas from four European governments, Great Britain, the Netherlands, Belgium, and France. Of the 288 passengers Great Britain admitted, all but one survived World War II. Of the 620 passengers who returned to the continent, eighty-seven (14 percent) were able to emigrate before Germany invaded Western Europe in May 1940. However 532 remained trapped on the continent. Over half (279) of these survived the Holocaust, but 254 died at the hands of the Nazis (Miller and Ogilvie 2006; Holocaust Encyclopedia 2012; Rosenberg 2013).

Thus, the immigration laws eugenicists inspired and influenced had a direct influence on the death of millions of Europeans. As Stephen Gould (1996, 262) has stated: "The eugenicists battled and won one of the greatest victories of scientific racism in American history." One could add that they aided and abetted one of the greatest tragedies in world history. Nash (1999, 155) summarizes: "Hitler had done exactly what Madison Grant had suggested: a eugenics program to sterilize 'defectives' followed by the extermination of 'inferior' peoples."

4

Eugenics and the Nazis

American immigration restriction legislation that was framed during the last quarter of the nineteenth and the first quarter of the twentieth century was directly influenced by Gobineau's "reasoning." Similarly, there was a direct linear connection between Gobineau's reasoning and the verbiage of Adolf Hitler's *Mein Kampf* of 1925 (Biddiss 1970, 258; Brace 2005, 122). Thus, the "center of gravity" of the ideology of the eugenics movement, which began in Europe with early polygenic theories such as those of Hume and Kant, moved to the United States with the Mortonites and their fight to preserve slavery, then back to Europe with Sir Francis Galton soon after Darwin's *Origin* was published. By the end of the nineteenth century, the Frenchman Gobineau's version of eugenics was dominating Europe and was being popularized by Chamberlain and Haeckel in Europe and by Ripley and Shaler in the United States.

By the early twentieth century, although eugenics activism, driven by polygenic-like ideology, was still popular in Europe, it was once again dominated by the United States. The leaders of the American eugenics movement—Davenport, Laughlin, Osborn, and Grant—had the backing of powerful academic and medical institutions. In fact, eugenics became a dominant scientific paradigm in American academia—conventional wisdom, so to speak. The American eugenics movement was amply supported by major financial barons and giant corporations, and its agenda was promoted and turned into policy and law by a large number of powerful politicians and lawmakers. However, American democracy and a large contingent of minorities in the United States forced the movement to move slowly and initiated a number of roadblocks and failures for the movement (see Chapter 3).

In the 1870s, Germany had "established an impressive range of liberal rights and Germany seemed to be on the brink of an age of enlightened liberal freedoms and tolerance" (Weindling 1989, 49). However, economic depression during the mid-1870s, defeat in World War I, and the Great Depression of 1929 changed that trajectory. During this period, anthropology in Germany changed from a public and participatory science to a discipline that served political requirements for racial classification. Weindling (1989, 307) has described postwar Germany: "There was an outburst of theorizing on the social organism and of interest in eugenic social policies. At the same time mass poverty, political polarization, and the shift towards socialization were all signs that at the very time that democracy became a reality, liberal and democratic values were in jeopardy. The defeat gave eugenics relevance with regards to national reconstruction. Virtually every aspect of eugenic thought and practice—from 'euthanasia' of the unfit and compulsory sterilization to positive welfare—was developed during the turmoil of the crucial years between 1918 and 1924."

When Hitler rose to power in Germany in 1933, he was hindered by the Weimar constitution. However, the burning of the German parliament (Reichstag) building shortly thereafter was blamed on communists and radicals and this enabled Hitler to demand emergency powers by decree. This swept away any vestiges of parliamentary control. It was later determined that the fire most likely was set by the Nazis (Bahar and Kugel 2001). Hitler's eugenic ideals, fueled by the writings and theories of Gobineau, Chamberlain, and Grant; the "science" of Davenport and Laughlin; and the German racial hygienists Ploetz, Baur, Fischer, and Lenz then could be carried out without the restrictions imposed by a democracy (Weiss 2010). He and his Nazi Party's racism and ethnic cleansing could proceed unabated. The focal points of the Nazi Party (the National Socialist German Worker's Party, or NSDAP) were "race" and "heredity," and its goal was to develop a genetically healthy and racially pure German national community, or *Volk* (Weiss 2010). Thus, Nazi Germany moved into dominance of the eugenics movement in the 1930s. In fact, Nazism can be seen as the logical culmination of this sinister ideology. As Black (2003, 318) states, "The war against the weak had graduated from America's slogans, index cards and surgical blades to Nazi decrees, ghettos and gas chambers."

As Weindling (1989, 337) has explained: "Eugenics was attractive to a broad range of medical specialisms such as hygiene and psychiatry, to biologists, to anthropologists and social scientists. It gave them a means of arguing for the social relevance of their disciplines, and fitted the mood of the times with the yearning for social order based on secure natural values." There was a symbiotic relationship between eugenic science and the Nazi worldview (Weiss 2010).

American eugenicists, who realized that their policies would never move as fast or be carried out as thoroughly as was being done in Germany, envied the Nazis and became cheerleaders for Hitler. For example, Leon F. Whitley of the AES wrote that "many far-sighted men and women . . . have long been working toward something very like what Hitler has now made compulsory. . . . And this represents but a small beginning we are told!" He continued: "While we are pussyfooting around," the Nazis were accomplishing great things. Dr. Joseph DeJarnette (who had testified in the Carrie Buck case) exclaimed that the "Germans are beating us at our own game!" (quotes in Spiro 2009, 364). Laughlin told Grant that "he was thrilled that the speeches of Nazi leaders . . . 'sound exactly as though spoken by a perfectly good American eugenist,' but he admitted that he was jealous that he and Grant were only humble researchers," while in the German dictatorship, scientists were actually getting things done (quoted in Spiro 2009, 364). Army tester and primatologist Robert Yerkes, who thought the Nazis were upstaging Americans by properly using mental testing for military purposes, stated: "Germany has long led in the development of military psychology. . . . The Nazis have achieved something that is entirely without parallel in military history. . . . What has happened in Germany is the logical sequel to the psychological and personnel services in our own Army during 1917–1918" (Yerkes 1941, 209). As Spiro (2009) has pointed out, Frederick Osborn believed that the program of the Nazis was an excellent one and that "taken together, recent developments in Germany constitute perhaps the most important social experiment which has ever been tried" (quoted in Spiro 2009, 365). Osborn emphasized that "Germany's rapidity of change with respect to eugenics was possible only under a dictator" (quoted in Spiro 2009, 368). Although many would later deny knowledge of the Nazi atrocities, Germany was quite proud of its treatment of Jews and others, and the Nazi fascist program was widely reported in the American media. American eugenicists kept tabs on the Nazi program (Black 2003, 299).

Thus, from the early beginnings of Nazism in Germany through the atrocities carried out by Hitler and his followers, even to the end of World War II, many American eugenicists supported Hitler and Nazism intellectually, with moral (or we might say immoral), political, and financial support. After all, American eugenicists had essentially written Nazi ideology and policy. As Dr. Heinz Kürten, the Nazi physician in charge of training other doctors in Nazi medicine, pointed out in 1933, U.S. legislation was used as the model for the new Germany, and the "American pathfinders" were Madison Grant and Lothrop Stoddard. Kürten was proud that although "U.S. legislators had been reluctant to endorse some of Grant's more drastic proposals, the Third Reich was actually carrying them out" (quoted in Spiro 2009, 362). Here I will describe only a few of the many interactions and interconnections among the American eugenicists and the Nazis.

As Weiss (1990, 14) has noted, German eugenicists with expertise in medicine proposed a national comprehensive eugenics program to improve national health. Toward the end of World War I, around 1917, the leading racial hygienists in Germany began to ally themselves with the political right. They began to use science and medicine as weapons for racial purity (Weindling 1989). Using a "selectionist" brand of Social Darwinism allowed German eugenicists to convert social and economic problems into a scientific crisis. The depression brought home to the elite classes the high costs of social welfare, and eugenics was recognized as a means of cutting these costs. Defeat in World War I and middle-class impoverishment were central factors in the rise of German racism. The eugenicists reached a high point of influence in Germany from 1929 to 1932, during the economic crisis. By 1932, only approximately one-third of the population of Germany was working (Weindling 1989; Weiss 2010). Eugenicists asserted that only eugenics-based elimination of the unfit, or "rational selection," could remedy the growing crisis. A eugenics ideal in Germany also made it possible for "science" to keep "a large and militant working class true to the state—and boost national efficiency" (Weiss 2010, 25–26).

No other country used science and medicine, under the guise of racism and eugenics, with the degree of wanton disregard for human life as did Nazi Germany. As Poliakov (1971, 327) has pointed out: "During the Renaissance, the Spanish statuses relating to 'purity of blood' brought about a type of segregation similar to that of the racial laws promulgated in the twentieth century by the Nazis and Fascists. . . . Even so, the 'inferior race' of the *conversos* . . . was persecuted by the Inquisition . . . it was not eliminated." In Nazi Germany, at first, emigration ("self-deportation") of those designated as "racial degenerates," such as Jews, was encouraged (Weindling 1989). However, as emigration became increasingly impossible in the years preceding and during World War II, Nazi policies resulted in the death of millions. The Reich's biological enemies were simply deemed genetically subhuman. "There was never the degree of wanton disregard for human life that existed in Nazi Germany. Such practices were also never seen as eugenic measures" (Weiss 2010, 267). While masses were being eliminated, patients from "racially elite" groups were receiving the highest-quality medical care. As Weindling (1989) has noted: "The Nazis rapaciously plundered elements from pre-existing movements for motherhood, family welfare, and national health, and incorporated these into their distinctive ideology of racial purity. . . . Nazi concepts of health meant that Jews, gypsies and homosexuals were stigmatized as 'alien parasites' or as 'cancerous growths' in the German body politic. Health care was selectively to promote a Nordic elite, which would lead a purged but fitter and healthier nation to military victories. . . . Social, economic and cultural factors were reduced to categories of evolutionary biology" (489–490, 580).

One individual who was particularly distinguishable as a catalyst for the German eugenics movement was Alfred Ploetz (Weiss 1990). Ploetz (1860–1940), a German physician and biologist, was one of the earliest proponents of the German eugenics movement; it was he who coined the term *Rassenhygiene,* or racial hygiene (Weindling 1989; Weiss 2010). He was fascinated with Darwinism and was profoundly influenced by Haeckel and his monist philosophy, from which he drew the idea of the need to elevate the status of the nation's biological elite. He believed "that if no more weaklings were bred, then they need not be exterminated" (Weindling 1989, 131). In 1895, he wrote *Grunlinien einer Rassenhygiene (Racial Hygiene Basics),* which helped establish eugenics as a science in Germany. He has been described as the founder and spiritual leader of the German eugenics movement (Weindling 1989; Kühl 1994; Weiss 2010). In this volume he emphasized that Galton and Weismann's refutation of Lamarckism meant that education and economic and social progress as means of assisting human evolution were no longer tenable. In 1904, he established the periodical *Archiv für Rassen- und Gesellschafts-Biologie (Archive for Racial and Social Biology);* he asked Fritz Lenz to join the editorial board in 1913 (Weindling 1989; Müller-Hill 1998). This was the first eugenics journal in the world. Its early articles linked racial hygiene with Mendelism (Weindling 1989; Weiss 2010). In 1905, along with Ernst Rüdin (his close friend and brother-in-law), he founded the Deutsche Gesellschaft für Rassenhygiene (German Society for Racial Hygiene); Haeckel and Weismann were honorary chairs. This was the first professional eugenics organization (Weiss 2010). It paralleled the ERO and ERA in the United States and became the main clearinghouses for Nazi eugenics. Ploetz had helped lay the groundwork for the First International Congress for Eugenics at the International Hygiene Exhibition in Dresden in 1911. The 1911 meeting was organized by the International Society for Racial Hygiene, a group dominated by German racial hygienists (Kühl 1994).

Ploetz was given an honorary doctorate at the University of Munich in 1930. He welcomed the Nazis' seizure of power and wrote in 1933 that Hitler would bring racial hygiene from its previous marginality into the mainstream. In the same year, the Third Reich's minister of the interior, Wilhelm Frick, established an Expert Advisory Council for Population and Racial Policy comprising Ploetz, Lenz, Rüdin, and Hans Günther. This council provided advice on all legislation related to racial and eugenic issues and on enforcement and justification of national policies (Weiss 2010). In 1936, Ploetz was nominated, unsuccessfully, for the Nobel Peace Prize for his work on racial hygiene (Cornwell 2003). In the same year, Rüdin joined Ploetz in editing *Archiv,* but the journal was increasingly controlled by Nazi politics (Weindling 1989). That same year, Hitler appointed Ploetz, who was seventy-six, to a professorship, and Ploetz remained the "Great Old Man" of German eugenics until his death at seventy-nine in 1940 (Cornwell 2003). He was recognized as the founder of racial hygiene (Weindling 1989). In 1937, Ploetz was elected an honorary member of the German Society for Racial Research, which had been reconstituted from the Society of Physical Anthropology that year (Weindling 1989). Two years before his death, Rüdin said that Ploetz "by his meritorious services has helped to set up our Nazi ideology" (Rüdin 1938).

Ploetz's contacts with international and U.S. eugenicists were long and extensive. In the 1880s, he lived in the United States, where he studied U.S. utopians and practiced medicine in Massachusetts. He also bred chickens and began getting interested in the eugenics belief in breeding humans. He served as one of the vice-presidents of the First International Congress of Eugenics in 1912, and the next year, he became a member of the elite Permanent International Eugenics Committee, which later evolved into the International Federation of Eugenics Organizations (IFEO). Just after the London conference, Ploetz was quoted in a German newspaper as saying that the "United States was a bold leader" in advancing the cause of eugenics (Weiss 2010, 50). In 1930, Davenport and Paul Popenoe contributed congratulations to Ploetz on his seventieth birthday on behalf of the United States in a special tribute issue of *Archiv*; *EN* wrote that the issue was "a worthy tribute of esteem and affection for the genial and high-minded scholar whom it honors" (quoted in Black 2003, 295). In a letter written in 1932, shortly after Rüdin had taken over as president of the IFEO, Ploetz's friend Davenport confided to Rüdin that he was "very glad the Federation is now under leadership of a German" (quoted in Black 2003, 294).

Ernst Rüdin (1874–1952), a psychiatrist and geneticist, became one of the major architects of Hitler's eugenics medical program. From 1917 to 1945, Rüdin served as director of the German Psychiatric Institute's genealogical and demographic department, which became the Kaiser Wilhelm Institute for Psychiatry in 1926. He later became the director of that institute. In 1926, he began collecting family profiles similar to those Laughlin had collected for the ERA and ERO at Cold Spring Harbor. The Rockefeller Foundation financed a new building for the German Psychiatric Research Institute in 1928 and for Rüdin's research until 1935 (Weindling 1989). In the late 1920s, Rüdin began a professional and public campaign for compulsory sterilization (Weindling 1989). In 1933, he was chosen to lead Germany's racial purity program, and in that same year he helped draft the Nazi Sterilization Law and wrote an official commentary for Nazi racism at several of Germany's psychiatric hospitals and later for the general Nazi euthanasia policies (Weiss 2010). Interestingly, as the sterilization and later euthanasia policies began to be carried out, psychiatrists (such as Rüdin) and anthropologists (such as Fischer) competed over the definition of the "unfit, asocial individuals" who were to be eradicated (Müller-Hill 1998). As Weindling (1989, 552) has stated: "'Euthanasia' was the culmination of a deadly combination of subordination to the Führer, a war economy, and professionalism taken to an extreme of scientific authoritarianism."

In 1935, the Nuremberg Laws defined who was a Jew and deprived Jews of rights of citizenship, forbade their marriage with German citizens, and forbade extramarital sexual relations between Jews and Germans. Jews could no longer employ female citizens under the age of forty-five as domestic help. They were dismissed from all government positions. The following year, Jews were excluded from professional positions (for example, medical, education, and industrial administrative jobs). Other laws were soon passed that were directed against Jews, Gypsies, the offspring of German mothers and black French soldiers, homosexuals, and other social and political "deviants" (Paul 1995; Farber 2011). In 1936, one of Rüdin's main tasks as leader of the Institute for Racial Hygiene was to decide which German citizens had Jewish blood and how much they had. In 1938, he stated that "the importance of eugenics has only become known in Germany to all intelligent Germans through the political work of Adolf Hitler, and it is only through him that our more than thirty-year-old dream has become a reality and racial-hygienic principles have been translated into action" (quoted in Wistrich [1982] 1984, 260). In 1944, Rüdin similarly stated: "It will always remain the undying, historic achievement of Adolf Hitler and his followers that they dared to [put] into practice the theories and advances of Nordic racism [and won] the fight against parasitic alien races such as the Jews and the Gypsies" (quoted in Spiro 2009, 376). On his sixty-fifth birthday, in 1939, he was awarded the Goethe medal for art and science by Hitler, and Interior Minister Wilhelm Frick praised him as the "meritorious pioneer of the racial-hygienic measures of the Third Reich" (quoted in Wistrich [1982] 1984, 261). In 1944, when Rüdin was seventy, Hitler awarded him a bronze medal bearing the Nazi eagle, naming him the "pathfinder in the field of hereditary hygiene" (Wistrich [1982] 1984).

Rüdin began to correspond with Davenport in 1910 (Weiss 2010). As was Alfred Ploetz, Rüdin was active in the First International Eugenics Congress. He served there in order to boost the international prestige of Nazi legislation. Rüdin was very visible to U.S. eugenicists and established warm and lasting contacts with Davenport and Laughlin (Weindling 1989; Weiss 2010). Articles about or by him appeared in the United States for years. From 1922 to 1925, many articles were published in *EN* about his work on the inheritance of mental defects and on his family archives. In fact, by this time he had become the "star of German eugenics" to American racists (Black 2003, 286). In 1928, Rüdin hosted the IFEO in Munich, where the delegates were given a tour of the Institute for Psychiatry. The institute was selected for IFEO membership the following year. In 1929, *JAMA* published a lengthy report on his work. In 1930, *EN* reprinted a long Rüdin paper on sterilization and eugenics, stating that in any member of a family with a hereditary disease, "propagation ought to be renounced. . . . We must make the eugenic ideal a sacred tradition" (Rüdin 1930, quoted in Black 2003, 296). In 1932, Davenport, Rüdin's longtime friend and colleague, suggested that Rüdin succeed him as president of IFEO and Laughlin proudly nominated him. With this, Germany became a senior partner with the United States in the international eugenics movement.

The following year, Rockefeller awarded Rüdin's institute a generous ten-year grant. Shortly after the Nazi Sterilization Law was passed in 1933, *Birth Control Review* featured an article by Rüdin entitled "Eugenic Sterilization, An Urgent Need" (Rüdin 1933) and reprinted a pamphlet he had prepared for British eugenicists urging them to "act without delay" (Black 2003, 301). The first meeting of the IFEO during the National Socialist era was held in Zurich in 1934. Rüdin presided over this meeting and Ploetz opened the congress (Weindling 1989). Rüdin believed this would serve as an international referendum on Nazi racial policy (Weiss 2010). In 1935, Rüdin and Eugen Fischer (see later in this chapter) were made members of the ERA and were added to the advisory board of *EN*. In return, later that year, Davenport was invited to join the editorial boards of two influential German eugenics journals (Black 2003).

Rüdin was interned for his war crimes in 1945 but was released the next year with the aid of the Max Planck Institute. He claimed he was a scientist and not a politician and was denazified and classified as a nominal party member. He lost his job and was stripped of his Swiss citizenship, but he was never tried because it could not be proven that he had direct, hands-on involvement in the Nazi murders. Rüdin's ambitions in the field of "biological research" were continued at the Max Planck Institute for Psychiatry by his daughter, who worked on a project involving the behavior of twins. Research on twins had fascinated the Nazis, including the infamous Dr. Mengele of the SS, who conducted many deadly experiments on twins in Nazi concentration camps (see later in this chapter). Rüdin died in retirement in 1952 and is still considered by many to be the founder of psychiatric genetics (Eisenberg 1995).

Eugen Fischer (1874–1967) was a professor of medicine, genetics, and anthropology. He studied under geneticist August Weismann (Eng 2005). Although his doctoral thesis was in anatomy/primatology at the University of Jena (on orangutan female genitalia), Fischer established a reputation as the leading expert on miscegenation after his book *Die Rehobother Bastards und das Bastardierungsproblem beim Menschen (The Rehoboth Bastards and the Problem of Miscegenation among Humans)* was published in 1913. In this book, he declared that mixing white blood with that of blacks would bring about the demise of European culture (Black 2003; Cornwell 2003; Spiro 2009). He became a professor of anthropology at the University of Freiburg, the head of two German anthropological societies, and the editor of several anthropology journals. In 1908, Fischer established the Freiberg Society, a small elite group of academics that sought to establish eugenics and anthropology on the basis of Mendelism. A textbook by Baur, Fischer, and Lenz on heredity and genetics was planned in 1913 in discussions between Baur, Ploetz, Lenz, and the publisher Julius Lehmann (1864–1935). Lehmann was a major Nazi publisher of eugenic literature, a mobilizer of anti-Semitic and ultranationalist causes, and a gifted racial propagandist. Fischer was well versed in the eugenics movement in the United States and developed many of his ideas directly from his American colleagues. He was recruited as a "corresponding scientist" of the CIW by Davenport in 1904 and became a long-term collaborator with Davenport and Grant (Kühl 1994; Spiro 2009). Fischer was among the leading eugenicists of the Third Reich and became known as "the founder of human genetics" in Germany (Spiro 2009).

In 1921, Fischer coauthored, with Erwin Baur and Fritz Lenz, the two-volume work *Grundrisse der menschlichen Erblichkeitslehre und Rassenhygiene* (*Human Heredity*). The book was based primarily on references and examples from the literature of American eugenics and geneticists, including East and Jones and Davenport (Black 2003; Farber 2011). It emphasized how science could explain the rise and fall of civilization, cure the diseases of the "body politic," and provide a scientific basis for population control and racial politics. It also called for all medical schools to add racial hygiene to their curriculum (Weindling 1989). While writing *Mein Kampf* in prison, Hitler reinforced his views and derived support from biology for his view from this text (Müller-Hill 1998). The book emphasized eugenics as the cure for the "diseases in the body politic" and used the U.S. army intelligence tests as proof of the intellectual superiority of the Nordic race. Like other eugenicists in Germany, Fischer believed that blacks and Jews were inferior subspecies (Müller-Hill 1998; Black 2003). In fact, in a speech in 1941, he all but denied that "Jewish Bolsheviks" were part of the human species (Weiss 2010).

Fischer served as director of the Kaiser Wilhelm Institute for Anthropology, Human Heredity, and Eugenics, often called the Dahlem Institute or KWIA (Weiss 2010), from its inception in 1927. Fischer was a political chameleon. An anti-Semitic and ultraconservative eugenicist, he had nonetheless suppressed these leanings while serving within the centrist and Social Democrat–run government. At first, this caused some Nazis to suspect him, and Fischer had to work hard to keep control of his institute and to ingratiate himself with the Nazis. However, after Hitler came to power, Fischer totally aligned himself with the new regime and "'sold' his institute to the National Socialist state. . . . There was no center for human heredity and eugenics in Germany that did as much to provide the intellectual underpinnings for the Nazi regime's racial policy than did the Kaiser Wilhelm Institute for Anthropology" (Weiss 2010, 71). In fact, in 1932, it was the eugenicists from this institute who provided the single largest interest group for compulsory sterilization (Weindling 1989).

In 1933, Hitler appointed Fischer as rector of Frederick William University of Berlin (now Humboldt University). In his inaugural address, Fischer declared: "The old worship of culture . . . is past . . . the theory of the heritability of mental as well as physical traits has finally been vindicated. What Darwin was not able to do, genetics has achieved, it has destroyed the theory of the equality of man" (quoted in Proctor 1988, 345). Soon after the Nazi Sterilization Law was implemented in 1934, Fischer stressed that his institute had provided the "scientific underpinnings" of the laws (Weiss 2010) and he was recruited to quickly train a legion of race experts to identify those who should be sterilized. He, Rüdin, and Otmar von Verschuer took a leading role in organizing Spanish Inquisition–like tribunals. They used hereditary censuses and surveys to standardize indices of the ancestry and health of patients' families over four generations. These surveys were modeled after a criminal biology survey Rüdin had developed earlier. Using IBM equipment, data was sorted by family, race, religion, heredity, and cultural identity, and so forth, and later became the data that led to a precondition for extermination (Weindling 1989). In 1935, Fischer publicly thanked Hitler for giving eugenicists the opportunity, via the Nuremberg Laws, to make their research useful to the general public (Spiro 2009, 366). In 1937, he was elected into the prestigious Royal Prussian Academy of Sciences (Weiss 2010).

By 1938, SS officers often accompanied Fischer as he fulfilled his duties. In that year, national anti-Jewish riots and pogroms began. Over 100 synagogues were burned, and thousands of Jews were taken to concentration camps. The SS and the Gestapo now took over the Kaiser Wilhelm Institutes, the Society for Racial Hygiene, and German medicine generally. Fischer said in a lecture that "when a people wants . . . to preserve its own nature, it must reject alien racial elements . . . it must suppress them and eliminate them. The Jew is such an alien. . . . I reject Jewry with every means in my power, and without reserve, in order to preserve the hereditary endowment of my people" (quoted in Black 2003, 316–317). By 1941, Fischer was a major advocate of "a total solution to the Jewish question" (Black 2003, 347) and had declared that the Bolsheviks and Jews were a distinct and degenerate species (Weindling 1989). He was the honored guest at a March 1941 conference on the solution to the Jewish problem at which Nazi scientists developed the idea of eliminating Jews en masse. In 1941, Fischer instituted *Phänogenetik* (developmental genetics) which, at least as envisioned at KWIA, necessitated the establishment of "biobanks" of "human material." As we will see, this soon led to brutal human experimentation at the institute (Weiss 2010). In 1943, he urged the mass sterilization of Gypsies on the grounds that they were primitive and antisocial (Weindling 1989). Under Fischer a mutually beneficial symbiosis developed between the institute and the Nazi regime. In 1944, on his seventieth birthday, Fischer was honored by the renaming of his institute as the Eugen Fischer Institut (Weindling 1989). By the end of his career, Fischer had become "the undisputed academic spokesmen for racial science under the swastika" (Weiss 2010, 111).

U.S. eugenic journals had been following Fischer's research closely throughout the years. Davenport first contacted him in 1908. In 1914, the *Journal of Heredity* published an extensive review of Fischer's book on race crossing (Fischer 1914). The grand opening of Fischer's KWIA in 1927 took place during the Fifth International Genetics Congress in Berlin. Davenport was chairman of the program and an honorary president of the congress. He also proudly delivered one of the commemorating addresses and later provided a brief history of American eugenics to the institute's administration. That same year, Davenport began a campaign to identify mixed-race people and asked Fischer to join a committee of the IFEO on race-crossing in view of "the international nature of the problem" (Müller-Hill 1998; Black 2003, 280, 291). Davenport wrote to Fischer: "No one has greater experience in the field than you." In 1929, Davenport asked Fischer to assume the chairmanship of the committee, and together they sent out a worldwide questionnaire to identify not only cross-race individuals but all peoples of "color." Fischer modified this study to become a German national "anthropological survey," which became his tool for identifying undesirable populations (on the basis of so-called pathological lines of heredity) within the Third Reich (Black 2003). In 1930, the Rockefeller Foundation supported this survey with a five-year grant (Weindling 1989).

In 1934, in the preface for a special issue of *Zeitschrift für Morphologie und Anthropologie* honoring Fischer, two of his students, Aichel and von Verschuer (1934, vi), wrote: "For the first time in world history, the Fuhrer Adolf Hitler is putting into practice the insights about the biological foundations of the development of peoples—race, heredity, selection. . . .

German science provides the tools for the politician." Two American eugenicists, Raymond Pearl and Charles Davenport, contributed to this volume. Both were to become presidents of the American Association of Physical Anthropologists (AAPA). In 1935, Fischer served as president of the International Congress for Population Science, which met in Berlin. Laughlin, Pearl, and Clarence Campbell (all members of the U.S. Galton Society) served as vice presidents. At the meeting, Fischer gave thanks to Providence for having given Hitler to Germany (Spiro 2009, 371). In 1936, on the anniversary of Hitler's 1934 purge of Jews from the faculty of Heidelberg University, Laughlin was invited to an award ceremony at the university to receive an honorary doctorate for his work on the "science of racial cleansing." He was unable to attend the ceremony but remarked that the award symbolized the "common understanding of German and American scientists of the nature of eugenics" (Lombardo 2008, 211). In 1937, Madison Grant asked Fischer to write the foreword to the German publication of his new book, *Conquest of a Continent*. Fischer happily complied, writing that Grant was "no stranger to German readers" who were familiar with his *Passing of the Great Race* and that "no one today should pay more careful attention to Grant's work than the people of Germany, where racialist thinking has become the chief foundation of the population policies of the National Socialist State" (Fischer 1937, quoted in Spiro 2009, 359).

In 1942, Fischer retired and became professor emeritus at the University of Freiburg. In 1943, Fischer summarized his thoughts concerning the Nazi approach to practical eugenics by stating: "It is a rare and special good fortune for a theoretical science to flourish at a time when the prevailing ideology welcomes it, and its findings can immediately serve the policy of the state" (quoted in Müller-Hill 1998, 20). In 1945, he was denazified and thus was not tried as a war criminal. As is often the case, those who provided the justification and ideology, the leaders, were not punished for the actions they ordered. As Weiss (2010, 72) has stated: "In assessing the extent of Fischer's and his colleagues' moral culpability, we must consider the role that the director played in establishing a research program at Dahlem that made such crimes possible in the first place." In 1952, he became honorary president of the newly founded German Anthropological Society. He continued to publish, and in 1959 he published his memoirs (Fischer 1959), avoiding any mention of his work with the Nazis (Müller-Hill 1998). He died in 1967.

Erwin Baur (1875–1933) was senior author, with Fischer and Lenz, of *Grundriss der menschlichen Erblichkeitslehre und Rassenhygiene*

(Foundations of Human Heredity Teaching and Racial Hygiene) (1921), the textbook upon which Hitler had formulated his eugenics policies (Kühl 1994). Baur was a geneticist and botanist who worked primarily on plant genetics. In 1911, he published a classic textbook on heredity in which he linked Mendelian discoveries to humans and expressed the need for more comprehensive means of weeding out degenerative populations. He was appointed director of the Kaiser Wilhelm Institute for Plant Breeding Research in 1928 (since 1938, the Erwin Baur Institute). He served as the chairman of the German eugenics committee at the Fifth International Genetics Congress in Berlin in 1927.

After World War I, Baur's nationalist and eugenic convictions intensified, and he praised the research at Cold Spring Harbor and the achievements of the Americans in racial hygiene (Weindling 1989). In 1920, a year before publication of *Foundations of Human Heredity,* Baur corresponded with Davenport to ask for a summary of the American eugenic laws so he could prepare a review for a new Prussian government advisory commission for race hygiene. Much of the eugenics in *Foundations of Human Heredity* was influenced by Ploetz's 1895 volume *Grunlinien einer Rassenhygiene* and by Davenport's *Heredity* (Black 2003; Weiss 2010). In 1923, Davenport invited Baur to join the International Eugenics Commission (Kühl 1994). In 1924, *JAMA* published an article based on a lecture Baur gave to Berlin's eugenics society. Presenting his comments as proven medical knowledge, the article emphasized his view that "the attempts to elevate the negroes of the United States by giving them the same educational advantages the white population receives have necessarily failed. . . . Race suicide brought about the downfall of Greece and Rome, and Germany is confronted by the same peril" (Baur 1924, quoted in Black 2003, 281–282). The 1931 edition of *Foundations* included a discussion of the U.S. army intelligence tests and their assumed implications for racial differences (Paul 1995).

In 1933, Baur stated: "No one approves of the new sterilization laws more than I, but I must repeat over and over again, that they constitute only a beginning" (quoted in Spiro 2009, 365). Himmler's SS philosophy represented a politically radical form of Baur's eugenics (Weindling 1989). Baur died in 1933, two years before the Nuremberg Laws were passed and before euthanasia became Nazi policy. It was in 1939 that the transition was made from sterilization of the "unfit" to their liquidation as part of the "euthanasia" project (Weiss 2010).

Fritz Lenz (1887–1976), coauthor of *Foundations of Human Heredity Teaching and Racial Hygiene,* was a physician, geneticist, and racial hygienist and a student of Alfred Ploetz and August Weismann (Spiro 2009). He also studied with Eugen Fischer (Engs 2005). World War I reinforced Lenz's beliefs that a racially oriented eugenics was needed and that eugenics needed to be a racial religion for it to succeed. Following Gobineau, he believed that Germany was the last carrier of supreme Nordic racial values and that it was at the crossroad of either reaching great heights or extermination. Racial hygiene was the answer (Weindling 1989). He edited the journal *Archives for Racial and Social Biology* from 1913 to 1933 and received the first chair of eugenics at the University of Munich in 1923. At that time, Lenz criticized Germany for being backward in the domain of sterilization in comparison with the United States (Lifton 1986). He was a great admirer of Mussolini (Weindling 1989) and one of the earliest to give praise to Hitler in 1930 as "the first politician of truly great import who has taken racial hygiene as a serious element of state policy" (quoted in Proctor 1988, 47). Like most eugenicists, Lenz rejected environment as having any substantial role in human behavior, stating in 1930, in celebration of his mentor's seventieth birthday: "Ploetz recognized as unsatisfactory . . . the principle of the omnipotence of the environment. . . . The recognition that not all evil is determined by the environment, and that the roots of most evil lie instead in hereditary defects became the decisive motivating force in racial hygiene" (quoted in Proctor 1988, 49).

In 1933, Lenz established the first department specifically devoted to eugenics at the KWIA and later served as the co-director of the Anthropology Department with Fischer. In 1939, when German eugenics graduated from sterilization to euthanasia, Lenz helped draft the guidelines allowing "misfits" to be killed "by medical measures of which he (the victim) remains unaware" (Müller-Hill 1998, 15). He was one of Josef Mengele's mentors (see later in this chapter) (Lifton 1986). It seems, however, that Lenz was not particularly anti-Semitic; he claimed that the German term for anti-Semitism was pseudo-scientific, since there was no Semitic race (Weindling 1989, 553).

Lenz frequently corresponded about the future of eugenics with California eugenicist Paul Popenoe and with Davenport and Laughlin (Kühl 1994; Black 2003). He wanted Davenport to send him publications of the ERO so he could report on them in his journal, and by 1923 articles from

Archives were highlighted and summarized regularly in *EN*. Popenoe became Lenz's translator for articles in *EN* and the *Journal of Heredity*.

After the war, from 1946 to1957, Lenz became director of the Institute of Human Genetics at the University of Göttingen (Kühl 1994; Spiro 2009). He was the first of the old guard of German geneticists to attain a post after the war (Weindling 1989). In 1952, criticizing the UNESCO statement on race (see Introduction and Chapter 7), Lenz disputed the notion that all human beings were a single species (see Marks 2010b). He continued to publish into the 1970s. In 1972, on the occasion of his eighty-fifth birthday, Lenz was honored as the grandfather of racial hygiene in Germany by the neo-Nazi organization Neue Anthropologie. He died in 1976 still believing that eugenic theories of racial differences had been scientifically proven (Proctor 1988).

Wilhelm Frick (1877–1946) was appointed as minister of the interior of the Third Reich when Hitler came into power in 1933 and was one of three in Hitler's original cabinet. He created the first police system for the German Reich and became its supreme head. It was he who named Heinrich Himmler as Gestapo chief (Wistrich [1982] 1984). Frick had supreme authority over concentration camps and personally inspected these camps (Jewish Virtual Library 2013). Through his Medical Division, he controlled the medical institutions that performed sterilizations and euthanasia. Frick drafted and administered much of the Nazi racial legislation, including the Nuremberg Laws. He introduced the yellow star as a sign of stigmatization of the Jews. He denounced the welfare system and believed that hereditary defectives had to be prevented from procreating and that miscegenation had to cease (Weindling 1989). Frick believed that up to 20 percent of the German population was made up of "hereditary defectives, whose progeny is therefore undesirable . . . [and given this cataclysmal statistic, eugenic science] gives us the right and the moral obligation to eliminate hereditary defectives from procreation. No misinterpretation of charity, nor religious scruples, based on the dogmas of the past centuries, should prevent us from fulfilling this duty" (Frick 1934, quoted in Spiro 2009, 362).

In 1933, French eugenicist Count Georges Vacher Lapouge wrote to Grant, stating that "Frick has been with us since the beginning" (quoted in Spiro 2009, 362). In the same year, Frick appointed Ernst Rüdin as president of the German Society for Racial Hygiene (Kühl 1994). In 1934, Laughlin penned a letter to Grant, explaining how *EN* would counteract the negative publicity beginning to appear against the Third Reich in the

United States: "We propose devoting an early number of *Eugenical News* entirely to Germany and to make Dr. Frick's paper the leading article. Dr. Frick's address sounds exactly as though spoken by a perfectly good American eugenicist in reference to what 'ought to be done,' with this difference, that Dr. Frick, instead of being a mere scientist is a powerful Reichsminister in a dictatorial government which is getting things done" (quoted in Black 2003, 305). The next issue of *EN* (March–April 1934) was exactly as Laughlin had promised, with Frick's lecture translated by Grant (Kühl 1994). The issue was filled with exuberant praise for Nazi policies, with one article stating: "But may we be the first to thank this *one* man, Adolf Hitler, and to follow him on the way towards a biological salvation of humanity" (quoted in Black 2003, 305).

In 1935, ERA president Clarence Campbell attended the World Population Congress in Berlin and passionately proclaimed: "The leader of the German nation, Adolf Hitler, ably supported by Frick and guided by this nation's anthropologists, eugenists, and social philosophers, has been able to construct a comprehensive racial policy . . . that promises to be epochal in racial history. . . . It sets a pattern which other nations . . . must follow . . . in their accomplishments and in their prospects for survival" (quoted in Kühl 1994, 34). In 1939, Laughlin presented testimony to Congress that argued against allowing persecuted German Jews to immigrate into the United States. Supported by the Carnegie Institution, he wrote a 267-page report attempting to justify these restrictions that was published by the New York State Chamber of Commerce. In it he claimed that America would suffer conquest if it allowed Jews to settle and reproduce, very much like the case of the house rat that had traveled from Europe to the United States. Laughlin and the ERO proudly sent copies of this tome to Frick and to other leading Nazis (Rüdin, Ploetz, Lenz, von Verschuer).

Frick was tried and convicted before the International Military Tribunal at Nuremberg for planning, initiating, and waging a war of aggression, war crimes, and crimes against humanity. He also was accused of being the most highly placed person who bore responsibility for the existence of the concentration camps. He was sentenced to death and hanged in 1946.

Hans Friedich Karl Günther (1891–1968) was a professor at the universities of Jena, Berlin, and Freiburg, where he taught courses in race science, human biology, and rural ethnography. He was considered to have had a major influence on National Socialist racialist thought, and was often referred to as Race Günther (Rassegünther) or Race Pope (Rassenpapst). He was a student of Eugen Fischer and a friend and disciple of Madison Grant. However, it was Julius Lehmann who turned Günther "from a heroic and chivalrous-spirited nationalist into a biological racist. . . . Lehmann went on to publish a steady stream of Günther's books. These defined the Nordic race on the basis of Gobineau's classification, and established a stereotype of the Jewish race. . . . Günther and Lehmann did much to fuse racial hygiene with racist nationalism and anti-semitism" (Weindling 1989, 312).

Günther became the Nazis' main popularizer of Nordicism and a favorite of Hitler's. His early works were a regurgitation of the racial theories of Gobineau and Chamberlain (Wistrich [1982] 1984). At first the academic community was skeptical about Günther, but Lehmann eventually won Lenz, Ploetz, and Fischer's support for him (Weindling 1989). He was the author of seventeen books on racial science and an anthropologist whose work was read most widely in the period (Spiro 2009). (By 1945, an estimated 500,000 copies of his works had been sold [Weindling 1989].) In 1931, one of the first acts of Wilhelm Frick as Nazi minister, with the backing of Ploetz and Lehmann, was to appoint Günther as chair of racial theory at University of Jena, even though his appointment was vehemently opposed by a majority of the faculty senate (Müller-Hill 1998; Cornwell 2003). Hitler attended Günther's inauguration lecture. From 1940 to 1945, Günther served as professor of "racial science" at the University of Freiberg (Wistrich [1982] 1984; Engs 2005).

Günther received many honors. In 1935, he was designated the "pride of the NSDAP." That same year he received the Rudolph Virchow plaque, and in 1940, Hitler awarded him the Goethe Medal for Art and Science. He was guest of honor, with Fischer, at the conference where the obliteration of Jews was discussed. Günther claimed that he could identify a Jew from his posture alone. His most famous work was *Rassenkunde des deutschen Volkes (Racial Science of the German People),* published in 1922, which was an immediate bestseller. It went through numerous editions and was mandatory reading in Nazi Germany; even the students at the Dresden School of Dance needed proficiency in *Rassenkunde* to obtain a diploma (Spiro 2009).

Günther was greatly influenced by Grant's writings, both in form and content. In fact, Kühl (1994) claims that Günther's understanding of race was based on Grant's. His writings were basically a regurgitation of *The Passing of the Great Race,* which had been translated into German in 1925, as well as the standard fare of Gobineau, Chamberlain, and

Lapouge. As these earlier works had, Günther warned that miscegenation (in his case, especially with Jews) threatened to cause the fall of the master race and that negative eugenic measures needed to be implemented. In this book and in another recapitulation of *Passing, Rassenkunde Europas* (1927; issued in English as *The Racial Elements of European History*), Günther included photographs of Grant. He praised Grant's volume as a "remarkable success (which prepared the ground) for the reception of racial and eugenic theories" in the enlightened nations (Spiro 2009, 361). Grant reviewed this book for *EN* and returned the compliment, stating that Günther's books deserved "careful attention of all Americans who have the welfare of the country at heart" (Grant 1928, 120).

Grant and Günther shared a mutual admiration. In 1930, in a letter to Ernest Sevier Cox (see below), Grant said that Günther was "one of the most distinguished anthropologists of Germany" and "was in entire accord with all of our views" (quoted in Spiro 2009, 361). When Günther received his appointment at Jena, he praised Grant for his role in the U.S. Immigration Restriction Act of 1924. Soon after, Grant informed Davenport that he was very pleased that Günther was now in a powerful position. He then sent Günther a library of books by Grantian followers. In return, Günther arranged for the publication in German of Stoddard's eugenics tome, *The Revolt against Civilization*. Günther later referred to Grant and Stoddard as the "spiritual fathers" of immigration restriction and urged that their work serve as a model for Nazi population policies (Kühl 1994).

After the war, Günther was put into internment camps for three years until it was determined that although he was a part of the Nazi regime, he did not instigate criminal acts. The University of Freiburg came to his defense during the postwar trials. Günther was released and resumed ethnographic work in Freiburg. However, he adopted the pseudonyms of Heinrich Ackerman and Ludwig Winter (Billig 1979). He never revised his racist thinking and denied the Holocaust until his death (Wistrich [1982] 1984; Spiro 2009). In 1951, he wrote *Husband's Choice,* listing biological qualities of good marriage partners, and in 1959, he published another eugenics book claiming that unintelligent people reproduce too frequently and that state-sponsored family planning was needed. Günther continued to "warn of the impending demise of civilization, though this time the threat came not from degenerate Jews but from godless communists" (Spiro 2009, 380).

In 1958, Günther became one of the founding members of the Northern League, an international neo-Nazi organization, along with Roger Pearson. Originally inspired by Günther, Roger Pearson (b. 1927) is a Nazi sympathizer, white supremacist, anti-Semite, and neo-Nazi who is still active in the modern eugenic movements and organizations such as the Pioneer Fund and the journal *Mankind Quarterly* (see Chapter 8). Günther died in retirement in 1968.

Otmar Freiherr von Verschuer (1896–1969) was an MD who also had a PhD in genetics; he specialized in the study of twins. Von Verschuer was an ardent racist, anti-Semite, and eugenicist long before Hitler gained power. He was an early and strong supporter of sterilization laws (Weiss 2010). Shortly after World War I, he belonged to the Freikorps, a radically conservative paramilitary group that was involved in several political assassinations and was active in the Kapp Putsch, a coup attempt in March 1920 aimed at overthrowing the Weimar Republic. It was supported by ultraconservative and reactionary forces (Weindling 1989). In 1922, von Verschuer wrote: "The first and most important task of our internal politics is the population problem . . . a problem that can only be solved by biological-political measures" (quoted in Müller-Hill 1998, 504). Four years later, he specified what he considered this problem to be: "The German folk struggle is primarily against the Jews, because alien Jewish penetration is a special threat to the German race" (quoted in Black 2003, 338–339). In 1925, he became secretary of the Tübingen branch of Ploetz's German Society for Racial Hygiene and, with the assistance of his close friend Fritz Lenz, gained a position as assistant physician at the Tübingen Medical Clinic (Ehrenreich 2007; Weiss 2010). In 1927, he was appointed department head of the Human Heredity Department of the KWIA under the direction of his mentor, Eugen Fischer. One year later, using a typical conservative argument, he attacked the welfare state, asserting that social insurance was a burden to the economy and that it reduced people's capacity for work (Weindling 1989). He used his position at KWIA to establish the institute as one of the world centers for twin studies (Weiss 2010). In the festschrift for Fischer, Aichel and von Verschuer (1934, vi) wrote: "We stand upon the threshold of a new era. For the first time in world history, the Fuhrer Adolf Hitler is putting into practice the insights about the biological foundations of the development of peoples—race, heredity, selection." Von Verschuer worked very closely with Fischer to fulfill one of the major functions of the director of the Dahlem Institute, to inform about and legitimize eugenics. Von

Verschuer and Fischer also served together on the Hereditary Health Courts that were established to hear cases regarding the Sterilization Law (Weiss 2010).

In 1935, von Verschuer left the KWIA to open the University of Frankfurt's Institute for Hereditary Biology and Racial Hygiene. This institute became the largest of its kind in Germany and a center of activity for the SS (Weiss 2010). During the 1920s, scientific research in Germany was funded on the basis of a peer review system. However, by the mid-1930s, the SS controlled research funding, and it supported von Verschuer, Fischer, and Rüdin and their institutes (Weindling 1989). Besides conducting research on twins, race biology, and race politics, the Institute for Hereditary Biology and Racial Hygiene "held courses and lectures for the SS, Nazi Party members, public health and welfare officials, as well as medical instructors and doctors in general to indoctrinate them with scientific anti-Semitic and eugenic theory" (Black 2003, 140). Von Verschuer's institute essentially became the center of Hitler's racially based medical program. He did not merely legitimize National Socialist race policies through his writings: in 1937, he agreed to become a "racial expert" for the "Jewish question" (Ehrenreich 2007). In 1941, von Verschuer wrote: "Today every person in Germany has the greatest interest in an objective identification of his or her blood ancestry" (quoted in Weiss 2010, 102). This was obviously an exaggeration, since by that time, Jewish identity could spell death for those deemed by Nazi science to be "racially alien." In the 1941 edition of his *Leitfaden der Rassenhygiene (Textbook of Racial Hygiene)*, he asserted that a "complete solution to the Jewish Question" was necessary and boasted that the "National Socialist state had broken new ground" in solving this problem. He praised Hitler as the savior of the Nordic race and linked genetics to anti-Semitism and the eradication of hereditary diseases (Weindling 1989). In the 1945 edition of his textbook, von Verschuer wrote: "It is a decisive task for the future of our *Volk* to protect the genetic material which is the biological precondition for German culture, from degeneration and mixing with other races" (quoted in Ehrenreich 2007, 62).

In 1937, the most notorious medical butcher of the Third Reich, Josef Mengele (1911–1979) (later known as the "Angel of Death"), became a research assistant to von Verschuer at his institute. The two soon became close friends and formed an inseparable medical team for the remainder of their careers. The year before World War II began, in 1938, Mengele joined the SS, but he continued his association at the Frankfurt University

Institute as an assistant doctor (Müller-Hill 1998). Earlier he had refereed at prosecutions that were held under the Nuremberg Laws (Weindling 1989). In 1942, after receiving two Iron Crosses and two combat medic awards for his work in the Medical Corps of the SS in battle, with von Verschuer's recommendation, Mengele was transferred to the SS Race and Settlement Office in Berlin.

In 1942, Fischer chose von Verschuer to replace him as director of the KWIA in Berlin. Fischer and von Verschuer were very close friends, and Fischer had groomed his student for this position for many years (Weindling 1989; Müller-Hill 1998; Weiss 2010). By this time, Fischer had called for a total solution to the Jewish problem and believed that Jews were a dangerous and inferior subspecies. Von Verschuer (1941) wrote that the war would lead to the "total solution to the Jewish problem. . . . [In my invitation to succeed Fischer] all of my requests were granted with respect to the importance and authority of the institute" (quoted in Proctor 1988, 211). Mengele continued to be von Verschuer's assistant, and in 1943, Mengele arrived at Auschwitz. Von Verschuer wrote to the German Research Society: "My assistant, Dr. Mengele, has joined this part of the research as a collaborator. . . . He is employed as an SS-captain and camp doctor in the concentration camp of Auschwitz. With the approval of the Reichsfuhrer-SS, anthropological studies have been carried out on the very diverse racial groups in this camp" (quoted in Müller-Hill 1998, 20). Now, under the direction of von Verschuer, who had a personal and professional connection to an SS physician, and with the knowledge of Fischer and many other eugenic scientists, cruel and inhuman eugenics research and experimentation reached its culmination. Mengele conducted a variety of horrific experiments on people of all ages (one of his main subjects was twin children) and on a "racially diverse" population at the concentration camps (Weiss 2010). Mengele's research methods drew on von Verschuer's twin studies as a basis for studying hereditary pathology (Weindling 1989). These included, among others, such atrocious things as injecting color dye into the eyes of twins; transplanting bone, muscle, and nerves between individuals; experiments involving head trauma and freezing; injecting malaria or other diseases and toxic chemicals; and experiments involving sterilization (Lifton 1986; Lagnado and Dekel 1991). He then murdered his victims and mailed the harvested tissue (including eyes, internal organs, skeletons, and blood samples of inmates infested with typhoid) in packages stamped "War Material—Urgent" to his superior, von Verschuer, at the KWIA for further research (Lifton 1986; Spiro 2009;

Weiss 2010). In 1944, von Verschuer expressed satisfaction that the hereditary danger Jews and Gypsies posed to the Germans had been "eliminated through the racial-political measures of recent years" (quoted in Proctor, 1988, 211). Since Gypsies and Jews were considered distinct degenerate species, like animals they could be used for Nazi experimentation (Weindling 1989). As Weiss (2010, 112) has stated: "Although there was certainly plenty of morally problematic research and 'service to the state' undertaken under Fischer's watch—more than enough to condemn the Institute in the eyes of future generations even had it closed its doors in 1942—scholars have naturally tended to focus on the unethical research and heinous medical crimes that occurred during the von Verschuer years."

By the mid-1920s, von Verschuer was already becoming a celebrity to American eugenicists. Articles by him and about him appeared regularly in *EN, JAMA, Journal of Heredity,* and the *American Journal of Obstetrics and Gynecology.* These continued to appear even after the Nazis assumed power in 1933. In 1934, *EN* listed von Verschuer as one of the eminent eugenicists in Germany, and the next year it ran a feature article on the opening of "von Verschuer's Institute." The glowing article emphasized that the mission of the facility was the betterment of "mankind" and a "culmination of decades of research by Mendel, Count Gobineau, Ploetz and even Galton himself" (Black 2003, 341). The *EN* article ended by wishing von Verschuer success in his new favorable environment. Articles continued to appear in *EN* and *JAMA* even after the plight of the Jews and thousands of other European refugees had become widely recognized as a world crisis. In 1936, *EN* printed a rave review (almost certainly written by Laughlin) of von Verschuer's book on genetic pathology, which was a clear precursor to the Nazi policy of stripping Jews of their citizenship (Weiss 2010). In 1937, Davenport asked von Verschuer for a special summary of his institute's work for *EN* so he could keep its readers informed.

American eugenicists began to regularly correspond with von Verschuer, especially after his institute opened. For example, in 1936, Stanford anatomist C. H. Danforth offered to translate articles for von Verschuer, and Goddard sent him several publications. Popenoe and his colleagues at the Human Betterment Foundation asked von Verschuer for information that would help them in their efforts to rebut negative publicity about Nazi policies and sent him the latest information on the California sterilization program.

Laughlin sent von Verschuer letters congratulating him on his new institute, offering to keep in touch with eugenic developments in the United

States and volunteering to send reprints of his articles. Von Verschuer replied with appreciation, congratulating Laughlin on his recent honorary degree from the University of Heidelberg. Von Verschuer stated: "You have not only given me pleasure, but have also provided valuable support and stimulus for our work here. . . . You have surely concluded from this [honorary degree] that we German hereditarians and race hygienists value the pioneering work done by our American colleagues and hope that our joint projects will continue to progress in friendly cooperation" (quoted in Black 2003, 342).

In 1937, von Verschuer's popularity among American eugenicists was at its peak. The president of the ERA at the time, Charles M. Goethe (1875–1966), wrote to von Verschuer asking if he could visit the institute, stating that he wanted to counter the anti-German propaganda in the United States. Goethe, a close friend of Madison Grant's, was an entrepreneur, a conservationist, and the founder of the Eugenics Society of Northern California. Although he had visited Germany a few years earlier (see later in this chapter), he was unable to visit in 1937 and sent an apologetic letter to von Verschuer, stating: "Dr. Davenport and Dr. Laughlin . . . have told me of your marvelous work. . . . I feel passionately that you are leading all of mankind herein. . . . America is flooded with anti-German propaganda. It is abundantly financed and originates from a quarter that you know only too well [Jews]. . . . However, this ought to not blind us to the fact that Germany is advancing more rapidly in Erbbiologie [genetics] than all the rest of mankind" (quoted in Black 2003, 343). Goethe continued to correspond with von Verschuer throughout the war and to praise his agenda and the Nazi agenda.

In 1938, a Harvard medical school researcher, Clyde Keeler, visited von Verschuer's institute, which by that time was bedecked with swastikas and devoted to its anti-Semitic and Aryan purity program. Upon his return, Keeler gave his fellow U.S. eugenicists a glowing report. After hearing this report, Danforth wrote his friend von Verschuer in 1939, telling him that Keeler "thinks that you have by all means the best equipped and most effective establishment. . . . May I extend my congratulations and express hope that your group will long continue to put out the same excellent work that has already lent it distinction" (quoted in Black 2003, 344). This was after the Nazi concentration camps and the brutality being carried out in them were well known to the world. In 1938, von Verschuer had agreed to join the advisory committee of the *EN* at the invitation of Davenport and was becoming an essential link between American eugenics and Nazi policy. In 1940, he was welcomed as a foreign member of AES (Weiss 2010). Von Verschuer considered genes in a holistic context as units striving for unity, order, and higher ideals in service to the state, or *Volk* (Weindling 1989). Interestingly, his theory has some similarities to the modern theory of the selfish gene (Weiss 2010).

In January 1945, Auschwitz was liberated and Mengele was listed as a war criminal. However, he evaded an Allied manhunt and escaped to South America, where he lived in hiding until his death in 1979. Tragically, he was never brought to justice. In February 1945, von Verschuer managed to ship equipment, his library, and boxes of records to his home in Solz (Müller-Hill 1998). He had other documents destroyed, including all of his correspondence with Mengele, before his institute was taken over by the "enemy." In March 1945, Hitler killed himself. The Third Reich and its eugenic horrors came to an end, and Nazi doctors and scientists began to reinvent the past. Von Verschuer denied that he and Mengele had complied with the Nazis in any way (Lifton 1986). He also sought support from his American eugenics colleagues, even attempting to obtain a faculty position in an American university. He was given moral but not practical support (Black 2003).

In Germany, in late 1945 or early 1946, a committee of scientists ruled that von Verschuer had indeed engaged in despicable acts in concert with Mengele at Auschwitz, but this report was kept secret for the next fifteen years. In fact, von Verschuer claimed that Mengele had opposed fascist cruelty (Weindling 1989). In 1949, another board of scientists ruled that he had committed no transgressions and soon after, in 1950, he was given a position at the University of Münster's Institute of Human Genetics (Müller-Hill 1998). He later became a dean of that institute and continued to publish. He also became an honorary member of a number of German and international scientific societies and president of the German Anthropological Association (Spiro 2009). It seems that after the war the scientific community attempted to forget von Verschuer's involvement in Nazi racist policy and even in negative eugenic science. For example, in 1956, the Italian medical journal *Acta Geneticae Medicae et Gemellologiae* dedicated a special edition in honor of von Verschuer's sixtieth birthday. This included an extensive bibliography of his work from 1923 to 1956. However, it is noteworthy that his 1924 pamphlet *Rasse,* his article on the "Racial Biology of the Jews" (1938), and his *Textbook of Racial Hygiene* (1941) were absent from this issue (Ehrenreich 2007). During the war, von Verschuer had been accepted as a member of the AES, a position he kept throughout his life. He continued his work as a member of the board of editors of the eugenicist, racist, and Nazi-sympathizing journal *Mankind Quarterly* until his death in 1970 (see Chapter 8).

Von Verschuer was never prosecuted for his crimes and was able to advance his research career through four distinct historical periods, the Weimar Republic, the Third Reich, the Allied occupation, and the Federal Republic of Germany (Weiss 2010). As Black has stated (2003, 380), "The legacy of his torturous medicine, twisted eugenics, and conscious war crimes lives on." When von Verschuer died in 1969 (in an automobile accident), obituaries in German scientific journals made no mention of his Nazi involvement.

Thus, these and other biomedical researchers had a major role in the overall structure and functioning of the Third Reich. As Weiss (2010, 263) has summarized: "They produced, legitimized, and, to a certain extent, popularized—both nationally and internationally—the human genetics knowledge required to make Germany the only fascist state where 'race' became the criterion for citizenship and where human genetics fostered genocide." However, as Paul (1995, 91) notes of the Nazi doctors and scientists: "After the war, nearly all would claim that they too had been victims and suffered greatly under the regime. They knew nothing of the mass murders of mental patients and Jews. . . . Some of their best friends were Jews" (Müller-Hill 1988). Most of the German academic, scientific, and professional institutions survived World War II unscathed or were reconstituted so that their social interests were left intact. Most university teachers and researchers in racial hygiene, public service personnel, and leading racial hygienists were back in office by the mid-1950s. It appears that "social stabilization and strategic advantage were preferred to justice. . . . While the Nazi political elite was removed, professional and administrative structures persisted" (Weindling 1989, 570).

Further Interactions between American Eugenicists and Nazis

As we have seen, American eugenicists were very impressed with their German counterparts and wanted to keep in touch with them and publicize the progress they were making in eugenic research and politics. In fact, as early as 1923, Davenport and Laughlin had renamed *Eugenical News,* adding the subtitle *Current Record of Racial Hygiene,* thus incorporating the German term *race hygiene*. This emphasized the racial component of the eugenics movement. At this time, the journal regularly began to highlight articles from German journals (Allen 1986), as did other eugenics-oriented journals such as the *Journal of Heredity*. Coverage of German publications was not limited to the eugenic press. *JAMA* employed a German correspondent who regularly reported on German eugenicists' medical research. The anti-Semitic slant of Nazi science began to become commonplace in these American journals (Black 2003; Spiro 2009).

By 1931–1932, Hitler's calls for fascist repression, territorial expansion, and persecution of non-Aryan (replacing the traditional term *Nordic*) races were prominent in newspapers and were being heard on American radio and in newsreels. This did not diminish American eugenicists' support of their German counterparts (Allen 1986). Just before Hitler's takeover, *EN* carried a long article praising his eugenic views and pointing out how he had been guided by American eugenicists Lothrop Stoddard and Madison Grant. It stated that "When they [Hitler and the Nazis] take over the government in Germany, in a short time there may be expected new race hygienic laws and a conscious Nordic culture and 'foreign policy' " (quoted in Black 2003, 298). As Nazi policies of racial repression and ethnic cleansing intensified, news of their atrocities was not kept secret. They were "chronicled daily on the pages of America's newspapers, by wire services, radio broadcasts, weekly newsreels, and national magazines. Germany bragged about its anti-Jewish measures and eugenic accomplishments. . . . Simultaneously, American eugenicists kept day-to-day tabs on the Nazi eugenic program" (Black 2003, 299). Much of the world was beginning to understand the atrocities the Nazis were committing and was repulsed by them. Most of the world's media was now focusing on the inhumane aspects of Hitler's regime.

As Germany passed its mandatory sterilization laws in 1933, American eugenics journals and other publications continued to cover Nazi policies with pride and excitement. *EN, JAMA, Birth Control Review,* and *American Journal of Public Health* all contained articles praising the new laws and other Third Reich policies and racist research. Black (2003, 303) describes the mid-1930s: "With each passing day, the world was flooded with more Jewish refugees, more noisy anti-Nazi boycotts and protest marches against any scientific or commercial exchanges with Germany, more public demands to isolate the Reich, and more shocking headlines of Nazi atrocities and anti-Jewish legislation. Still none of this gave pause to America's eugenicists. Correspondence on joint research flowed freely across the Atlantic. American eugenicists, and their many organizations and committees . . . maintained and multiplied their contacts. . . . Eager and cooperative letters, reports, telegrams and memoranda did not number in the hundreds, but in the thousands of pages per month."

Things began to change only after the negative publicity surrounding the Nuremberg Laws began to spread in 1935. It was an attempt to counter negative publicity in the United States that stimulated German authorities to publicly recognize Harry Laughlin's contribution to Nazi policy by conferring him an honorary degree (Black 2003; Weiss 2010). In fact, the awarding of many honorary degrees to U.S. recipients and universities was politically motivated (Weiss 2010). Even after the Nuremberg Laws were passed, Laughlin and Davenport continued to publish regularly in German journals.

From 1936 to 1939, Nazi Germany began attempting to appropriate territory from its European neighbors and preparing for war. News of gruesome treatment of eastern Europeans in concentration camps began to fill the world's media, and refugees began to flood the world. American eugenicists soon began to distance themselves from Nazi policies. "Genuine revulsion with Nazified eugenics was beginning to sweep over the ranks of previously staunch hereditarians who could no longer identify with a movement so intertwined with the race policies of the Third Reich. . . . Longtime eugenicists and geneticists spoke of a resolution to dissociate eugenics from issues of race" (Black 2003, 313). In 1936, journals such as *JAMA* finally began to criticize Nazi medical policies and research, and American funding of the Nazis began to dissipate (see Chapter 6). However, die-hard American eugenicists continued to admire Hitler's regime and defend his policies.

In late 1935, ERA president Clarence Campbell attended the World Population Congress in Berlin and stated that Hitler and the German nation "sets a pattern which other nations and other racial groups must follow if they do not wish to fall behind in their racial quality, in their racial accomplishments and in their prospects for survival" (quoted in Black 2003, 314). Campbell, Laughlin, and Raymond Pearl all served as vice-presidents at that congress. Charles M. Goethe, the California eugenicist and successor to Campbell as ERA president, was able to tour Germany in 1935 and was thrilled by their program. In his presidential address to the ERA in 1936, Goethe pointed out how Germany in only two years had outpaced even California in its sterilization operations. During that same year, eugenicist Marie Kopp toured Nazi Germany and witnessed its Heredity Courts. She wrote articles throughout that year praising the "fairness" of the Nazi program. In 1936, the Nazi regime held a major propaganda extravaganza at the 550th anniversary of Heidelberg University. This was the occasion at which Laughlin was to receive his honorary degree. Although the *New York Times* called for American scholars to boycott the event (and many educators from Europe refused to attend), representatives from Yale, Cornell, Columbia, Vassar, Johns Hopkins, Michigan, and Harvard were there. During the same year, Madison Grant's brother, De Forest, spent two weeks in Germany, and W. A. Plecker delivered a paper in Germany on Virginia's miscegenation laws and their success in halting the "spread of the mongrel races." His ticket was paid for by Prescott Bush, father of the forty-second president of the United States and grandfather of the forty-fourth (Spiro 2009).

In 1937, Hermann Goering was commander in chief of the Luftwaffe, president of the Reichstag, prime minister of Prussia, and Hitler's designated successor. He was also, like Madison Grant, a conservationist and hunter, and Reich Master of the hunt in Germany. In this capacity, he planned and hosted a massive international hunting exposition in Berlin. The three-week-long festival was to promote comradeship, goodwill, and understanding among the international hunting fraternity. As his final undertaking with the Nazis, Madison Grant helped organize this event and was to be one of Goering's honored guests. Although Grant did not live long enough to attend the hunt, his last project incorporated his two main life passions, hunting and Nazi-like eugenics (Spiro 2009).

In 1937, Frederick Osborn attended an AES conference in Berlin and stated that "recent developments in Germany constitute perhaps the most important experiment which had ever been tried" (quoted in Spiro 2009, 364). Three years before that, his father, Henry Fairbanks Osborn, had traveled to Frankfurt to receive an honorary doctorate and had praised the Hitler regime, stating that "the metempsychosis of Germany is one of the most extraordinary phenomena of modern times" (quoted in Spiro 2009, 371). Also in 1937, Laughlin, the ERO, the ERA, and the Pioneer Fund (see below) distributed Nazi eugenic propaganda films (such as *The Hereditarily Diseased*) to American churches, clubs, children's welfare agencies, and high schools. Even into 1938, when Germany had begun to seize all Jewish property and threatened to devour neighboring countries, Laughlin's *EN* and ERO still applauded Hitler's campaign. Finally, however, in late 1938, the Carnegie Institution disengaged from *EN,* and on December 31, 1939, it formalized Laughlin's retirement and closed the ERO. However, just before his forced retirement, Laughlin was allowed to deliver testimony to the U.S. Senate Immigration Committee and to a

Special Committee on Immigration and Naturalization in New York concerning limiting immigration of Jewish refugees into the United States. In this, he repeated a great deal of material directly from Hitler's *Mein Kampf*. His report was published as a 267-page document entitled *Immigration and Conquest* and contained a statement comparing Jewish immigration to rodent infestations.

In 1939, Lothrop Stoddard (1883–1950), a Grant protégé and Harvard PhD, visited Nazi Germany. Earlier, in 1920, he had written the racist diatribe *The Rising Tide of Color: Against White World-Supremacy,* which Grant wrote an introduction for. In it, Stoddard parroted the views of earlier American eugenicists on polygenics, the dangers of mixed breeding, and genetic determinism almost verbatim. He toured the country for four months, visiting American eugenicists, including Fischer, Günther, Frick, and Lenz (Black 2003; Engs 2005; Spiro 2009). He also visited many of the top Nazi political and military leaders, including Hitler himself. In fact, Stoddard was the first foreigner allowed an audience with Hitler since the beginning of the war. He was granted the visit when Stoddard assured the foreign office that he wanted to present Hitler's "human side" to his readers. Stoddard published his observations in *Into the Darkness* (1940). In it, Stoddard presented Germany as "dark" not because of its atrocious activities but rather because Stoddard believed that its "pragmatic" policies were being hidden (kept dark) from Western view. Nazi laws, Stoddard (1940, 147, 189) stated, are "weeding out the worst strains in the Germanic stock in a scientific and truly humanitarian way . . . [and the Jewish problem would soon be settled] by the physical elimination of the Jews themselves from the Third Reich."

After the Carnegie Institution dropped *EN,* the journal continued to be published by the AES. When Laughlin died in 1943, Davenport eulogized him in the new *EN,* stating that his views were founded more on a thorough analysis of fact than on sentiment and that within a generation his work would be widely appreciated (Black 2003). However, Laughlin left no children and left no meaningful legacy. Interestingly, he also died of a hereditary defect, epilepsy, a disease that he believed should be eliminated from society by sterilization or the imprisonment of the diseased in segregated camps, just like the feeble-minded, the "blind" (of various degrees), blacks, Jews, and other "inferior, unfit" races. Davenport continued to do research and be the active elder statesman of eugenics until his death, from pneumonia, in 1944. He never wavered in his support for and his scientific solidarity with Nazi science or policies, and he defended Nazism as late as 1943 in *EN*.

After the United States entered the war in 1941, fewer eugenicists were willing to voice their strong support for the Nazis or even eugenics, in speeches or in writing. However, eugenics did not disappear. As I will discuss in later chapters, eugenics has resurfaced over the years both literally and theoretically, and some of the same individual eugenicists, both German and American, were involved in keeping it alive.

Funding of the Nazis by American Institutions and Businesses

From its beginnings, eugenics in the United States was an elitist movement and was supported and funded by a number of very rich individuals, institutions, and corporations.

One of the earlier supporters of Adolf Hitler and the Nazi regime was the American millionaire Henry Ford (1863–1947), inventor of the assembly line production of automobiles, manufacturer of the Ford motor car, and founder of Ford Motor Company. Ford was a eugenicist and virulent anti-Semite who believed that Jews had a secret plot to take over world finances and politics by deceit and subterfuge. In 1920, Ford began publishing a series of ninety articles entitled "International Jew: The World's Foremost Problem" (later published as a set of booklets with the same title) in his own newspaper, the *Dearborn Independent*. His evidence for this worldwide plot presumably came from *Protocols of the Meetings of the Learned Elders of Zion,* an anti-Semitic book that was ultimately derived from a series of novels traced back to the 1840s or even earlier in the eighteenth century (Brace 2005). Ford's *International Jew* was translated into German soon after its original publication and became very popular in Germany. Hitler read both it and Ford's autobiography and was extremely impressed. In fact, some passages of *Mein Kampf* are essentially identical to what appears in *International Jew,* and it appears that Hitler copied directly from it (Pool and Pool 1979). Hitler also was influenced by Ford's insistence that Jewishness was a "race" and not a religion.

Ford disliked the British, and as World War II approached, he refused to provide aircraft engines to Britain. He continued to sell engines and vehicles to the Nazis, however, until 1941. He opposed the U.S. entry into the war. Hitler boasted of Ford's support and had a large photograph of Ford on his office wall (Brace 2005). He awarded Ford by presenting him with the Grand Service Cross of the Supreme Order of the German Eagle on Ford's seventy-fifth birthday, July 30, 1938. The Italian fascist dictator Mussolini was the only other non-German to receive this medal.

Other U.S. industrialists who were decorated by Hitler included Thomas Watson of IBM and James D. Mooney of General Motors. Watson's punch card technology had enabled Hitler to establish his rearmament program and to carry out his identification and attempted elimination of Jews. Watson had supported Nazi Germany against his anti-Nazi countrymen in the United States, had made it possible for Hitler to conduct commercial interactions despite international boycotts, and had brought an international summit to Berlin. He was also a valued supporter and crony of Franklin D. Roosevelt (Black 2001). Mooney was an avid Nazi supporter, and General Motors provided moral and financial support to Hitler. It also provided trucks, armored cars, tanks, jet fighter planes, and other war equipment to the Nazis well into the war against the United States and its allies (Higham 1983).

The Kaiser Wilhelm Institutes became the center of German eugenics medical research early in Hitler's regime. Three of the most prominent were the Institute for Psychiatry; the Institute for Anthropology, Human Heredity, and Eugenics; and the Institute for Brain Research. All of these received American funding, especially from the Rockefeller Foundation (Higham 1983; Black 2003). Using the model of American-style philanthropy, Germany had attempted to compete with American science and medicine by forming the Kaiser Wilhelm Institutes in 1914. German medicine and scientific research, which had been world class, had been paralyzed by defeat in World War I and by crippling inflation.

In an attempt to save German science and medicine, the Rockefeller Foundation decided to fund German government research institutions. Beginning in 1922, the foundation awarded hundreds of fellowships to German scientists and institutions. Since U.S. and German relations were still uneasy at that time, most of these were administered through the foundation's Paris office (Weindling 1989). By 1926, the foundation had given the equivalent of $4 million to German research. Even some German scientists were complaining about the Rockefeller Foundation's control over German science (Weindling 1988). In 1926, the foundation began supporting the newly formulated Kaiser Wilhelm Institute for Psychiatry under the direction of Ernst Rüdin, who had just begun collecting records about German families, similar to those of the ERO. However, Rockefeller funding for Rüdin ended in 1937 (Weindling 1989; Weiss 2010).

In 1927, Germany added the Kaiser Wilhelm Institute for Anthropology, Human Heredity, and Eugenics, with Eugen Fischer as its director. This institute was funded by the German government and the Rockefeller Foundation (Spiro 2009). Fischer was a longtime friend and collaborator of Davenport and an associate of the Carnegie Institution. As mentioned earlier in this chapter, the Carnegie Institution funded a large contingent of scientists and officials to attend the grand opening of this institute, which was held during the Fifth International Genetics Congress in Berlin. In 1929, the Rockefeller Foundation funded the renovation of the Kaiser Wilhelm Institute for Brain Research, which then became the center of German race biology. By 1936, under increasing criticism, the Rockefeller Foundation had become increasingly reluctant to fund Nazi German science. However, some funding continued until at least 1939 (Black 2001). The rationale behind the foundation's funding for Nazi eugenic science was well articulated by the director of its Division of Natural Sciences, Warren Weaver, who stated in a report to the trustees: "Work in human genetics should receive special consideration as rapidly as sound possibilities present themselves. The attack planned . . . is a basic and long-ranged one." A year later he specified that the long-range plan was to "develop so sound and extensive a genetics that we can hope to breed, in the future, superior men" (quoted in Black 2003, 370). The Rockefeller family also contributed to the Nazi war effort through its business interests. For example, Rockefeller-owned Standard Oil of New Jersey shipped fuel to Nazi Germany through neutral Switzerland after the United States had entered the war. Rockefeller-owned Chase Bank was doing millions of dollars of business in Nazi-occupied Paris after Pearl Harbor, with the full knowledge of its Manhattan office (Higham 1983).

International Business Machines (IBM) began contributing to Davenports' eugenics research during his research on race-crossing in Jamaica in 1927. Davenport was worried that mixed breeding would lower the standards of Western civilization and wanted to systematically survey the world for mixed-race populations. He believed, for example, that "well established races" were "harmoniously adjusted." He wondered about such questions as this: If a Scot, a member of a tall race, mixed with an Italian of shorter stock, would it yield offspring with inadequate organs, since the Scot's organs were adapted for a large frame and the Italian's organs would be too large for a small frame (Davenport 1928; Paul 1995)? Multimillion-aire Wickliffe Draper gave Davenport the funds for a demonstration project in Jamaica. Davenport collected his data using a new system created by IBM—punched data cards processed by Hollerith processing machines. These punch cards could store an unlimited amount of information about people, including family trees, location, specific traits, religion, "race," maladies, bank accounts, and so forth. The machines could process 25,000 cards per hour, alphabetize information, and print it out. They were originally developed for the U.S. Bureau of the Census but were ideal for tracking personal information for eugenics research. IBM monopolized this type of data processing and machinery worldwide (Black 2001).

The Jamaica race-crossing investigation was the first time IBM had custom designed a system to track and report on racial characteristics for a eugenics project. The Jamaican project was considered a success, and the Carnegie Institution published the results (Davenport and Steggerda 1929). Davenport, who wanted to expand this research worldwide, developed and chaired a Committee on Race Crossing within the IFEO. He asked Eugen Fischer to join this committee and be a partner in his cross-breeding research project. In 1929, the project went global and Fischer was asked to assume chairmanship of the committee. In a joint memo, Davenport and Fischer reported that eventually they would identify not only race-crossed individuals worldwide "but entirely foreign people, that is the so-called colored ones" (quoted in Black 2003, 294). Later that same year, using IBM punch card techniques, Fischer began a national "anthropological survey" in Germany to locate and identify mainly the 600,000 German Jews (Black 2001). German officials stated: "In this way it is hoped to find new solutions about the appearance of certain signs of degeneration, especially distribution of hereditary pathological attributes" (quoted in Black 2003, 294). The project was funded by the Rockefeller Foundation.

By 1933, IBM had adapted the technology used in the Jamaica study for general use by Hitler's Reich, and in 1934 IBM opened a million-dollar factory in Berlin. At the factory opening, the manager of the German subsidiary, Willi Heidinger, standing next to a representative of T. J. Watson, the president of IBM, "emotionally declared that population statistics were key to eradicating the unhealthy, inferior segments of German society" (Black 2003, 309). IBM custom-designed the systems that created national identification card files for people designated as unfit and helped locate European Jews and other "undesirables." The punch card programs enabled the Nazis to trace family trees, index bank accounts and other property, organize eugenic campaigns, and even manage extermination in death camps (Black 2001). The punch cards the SS Race Office used were almost identical to those used in the Jamaica study (Black 2003). As Black (2009) has stated, "With IBM as a partner, the Hitler regime was able to substantially automate and accelerate all six phases of the 12-year Holocaust: identification, exclusion, confiscation, ghettoization, deportation, and even extermination. . . . There was an IBM customer site in every concentration camp." Indeed, IBM data processing became the key to Nazi persecution of "unfit" people and "races." And it should not be forgotten that IBM generated massive profits as it took over the organization and systematization of the Nazi eugenics programs. IBM New York understood from the beginning in 1933 "that it was courting and doing business with the upper echelon of the Nazi Party. The company leveraged its Nazi Party connections to continuously enhance its business relationship with Hitler's Reich, in Germany and throughout Nazi-dominated Europe" (Black 2001, 9). IBM's U.S. top officials were aware of what was happening during the twelve-year reign of the Third Reich and constantly monitored the subsidiary activities in Germany and Geneva. The company owner, Thomas Watson, was a highly decorated businessman in the United States and was a financial supporter and close consultant of President Roosevelt. However, his business interests in Nazi Germany were becoming extremely profitable, and he was very sympathetic to the Nazi racist goals. He was a staunch supporter of both Mussolini and Hitler, and although he was not particularly driven by fascism, he recognized its profitability. As Black (2001, 69) has stated: "Thomas Watson and IBM had separately and jointly spent decades making money any way they could. . . . To a supranational, making money is equal parts commercial Darwinism, corporate ecclesiastics, dynamic chauvinism, and solipsistic greed."

Many other U.S. companies, while serving the U.S. war effort, also contributed to Nazi Germany immediately before and during World War II. These companies included banks, merchants of diamonds and other raw materials, automobile makers, armaments and machinery manufacturers, and telecommunication, filmmaking, oil, chemical, drug, and food companies. Although we cannot assume that the executives and owners of all of these companies were eugenicists, we know that some of them, such as Ford, Rockefeller, Carnegie, and some of the IBM executives, were. Most likely others were more interested in making a profit than in politics or human rights. As Charles Higham (1983, xiii) stated in his book *Trading with the Enemy:* "a number of financial and industrial figures of World War II and several members of the government served the cause of money before the cause of patriotism. . . . While aiding the United States' war effort, they also aided Nazi Germany's."

As Higham (1983) has pointed out, a number of major American corporate leaders formed interlocking directorates and financial connections with Nazi corporations, including IG Farben, the massive Nazi industrial trust that was critical to the development of the German war machine leading up to World War II and that helped create Auschwitz. In fact, Henry Ford merged his German assets with the IG Farben chemical cartel in Germany (Sutton 1976). IG Farben had business contracts with companies around the world, including Standard Oil, DuPont, Alcoa, Bayer, and Dow Chemical.

Higham (1983) labeled this interconnected group of international corporate giants The Fraternity. These included First National Bank (affiliated with J. P. Morgan), National City Bank, and Rockefeller's Chase National Bank, as mentioned above. These banks had connections with Hitler's Reichsbank and the Bank for International Settlements of Switzerland. For example, in 1944, T. H. McKittrick, the American president of the bank, held an annual meeting of bankers from Germany, Japan, Italy, Britain, and America to discuss millions of dollars' worth of gold that had been sent to the bank by the Nazi government for use by Nazi leaders during and after the war. McKittrick, a Nazi supporter, was enamored of both Hitler and Mussolini and, interestingly, like many of the eugenicists of the time, was a 1911 graduate of Harvard. The gold had been looted from European national banks after German takeovers or was Reichsbank gold that had been melted down from teeth fillings, eyeglass frames, cigarette cases and lighters, and wedding rings of murdered hospital and concentration camp victims. In 1943 and 1944, members of Congress and the Senate called for an investigation of the bank and its American president because (1) the bank was being used to further the war efforts of the Nazis; and (2) American money was still flowing into the bank— essentially, the Bank for International Settlements was aiding and abetting the enemy (Higham 1983). The U.S. government called for a liquidation of the bank, but it quietly ignored the American resolution. After World War II, the Bank for International Settlements reemerged as the main clearinghouse for European currencies (Epstein 1983).

J. P. Morgan and Union Bank of New York were owned by Brown Brothers and the Harriman family. Union Bank was intimately linked to the German industrial empire of steel magnate Fritz Thyssen, who had helped Hitler rise to power. The bank was managed by Prescott Bush, father and grandfather of the two Bush presidents. Prescott Bush was a strong supporter of Hitler, funneled money to him through Thyssen, and made considerable profits in dealings with Nazi Germany (Pauwels 2003).

Shortly before the war, the Nazi government, through Chase National Bank, offered Nazis in America the opportunity to buy German currency (Deutsche Marks) with dollars at a discount. Investors were told that the money would be used for Nazi interests and that the German Marks would greatly increase in value after victory in the impending war. "The bankers agreed that special attention should be focused on shopkeepers, factory workers, and others with little money but great potential for Germany. . . . They should be able-bodied young men and women of pure Aryan stock" (Higham 1983, 21).

The Fraternity also included Standard Oil, Davis Oil Company, Texas Company (Texaco), Ford Motor Company, General Motors, and Chrysler. In fact, GM "brought mass productions [of automobiles] to the Reich, converting it from a horse-drawn threat to a motorized powerhouse" (Black 2003). GM also built thousands of bombers and jet propulsion systems for the Luftwaffe at the same time that it was producing aircraft engines for the U.S. Army (Price 2001). Germany obtained raw materials such as rubber and nonferrous metals through the American Ford Company, and Ford-Werke was instrumental in German armament production after the United States entered the war, with full knowledge of its American executives (Sutton 1976). When U.S. troops landed in Normandy in 1944 and captured German trucks, they were amazed to discover that the engines were powered by GM and Ford (Pauwels 2003).

The American conglomerate International Telephone & Telegraph helped Hitler improve his communication systems and develop bombs that were dropped on Britain and on U.S. troops (Higham 1983). The Fraternity also included New York diamond merchants who smuggled commercial and industrial diamonds to Germany through South America. Other well-known American companies with strong interconnections with Nazi Germany were DuPont, Union Carbide, Westinghouse, General Electric, Gillette, Goodrich, Singer, Eastman Kodak, and Coca Cola (Pauwels 2003). American companies profited greatly by dealing with Nazi Germany because Hitler had cut labor costs tremendously. He had dissolved labor unions and had thrown many communist and socialist sympathizers and left-wing "liberal" supporters (political prisoners) into prisons and concentration camps, thus emasculating the German working class. German workers "were little more than serfs forbidden not only to strike, but to change jobs," driven "to work harder [and] faster" while their wages "were deliberately set quite low" (Pendergrast 1993, 221). Companies in Nazi Germany also used prisoners as laborers (forced laborers, or *Fremdarbeiter*), who were supervised directly by the Gestapo. Coca Cola, Ford, and GM used these laborers. Because of these tactics, wages declined rapidly and profits increased precipitously, even during the Great Depression. The war was so profitable to many of these companies that a number of their executives thought that the longer the war lasted, the better (Black 2001; Pauwels 2003).

The Fraternity had set itself up to maintain financial, industrial, and political autonomy no matter who won the war (Higham 1983). The American ambassador to Germany, William E. Dodd, was aware of the connections between American businessmen and other wealthy Americans and Nazi Germany. In 1937, he wrote to President Roosevelt: "A clique of U.S. industrialists is hell-bent to bring a fascist state to supplant our democratic government and is working closely with the fascist regime in Germany and Italy. I have had plenty of opportunity in my post in Berlin to witness how close some of our American ruling families are to the Nazi regime. . . . Certain American industrialists had a great deal to do with bringing fascist regimes into being in both Germany and Italy. They extended aid to help fascism occupy the seat of power, and they are helping to keep them there" (quoted in Higham 1983, 167).

It appears that the goal of many of these wealthy and powerful individuals was to maximize profits by supplying both sides with materials to conduct the war. Did the racial hatred of Hitler bother these industrialists and businessmen? "Apparently not much, if at all. . . . After all, racism against non-Whites remained systematic throughout the U.S. and anti-Semitism was rife in the corporate class" (Pauwels 2003). The anti-Semitism of corporate America soon evolved into anti-socialism, anti-Marxism, and anti-communism. Many American businessmen denounced Roosevelt's New Deal as "socialistic." Anti-Semites of corporate America considered Roosevelt to be crypto-Communist and an agent of Jewish interests, and his New Deal was vilified as the "Jew Deal" (Pauwels 2003). As Pauwels (2003) has emphasized, many of the owners and managers of American corporations admired the Third Reich because unconditional collaboration with Hitler made it possible for them to make profits like never before.

As Higham (1983, xiv) has noted: "When it was clear that Germany was losing the war the businessmen became notably more 'loyal.' Then, when war was over, the survivors pushed into Germany, protected their assets, restored Nazi friends to high office, helped provoke the Cold War, and insured the permanent future of The Fraternity" (see Martin 1950 for more on this subject).

5

The Antidote: Boas and the Anthropological Concept of Culture

How did the eugenics paradigm lose its tight grip over U.S. and Western European science? Although we can't say that it has vanished entirely from the scene, as I will soon show, we can say that it no longer dominated the majority of science after World War II. What were the basic tenets of polygenics and eugenics from the sixteenth century right up until the 1930s and 1940s? They were: (1) that certain complex behaviors and morphological traits were fixed and immutable; and (2) that these traits were created either by God and/or biologically determined through simple Mendelian inheritance. These traits included such things as the size and shape of the head and brain, intelligence, and the ability to create, to meaningfully participate in, and to contribute to "civilization." Furthermore, the polygenicists and eugenicists believed that human beings were biologically divided into races, and even economic classes, which were dealt different inherent abilities and that environmental or behavioral interventions could not change these abilities. They also believed that these races and classes could be put into hierarchical categories according to their abilities. Finally, they believed that some races, especially Nordic/Aryan races, were highly superior to others and were the "genetic carriers" of civilization and of the biological traits that made civilization possible.

Theoretical Background

What *scientific* evidence was brought to bear against this mythology that was over 400 years old? Just around the time when eugenics was rampant in the United States and sterilization laws, anti-miscegenation laws, and

146

restricted immigration policies were being imposed in the United States, Franz Boas (1858–1942), a German-Jewish immigrant to the United States, was beginning his lifelong fight against the basic "scientific" concepts of eugenics. In anthropology, he developed the profoundly important anthropological concept of culture. He also began to counter the adherence in physical anthropology to strict static anthropometric measurement and fixed racial typological thinking. He began to change this discipline from a static, one-dimensional field of study based on typology to one that was problem-oriented, dynamic, and focused on process (Sussman 1999; Little and Kennedy 2010). In fact, his work was just the beginning of an avalanche of scientific proof that was to show that eugenics was pure fallacy and was driven by ideology.

In 1911, Boas wrote two books that were pivotal to the eventual collapse of the acceptance of eugenics as conventional wisdom, *Changes in Bodily Form of Descendants of Immigrants* and *The Mind of Primitive Man*. These volumes appeared the year after Davenport had hired Laughlin to head the ERO, the same year that Davenport's textbook *Heredity in Relation to Eugenics* appeared, and the year before the First International Eugenics Congress in London.

Boas was born in Minden, Germany. His grandparents were observant Jews, but his parents were not. They were educated, well-to-do, and liberal and opposed dogma of any kind. Thus, Boas was given the independence to think for himself and pursue his own interests. However, although he was not a practicing Jew, Boas was sensitive about his Jewish ancestry and vocally opposed anti-Semitism (Cole 1999). In an autobiographical sketch, Boas ([1938] 1974, 41) wrote, "The background of my early thinking was a German home in which the ideals of the revolution of 1848 were a living force. My father, liberal, but not active in public affairs; my mother, idealistic, with a lively interest in public matters; the founder about 1854 of the kindergarten in my home town, devoted to science. My parents had broken through the shackles of dogma. My father had retained an emotional affection for the ceremonial of his parental home, without allowing it to influence his intellectual freedom. Thus I was spared the struggle against religious dogma that besets the lives of so many young people."

From an early age Boas was interested in natural history and geography. However, when he entered Heidelberg University, he studied mathematics and physics, enrolling in the University of Kiel for his final degree. His first choice for a thesis topic was related to statistical probability and random error. However, his dissertation was on a physics topic, the optical properties of water. He received his PhD in 1881. This subject was not his passion, and by patching together a number of jobs after receiving his degree, Boas was able to begin work on his true academic interest: human geography, specifically, the relationship between people and their physical environment, with a concentration on the Eskimo. Doing research in the Arctic had been a boyhood dream (Cole 1999; Baehre 2008).

In 1882, he moved to Berlin and began working with the prominent ethnographer Adolf Bastian at the Berlin Museum of Ethnology, which had collections of Eskimo artifacts. Bastian's recommendation made it possible for Boas to conduct his first field research, as a member of the German Polar Commission mission to Baffin Island. While at the museum, Boas also learned anthropological measurement techniques (anthropometry) from Rudolf Virchow, who was responsible for the physical anthropology section of the museum.

Adolf Bastian (1826–1905) was the reigning patriarch of German ethnography. He was a prodigious scholar who first studied law and then medicine. But while traveling around the world as a ship's doctor, he turned to ethnography. He founded the journal *Zeitschrift für Ethnologie* and, with Rudolf Virchow, organized the Berlin Society for Anthropology, Ethnology, and Prehistory. Bastian also was the head of the Royal Geographical Society of Germany (Köpping 2005). His major interest was in documenting various aspects of peoples around the world before their "civilizations" vanished. He developed the idea of the "psychic unity of mankind"—that is, that all humans shared a basic mental framework. Haeckel, a contemporary German scientist and eugenicist, diametrically opposed this idea (Marks 2010b; 2012). As we shall see, Boas accepted the idea of human psychic unity from Bastian but in a modified form.

Rudolf Virchow (1821–1902) was a doctor, pathologist, anatomist, biologist, prehistorian, and what we might call today a biological anthropologist. He made contributions to medicine, anthropometry, cell biology, human adaptability, archaeology, ethnology, and epidemiology (Baehre 2008; Marks 2010a). Among his many accomplishments in medicine and anatomy, he conducted a craniometric and ethnological study of Eskimo families that had been brought to Germany. He asked questions about their diet, their ability to count, and their color perception and compared their tools, clothes and tattoos. His findings "contradicted anthropological claims of innate Inuit racial inferiority and raised questions about the overriding role of geography in determining their characteristics" (Baehre 2008, 21). His conclusions thus contradicted contemporary scientific racist theories about the "Aryan race," leading him to denounce "Nordic mysticism" and the idea that any "European race" was superior to any other (Virchow 1880; Weindling 1989; Orsucci 1998; Baehre 2008). Virchow was also a defender of a tolerant, pacifist, humanistic vision of human society; a leader of the liberal political party in Germany; and a political reformer. In contrast to Haeckel, he rejected the idea that there were "higher" or "lower" races of mankind. His ideas reflected an anthropology of scientific liberalism (Weindling 1989).

Interestingly, both Bastian and Virchow were extreme empiricists who questioned Darwin's theory of natural selection because of the lack of observable empirical evidence for it. Also, as Marks (2010a, 2012) has noted, these scholars were obliged to strongly reject Darwinism in opposition to the growing use of Darwin's theories in support of Haeckelian, racially intolerant, proto-Nazi political ideologies in Germany.

Thus, Boas was receiving training and support for his liberal ideology from leading men in their fields at the time. Boas was greatly influenced by these two men and continued to correspond with them for the remainder of their lives. Of Virchow, Boas would later state he was the model of the natural scientist, "the ice-cold flame of the passion for seeking the truth for truth's sake" (Boas 1945, 1; Stocking 1974, 22; Baehre 2008).

During his first field expedition to Baffin Island in 1883, Boas concentrated his research on the language, myth and songs, customs, habits, material life, and seasonal cycle of the Netsilik Eskimo. He also studied their migration patterns, relationships between groups, the configuration of the land, and its food supply (Cole 1999). His goal was to study "the interaction between the organic and the inorganic, above all between the life of a people and their physical environment" (Stocking 1974, 43–44, quoting the draft of a letter to A. Jacobi). He also began to worry about how to conserve indigenous peoples and the urgent need to obtain knowledge from and about them, since he foresaw the rapid disappearance of their way of life. He lived among the people, completely adopting their mode of life (Cole 1999). In this first experience, Boas was already expressing his view of the worth of these people. He believed in the equality of virtue among peoples and that their inner character *(Herzensbildung)* was far more important than the veneer of "civilization" or learning (Baehre 2008). "I often ask myself what advantages our 'good society' possesses over that of the 'savages' and find, the more I see of their customs, that we have no right to look down upon them. . . . We have no right to blame them for their forms and superstitions which may seem ridiculous to us. We 'highly educated people,' relatively speaking, are much worse" (Boas diary 1883, quoted in Cole 1999, 79).

Boas's interests were shifting from geography to anthropology. In fact, he was becoming an anthropologist. However, his approach to anthropology was different from those who had worked in this field before him. Boas was beginning to combine the various fields of research on humans into an integrative anthropology, one we would now call the four-field approach, one that includes ethnography, archaeology, biological anthropology, and linguistics (Boas 1940). To this he added a scientific background of inductive reasoning, an interest and knowledge in statistics, and a focus on dynamic processes as opposed to a descriptive, typological, static approach. He also emphasized a historical and relativistic view of human behavior and morphology (Baehre 2008).

Basically, Boas was developing modern scientific anthropology, while others who considered themselves anthropologists at that time were doing nothing that would be considered professional anthropology today. In fact, with his training and research in ethnography, geography and geology, anatomy and anthropometry, statistics, language, and historical content and processes, Boas essentially formulated and developed what would now be considered modern anthropology, at least by most contemporary anthropologists.

After returning to Germany and doing more work and training with Bastian and Virchow, Boas migrated permanently to the United States in 1886. He had become disillusioned with German politics, academics, and academia. In addition, his fiancé and future wife lived in New York. In the United States, Boas began to work in earnest on further field and museum work as an anthropologist. Although he had to move around in various professional positions (he was an assistant editor for the journal *Science* and an assistant professor at Clark University, and he did museum work at the Smithsonian Institution, the Field Museum, and the American Museum of Natural History), Boas finally landed a permanent position at the American Museum of Natural History as assistant curator of ethnology and somatology in 1896, with a joint appointment at Columbia University in 1897. Two years later, in 1899 at the age of forty-one, he became a professor at Columbia, where he remained until he retired.

From the time of his immigration to the United States until he published his two major works in 1911, Boas continued to expand his work on a new and dynamic anthropology. In what would now be considered physical anthropology, he collected massive data on the anthropometry (physical measurements) of Native Americans (Little 2010). Given the current belief that skull shape revealed racial characteristics and such traits were fixed, he was surprised by the amount of variation he encountered in the measurements of different tribes, even those that spoke the same language (Cole 1999). He also had done studies of tribal mixture and found no evidence of any ill effects from miscegenation among different peoples (Boas [1894a] 1940). At first, Boas believed that geography was the major determinant of human life (Cole 1999). However, although he recognized that humans were indeed dependent upon their environment, he came to the conclusion that the links between the external world and complex internal and social behavior were superficial (Boas [1936] 1940). As he later stated, geographic influences were "patent" and "so shallow they did not throw any light on the driving forces that mold behavior" (quoted in Stocking 1974, 42). He abandoned environmental determinism; for Boas, ethnology was essentially concerned with the mental life, the psychology of primitive groups and was not an environmental science (Cole 1999).

So what then? Was human psychology genetically determined and fixed? Boas had been trained in statistics, physics, and geography, and his interest in the methodology of science bordered physical science and history. At the time, Boas wrote that the physical sciences were focused on developing all-encompassing laws. "Single facts were unimportant; the stress was upon their accumulation to demonstrate a general law. To the historian, however, facts themselves were interesting and important." (Cole 1999, 122). Boas came to the conclusion that in anthropology, the same subject could be studied from either standpoint, the generalizing method of the physicist or the individualizing method of the historian. However, to get to the former, general laws, one had to pay particular attention to the latter, the particular facts. Generalizing without a deep understanding of the underlying facts just led to false generalizations. This approach led Boas to lean toward history in his own research. This was, in essence, a Darwinian approach. As Boas wrote, historical aspects of nature had taken hold of investigations in the whole of science, revolutionizing the methods of both the natural and human sciences (Stocking 1974). To Boas ([1889] 1940), every living being and every society was the result of its historical development: "Historical factors are of greater importance than the surroundings. . . . The longer I studied, the more convinced I became that customs, traditions and migrations are far too complex in their origin to enable us to study their psychological causes without a thorough knowledge of their history. . . . No event in the life of a people passes without leaving its effect upon later generations" (letter to Powell, 1885, quoted in Cole 1999, 126).

Boas believed that the value and nature of anything could be entirely comprehended by its development and that histories were so complex that no single comprehensive plan could be envisioned. He put "emphasis upon complexity: upon how 'complicated' the elements were affecting the human mind, upon the 'intricacy' of the causes that lay behind ethnological phenomena, upon the 'extreme complexity' of the elements behind the character of a people: his position challenged the assumption of a simple teleological plan" (Cole 1999, 126), an assumption that was common in his day. The common belief at that time, for example, was that similar ethnological phenomena in different areas could be explained by either a common source of ideas (for example, peoples attaining similar levels of material culture and thought processes as they separately passed through similar stages in cultural evolution) or by independent invention in response to similar environmental or social circumstances.

However, Boas stressed that any similarities might be simply accidental and that one could not assume that apparently similar phenomena were comparable. Two rattles from different areas might look the same, but one might have important religious functions and the other might simply be a child's toy. Although they might appear identical, the chain of events leading to their use and function—the historical causes—could be quite different. Similar effects do not necessarily mean similar causes. This approach drove Boas's anthropology: before one can discover general laws, it is necessary to prove that similar causes always have similar results and that similar results are indeed similar. "Identical developments could arise from very different environments, just as identical environments could produce very different phenomena" (Cole 1999, 127; see also Boas [1896] 1940; Stocking 1968).

Furthermore, Boas's physical scientific training contributed to his historical approach: "every theoretical deduction needed to be scrutinized by empirical induction" (Cole 1999, 128). Although he is often accused of being a strict empiricist and of being opposed to theory, this was not the case. Boas was not critical of theory in itself, but he believed that a researcher, either a physical scientist or a historian, needed to use empirical induction, the only proper scientific method for either approach. Boas was not content with deduction and arguments from analogy. Instead, for him, science demanded inductive methods for drawing conclusions from facts; seeking facts to support a priori theoretical conclusions was not good enough (Cole 1999). The latter approach was the dominant approach in the eugenic science of the time and is not that uncommon today.

Boas, however, did not want to ignore the complexity of the history or the development of any organism or any society. He believed that "the physiological and psychological state of an organism at a certain moment is a function of its whole history" and that "the character of a biological phenomenon is not expressed by the state in which it *is,* but by its whole history" (Boas 1887b, 589).

Boas believed that only historical investigation of particular phenomena could provide meaningful material for general laws of development. Historical research furnished the inductive building blocks for the discovery of general laws that governed the development of humans and their societies. To Boas, then, ultimately, "the purely historical method without a comparative study will be incomplete" (Stocking 1974, 68). He argued that it was necessary to work upward to develop theory from the data rather than working downward, imposing theory on the data. To Boas, ethnological knowledge at the time was too preliminary to ascertain any general laws with certainty (Boas [1932] 1940).

To his call for empirical research, Boas added a cultural relativist approach. As biographer Douglas Cole (1999, 132–133) has summarized Boas's thought: "If all knowledge was historically determined, then the judgments establishing this knowledge were themselves subject to the historical conditions under which they were made. To many this was a liberating conception, opening the mind to possibilities, dispensing with dogmatism, overthrowing tradition, and enlarging freedom of will. Such relativity went hand in hand with a relativity of ethnic groups. . . . This deep relativism and the liberating value he ascribed to it distinguished Boas's thought from his American contemporaries as much as his historicism and emphasis upon the inductive method."

To discover the history of a people, then, Boas used a historical, individualistic, and relativistic approach and collected empirical data on their physical characteristics, language, and ethnology. Furthermore, he understood that archaeological data could add to a peoples' history, and beginning in 1887, the archaeologist Frederic Ward Putnam served as his mentor and major supporter (it was Putnam who got Boas his appointment at Columbia [Patterson 2001]). Finally, Boas added biostatistics to his methodology. This

"Boasian" agenda (Boas 1940; Stocking 1968; Darnell 1971; Cole 1999) became the blueprint for modern anthropology.

Boas made his first public, general statement against the conventional wisdom of the eugenicists' Humeian approach to history and ideology in 1894 at the annual meeting of the AAAS in a paper entitled "Human Facility as Determined by Race." "Historical events appear to have been much more potent in leading races to civilization than faculty, and it follows that achievements of races do not warrant us to assume that one race is more highly gifted than the other. . . . No specific differences between lower and higher races can be found [in mental qualities] . . . no unquestionable fact which would prove beyond a doubt that it will be impossible for certain races to attain a higher civilization" (Boas [1894b] 1974, 308, 323, 317). The data available at that time still indicated that skull and brain sizes were larger among white than among "colored" peoples. However, Boas noted problems with these measurements and argued against the assumption that this provided any conclusive evidence that whites were superior. He noted that even though the cranial capacity of women was smaller than that of men, "the faculty of women is undoubtedly just as high as that of men" (Boas [1894b] 1974, 315). Early in his career, Boas began to work closely with W. E. B. Du Bois (1868–1963), the nationally prominent African American sociologist, on racial equality and integration, and he helped him establish the National Association for the Advancement of Colored People in 1909 (Barker 2010).

By 1900, Boas had conducted nine field trips to study indigenous peoples, in Canada, in Alaska, and on the northwest coast of the United States, often working with indigenous collaborators as he wrote his ethnologies. He had also worked with the major ethnographers and archaeologists of his time (Bastian, Virchow, Tyler, Jessup, Putnam, Powell) and had collected anthropometric data from 17,000 Native Americans and "half-bloods." He was impressed by the variability of his measurements. His statistical data demonstrated the extent of variability within groups in terms of growth rates by age, origin, social class, and between boys and girls. His statistics demonstrated quite clearly that morphological "type" was a constructed abstraction. He showed that the prevailing eugenics view that mixed-bloods were inferior was unfounded: instead of being infertile and physically inferior, they had more children, were taller, and showed more variation than either ancestral group (Boas [1894a] 1940; Cole 1999; Little 2010).

Boas emphasized variation and complexity. He recognized that these could be explained by social environment, geographical influences, selection, chance variation, or hereditary mixture (Stocking 1968). Boas was far ahead of his contemporaries in the field of anthropology in his understanding of the dynamics of anthropological and social process. He had set for his goal "to build up ethnology to a recognized discipline in the university" (Boas letter to von Andrian-Werberg, 1898, quoted in Cole 1999, 220). He had begun a new approach to anthropology and ethnography that focused on process and not on pure description, old theories, and static typological approaches. He set out to establish a new anthropology in the United States and to train students using his broad, four-field methodology: ethnography, physical anthropology, linguistics, and archaeology. "Almost all of the American ethnologists of the next generation entirely or partly will come from my school," he wrote (Boas's letter to his parents, 1899, quoted in Cole 1999, 221).

For Boas, anthropology was a science based on inductive methods and its aim was to understand the biological, psychological, and social history of mankind—"to trace the history and distribution of every single phenomenon related to the social life of man and to the physical appearance of tribes" (Boas 1887a, 231). He was interested in understanding all of the complexities that led to those phenomena for each group of people and the development of those phenomena within each group. Up to then, anthropologists were more interested in explaining the origins of "civilization" in general, the *generic* view of human nature and development. Boas was interested in the history of specific human groups and believed that generalizing should not precede an understanding of the dynamics of these specific histories.

Other ethnographers' views of the "psychic unity of mankind" were based on the idea that the human mind was like a machine, the notion that given the same input it will always come up with the same product. Thus, it was expected that there was parallel development everywhere. The stages of human social evolution, the progress from savagery to barbarism to civilization, were believed to be based on uniformity in the thoughts and actions, aims and methods as humans passed the same degrees of development throughout their history. Living groups of primitive peoples were seen to approximate the earlier stages of "civilized" peoples, which enabled researchers to reconstruct the past of the "more-developed" groups. It was posited that living "uncivilized" peoples were at an earlier stage of development and had more primitive, less "civilized" thought processes than "civilized" peoples. The idea that evolutionary developments are reflected in the ontogeny (or internal development) of modern peoples was an extension of

Ernst Haeckel's biogenetic law of recapitulation, an indication of Haeckel's widespread influence in numerous fields at that time (Haeckel 1874; see Gould 1977b; Baehre 2008).

Although Boas was not critical of the idea of an evolutionary development of social history, he did criticize two specific aspects of this approach: the first related to the problem of cause and effect and the second to the contemporary view of psychic unity. First, as mentioned above, Boas argued that similar phenomena were not always or even usually due to the same cause. In fact, his ethnographic research had illustrated that similar phenomena developed in a multitude of different ways. Boas used examples that ranged from totemic clans to decorative art to the use of masks in the northwest coast to show similar effects from very dissimilar causes. He illustrated the complexity of the origins of superficially similar phenomena and called for the slow, careful, and detailed study of the actual relationships among and between phenomena. This required the historical method (Cole 1999). Only with this approach could one determine "with considerable accuracy the historical causes that led to the formation of the customs in question and to the psychological processes that were at work in their development. . . . The application of this method is the indispensable condition of sound progress" (Boas [1896] 1940, 277, 279).

Boas believed that diversity, not uniformity, among human groups was the striking feature anthropology revealed (Cole 1999). This was the basis of the development of Boas's anthropological concept of culture (see below). Boas was opposed to the hypothetical and deductive theories of his day because he saw them as having little or no empirical basis. He rejected speculative theories not because they were theoretical but because they were speculative. In most cases, the method determined the results in research in human biology or behavior; the data were collected not to test the theory but to prove it. Although peoples had been classed into broad typological groups by race, language, and culture, Boas showed that there was incongruity among such classifications and that there was no necessary association between race, language, and culture. In each case, only the detailed history of the physical and social interactions of groups made explanation of the result possible.

The second problem Boas had with contemporary theory related to the concept of the "psychic unity of mankind." This had been interpreted to mean that people everywhere would make essentially the same inventions under the same circumstances. Much like the "racial" typologies and theories about the stages of human social evolution of his day, this was a static, typological way of looking at human thought. It assumed that given certain circumstances, people would always think in limited, predictable ways. In this approach, human thought was assumed to be similar among peoples at similar stages of human social evolution. For example, the thought processes of contemporary "primitive" people were similar to those of peoples who lived at earlier stages of human social evolution, but they differed from those of contemporary "civilized" peoples.

This was not what Boas meant when he spoke of psychic unity. To Boas, similarity of the human mind meant that all peoples in all societies shared similar mental *processes,* and one of the major foci of Boas's research agenda was to seek the laws of these shared human mental processes. To Boas, the organization of the human mind was essentially identical among all past and present human societies and among all races and mental activity followed the same laws everywhere. The manifestation of these laws was not dependent upon race or any genetic differences among peoples but was "a product of the diversity of the cultures that furnish the material with which the mind operates" (Boas 1904a, 243). Human minds developed not in a vacuum but within a particular society with a particular history, and each mind was enveloped in a cocoon of shared tradition, habit, learning experience, education, customs, and so forth. Mental manifestations "thus depended upon the individual experience as received through imitation and the association of ideas" within a social group (Cole 1999, 272). People developed their "world views" by associations among ideas, habits, and tradition; through normal socialization and acculturation; and by individual experience, and not necessarily consciously (Boas 1940).

Changes in Bodily Form, The Mind of Primitive Man, and the Concept of Culture

With this way of thinking, Boas was developing a critique that would help diminish the hold of eugenics in social science and society. He also was developing a new paradigm that would lead to an enlightened way of thinking about and understanding both the physical and social aspects of being human. By 1911, when his two seminal books were published, Boas was already fifty-three years old and had been developing his new anthropology for almost thirty years. These two books seemed to synthesize his previous work and stimulate his future endeavors. And although it took some time, these books were among the initial catalysts that ultimately led to the end of twentieth-century eugenics and Nazism, though, unfortunately, not to the complete elimination of racism or simple-minded biological determinism.

At the suggestion of a physician, Maurice Fishberg, who had done preliminary measurements on New York Jewish immigrants, Boas successfully applied for funds and was commissioned by the U.S. Immigration Commission to study the physical effects of immigration on migrants to New York City (Little 2010). From 1908 to 1910, he patiently measured nearly 18,000 immigrants and their children. The results were surprising even to him (Degler 1991; Little 2010). His results were published in his 1911 volume *Changes in Bodily Form of Descendants of Immigrants* (1911a). He found that within ten years of their mothers' immigration into the United States, the head shapes of the children of immigrants changed. Boas claimed that because the mother and father were both immigrants and presumably from the same genetic background, some aspect of the social or physical environment must have been the source of the change. Boas emphasized these changes in the shape of the head since the cephalic index (the ratio of the length to the width of the head) had long been used as the most reliable, unchanging measure for identifying and classifying human races. Thus, it was widely accepted that races could be identified by this measure, that it was stable through time, and that it could not be changed by environmental and social influences. Skull shape and size was one of the most identifiable symbols of the biological fixity of the races, one of the mainstays of polygenics and eugenics. The Immigration Commission dismissed Boas's claims (Barker 2010).

Even Boas had been convinced of these previous empirical findings of physical anthropology, stating in 1910, "I have always been so thoroughly convinced of the great stability of the cephalic index, that it has taken me a long time to get ready to accept the results of my own investigation" (Boas letter to Gustaf Retzius, May 3, 1910, quoted in Degler 1991, 66). Boas, however, was also aware of the great significance of his results. This seemingly unchanging biological feature, one that it was assumed could never be influenced by environment and that was used as one of the basic symbols of typological, polygenic, eugenic theory, could change within one generation. Boas (1911a, 5, 76) emphasized that

> not even those characteristics of a race which have proved to be most permanent in their old home remain the same under the new surroundings. . . . We are compelled to conclude that when these features of the body change, the bodily and mental make-up of the immigrant may change. . . . The evi-

dence is now in favor of a great plasticity of human types. . . . We are necessarily led to grant also a great plasticity of the mental make-up of human types. . . . [We must] conclude that the fundamental traits of mind [are also subject to change in a new environment]. . . . If we have succeeded in proving changes in the form of the body, the burden of proof will rest on those who, notwithstanding these changes, continue to claim the absolute permanence of other forms and functions of the body.

Boas was fundamentally recasting the discussion of the role of race in shaping the human mind. He stated in the report that "it is probably not too much to say that [the findings] indicate a discovery in anthropological science that is fundamental in importance" (quoted in Degler 1991, 65). He set out to make sure his results were well disseminated. He corresponded with professional colleagues, reported his results at national and international conferences, and presumably sent out copies of his report to a large number of people, since he asked the Immigration Commission for 1,000 copies. Even though Boas's findings on the cephalic index were extremely important empirical evidence against the conventional wisdom of the fixity of racial characteristics, this was only the tip of the iceberg in terms of his emphasis on the role of environment in human biology. It may have been a very important finding at that time for many who had been convinced by the conventional wisdom that environment could have no influence on human biology or behavior. However, to Boas, his study of head shapes was not needed to convince him of the power of the environment in shaping human biology and the irrelevance of racial influences (Degler 1991). His conclusions on this topic had been reached long before 1911. (Boas's results on changing skull shape have recently been questioned on empirical grounds [Sparks and Jantz 2002], but for an analysis of this controversy and excellent rebuttals to these questions, see Gravlee, Bernard, and Leonard 2003a, 2003b; Relethford 2004; and Little 2010.)

Boas's views on the changes in bodily form in relationship to environmental influences, thus, came from his research endeavors in anthropometry, or physical anthropology. However, his general views of environmental influences on human behavior were gained from his ethnological and linguistic research. He expressed these views in his second 1911 volume, *The Mind of Primitive Man* (1911b). In this book, Boas summarized, developed, and synthesized the views he had been developing for years. It presented his ideas in a coherent, readable manner to a wide academic and popular audience. In this volume, Boas's rejection of the traditional conventional wisdom was truly radical; it illustrated Boas's early commitment to racial equality and the separation of race and culture (Little 2010).

Boas argued that there was no difference in mental capacities or abilities between "savage" and "civilized" peoples or between "coloreds" and whites. The differences in physical appearance and in stage of social development did not correspond to any significant difference in mental ability or social function. What, then, led to the obvious differences that could be observed among different human populations? To Boas, the differences in behavior and society among peoples were the product of their different histories, not differences in their basic biology. His approach was historical and relativistic. The ideas, worldviews, technology, myths, religions, language, kinship patterns, art, reasoning, and so forth of a particular society were the products and were based on the influence of their predecessors and their historical environment. "Our ideas were relative to our environment and were not absolute across the spectrum of human experience," he wrote (quoted in Degler 1991, 67). Each society had its own history and, through this history, developed its own set of physical and mental contexts with which it dealt with the world. Although each society might approach the world in a different way, within each context, the approach was integrated and logical. Though the logic might be different in each of these social settings, it developed over the long history of each society and no group was absolutely right or wrong. To Boas, anthropologists should "study the human mind in its various historical . . . ethnic environments." The purpose of ethnology, he argued, was the "dissemination of the fact that civilization is not something absolute, but that it is relative, and that our ideas and conceptions are true only so far as our civilization goes. . . . How far each and every civilization [goes] is the outcome of its geographical and historical surroundings" (Boas 1887b, quoted in Degler 1991, 67).

In *The Mind of Primitive Man*, Boas summarized and synthesized his views of the importance of history over biology in the development of the mental, subsistence, technological and other major features of a coherent human population, social group, or society. Based on his ethnological research, he insisted that the differences in societies were not derived from differences in innate capacity; "primitive" peoples did not differ in mental capacities from "civilized" peoples. All peoples had similar mental capabilities, and the differences among them were the products of the development of their societies through their particular histories. He coined the anthropological use of the term *culture* to embody this aspect of human social existence. In doing so, Boas created a new and innovative paradigm in social science. As Degler (1991, 70–71) has described the new development: "Ideas expressed in the 1880s appeared almost word for word in his 1911 book. . . . The only serious exception to that is his use of the word 'culture,' the concept that later would come to be the contribution that Boas made to social science thinking about differences in human action." Before 1911, the term *culture* was always used with a singular meaning, as another term for high society or civilization such as "high culture." Those who listened to opera or attended art galleries might be considered to be "cultured." However, Boas began to use the term in a plural sense to express the idea of multiple cultures, as it is used in anthropology today, and assumed that every society exhibits a culture. Boas (1904b, 522) also recognized that "before we seek what is common to all culture, we must analyze each culture."

In contrast to many of his critics, both past and current, Boas did believe that heredity or biology was important in shaping human behavior; "it is readily seen that all the essential traits of man are due primarily to heredity," he wrote (Boas 1911b, 76). He realized that there were organic differences among individuals that related to differences in heartbeat, basal metabolism, hormonal and mental development, and mental behavior. However, he saw individual variation within each society or culture and within each race as being so wide as to overlap to a large extent the range of variation in every other culture or race. Basic and defining differences among societies were not the result of differences in individual biology or genetics but were due to the processes of history that led to underlying differences in their culture.

A classic definition of culture by the renowned cultural anthropologist Clifford Geertz (1973, 89) is this: "an historically transmitted pattern of meanings embodied in symbols, a system of inherited conceptions expressed in symbolic forms by means of which men communicate, perpetuate, and develop their knowledge about and attitudes toward life." A more recent definition speaks of a set of guidelines (both explicit and implicit) that individuals inherit as members of a particular society and that tells them how to *view* the world, how to experience it emotionally, and how to *behave* in it. Through enculturation, an individual slowly acquires the cultural "lens" of that society. Without such a shared perception of the world, human society would be impossible (adapted from Helman 1994). Of course, with a shared perception of the world comes all of the other ramifications of a culture that members of a society share through their history, such as subsistence techniques, technology, social structure, language, myths, and religion. This concept is (or should be) the basic paradigm of anthropology. It encapsulates the basic difference between human behavior and the behavior of other animals (Read 2012). An understanding of the term *culture* in humans, as Boas intended it to be used, identifies a unique and profound characteristic of human behavior. It is part of human nature (Sussman 2010). In this way, Boas was also addressing the popular concept of animals' culture. Although there are differences, for example, between different populations of chimpanzees across Africa, these differences are trivial when compared to differences between different societies, even those living sympatrically in the same valleys in Africa. Culture is more than the passing on of learned behavior or social learning. Many animals are capable of that. However, the learning abilities of humans and other animals are quite profound.

A chimpanzee population might have a different hand clasp or hunt for a termite in a different way than other populations. However, people living in different cultures, even if these people overlap in the same geographic area, can differ in essentially every major aspect of their existence. They can have differences in their diets, technology, language, kinship structure, religion, whom they see as friends and enemies, the way they measure time, the way they see colors or hear sounds, the way they classify every aspect of the world around them. Basically, they can approach the world in very different ways. They see the world differently—they have different worldviews. Just as one can say that humans are unique in being bipedal, they are also unique in having culture. Paraphrasing turn-of-the-twentieth-century sociologist Max Weber, Clifford Geertz (1973, 49) has said: "Man is an animal suspended in webs of significance he himself has spun." Geertz continues: "Without man, no culture, certainly; but equally, and more significantly, without culture no men." I would add: without culture, no anthropology. For I believe that a proper understanding of the concept of culture defines good anthropology and characterizes a good anthropologist.

Stocking (1968, 232–233) summarizes the importance of Boas's use of culture:

> Focusing only on those aspects of the change having specifically to do with the culture idea, one might say that it involved the rejection of simplistic models of biological and racial determinism, the rejection of ethnocentric standards of cultural evaluation, and a new appreciation of the role of unconscious social processes in the determination of human behavior. It implied a conception of man not as a *rational* so much as a *rationalizing* be-

ing. . . . Boas did not . . . offer a definition of anthropological "culture." But what he did do was to create an important portion of the content in which the word acquired its characteristic meaning. . . . By changing the relation of "culture" to man's evolutionary development, to the burden of tradition, and to the processes of human reason, transformed it into a tool quite different from what it had been before. In the process he helped to transform both anthropology and the anthropologist's world.

Around the same time that Boas wrote his two volumes in 1911, the eugenicists believed that environment could have little influence on human behavior. In 1912, Goddard wrote his book on the Kallikak family, presumably showing that heredity trumped any environmental influences on complex human behavior. The environmental alternative to genetic biological determinism was dead, at least in most of Europe and the United States. The eugenicists thought that they had won the over-500-year battle and that differences between human societies, between classes, and between "unfit" and socially acceptable individuals could be explained only by heredity and biology. Even as late as 1933, as mentioned above, Eugen Fischer could claim that "the old worship of culture . . . is past. . . . The theory of the heritability of mental as well as physical traits has finally been vindicated" (quoted in Proctor 1988, 345). Although Boas's concept of culture was a new idea in its infancy in 1911, it would eventually be the alternative paradigm, the one that better fit the empirical evidence, and the one that would ultimately trump the simple-minded genetics and biological determinism of the eugenicists and the Nazis.

Stocking (1968) examined the use of the term culture in the social science literature from 1890 to 1915. He found that its more expansive definition was not used before 1895 except by Boas. After 1910 this usage was common among cultural anthropologists and other social scientists. At that time, Boas had provided science with an alternative to the Lamarckism that had been disproven. Although eugenics had an upper hand in U.S. science into the 1920s and early 1930s, and Nazism had the upper hand into the 1940s, eventually the anthropological concept of culture and the insight it provided into human behavior would provide the evidence needed to show that the biological determinism and simplified genetics of eugenicists and Nazis was devoid of any scientific credibility. Unfortunately, some of this biological determinism and oversimplified genetics has crept back into biology and anthropology. Along with this, and related to it, the Boasian term culture has been misunderstood and misused, even by many anthropologists.

However, in their fight against eugenics, Nazism, and racism in the early part of the twentieth century, social scientists understood the importance of Boas's concept of culture and ultimately used this as a major tool. Boas embarked on a lifelong assault on the idea that race and heredity were the primary sources of differences found in the mental or social capacities of human groups. The concept of culture, which Boas, his students, and other social scientists tirelessly elaborated and defended, ultimately undermined the concept of racism. As Degler (1991, 61–62) has emphasized, in the twentieth century, the concept of culture "became not only an alternative to a racial explanation for human behavioral differences but also a central concept in social science. . . . Boas' influence upon American social scientists in matters of race can hardly be exaggerated. . . . He accomplished his mission largely through his ceaseless, almost relentless articulation of the concept of culture. . . . This assertion by Boas [that differences in human societies were due to differences in culture and not to innate mental capacities] . . . proved to be truly revolutionary for the development of anthropology and eventually of social science in general." The culture paradigm, as Boas conceived it, eventually was instrumental in overturning both the concept of degeneration and the concept of polygenics that had dominated the discussion of race and human behavioral differences for over 500 years.

This new view of culture provided the rebuttal of the concept of race and its underlying meaning in eugenics. Again, Degler (1991, 82) summarizes the impact of the new idea: "Certainly, the culture concept, by denying the influence or power of biology in the social sciences, offered a new discipline like anthropology a clearly defined intellectual tool that could identify and thereby justify the discipline in competition with the better established and more highly regarded biological sciences of the time."

6

Physical Anthropology in the Early Twentieth Century

The leaders of the eugenics movement understood the implications of Boas's views and the threat they posed to the eugenics agenda. They energetically went after him. In fact, Madison Grant had a long-standing "cold war" with Boas. As we have seen, Grant was a social Darwinist and anti-Semitic. He believed in the superiority of the Aryans and emphasized the dangers to civilization of the dilution of Aryan blood. Grant also idolized Ernst Haeckel, who was a staunch follower of Gobineau. Boas, on the other hand, was mentored by Rudolf Virchow, Haeckel's main nemesis in Germany (Weindling 1989; Marks 2012). But Grant did not use a scientific argument to criticize Boas's findings that the most "immutable" of hereditary traits, skull shape and cephalic index, were susceptible to environmental influence, findings that challenged the basis of eugenics theory.

Early German "anthropologists," like Virchow, were attempting to establish the idea that all people were cognitively comparable, to provide the validity of the concept of the psychic unity of mankind. Early German "Darwinists," like Haeckel, dehumanized non-Western Europeans in order to connect them more closely to the apes. Haeckel's form of Darwinism was at odds with Virchow's anthropology and ethnology, and it undermined the assumptions of the psychic unity view of mankind that would make Virchow's anthropology possible (Marks 2010a, 2012). Haeckel and, later, Nazi biologists actually emphasized the continuity of particular living and fossil apes with modern human "races" in their polygenic theories of human variation.

For example, around 1918, as a joke, Francois de Loys, a Swiss geologist and explorer working in Venezuela, took a picture of a recently deceased pet spider monkey that was propped up and posed to look like a

165

"primitive ape." Around 1929, in a series of scientific and popular papers, George Montandon, a Swiss/French physician, anthropologist, ethnologist, and Nazi sympathizer, perpetuated this "joke," giving the dead pet monkey the scientific name *Ameranthropoides loysi* and claiming that it verified the polygenic scheme of Haeckel and the Nazis in which each of the four presumed races of humans had independent evolutionary histories. In this scheme, "whites" were derived from chimpanzees, black Africans from gorillas, and Asians from orangutans. Montandon claimed that with the discovery of *A. loysi,* the "missing link" between Native Americans and the apes was identified, thus providing further evidence and support for this polygenic theory from the sixteenth century that had more recently been regurgitated by the Nazis (Urbani and Viloria 2008).

Grant's vs. Boas's Anthropology

The rivalry between Boas and Grant epitomizes the major differences in racial theory during the early 1900s. It also gives us an outline of the history of anthropology at that time, especially physical anthropology. At the beginning of the twentieth century, in the wake of the demise of Lamarckism and of the idea that environment could influence behavior, the followers of eugenics claimed victory. Grant (1916, 14), sounding very much like the authors of the past (and of the more recent tome *The Bell Curve* [Herrnstein and Murray 1994]), claimed, for example: "It has taken us fifty years to learn that speaking English, wearing good clothes, and going to school and to church, does not transform a negro into a white man." When Boas's *Changes in Bodily Form* appeared, Grant countered that it was motivated by immigrant Jews like Boas who were attempting to overemphasize the power of environmental influences in order to downplay the reality of their inferior heredity. In *Passing of the Great Race* (1916), Grant emphasized the inferiority of blacks and Jews and the dangers of miscegenation between inferior races and Nordics.

It is interesting to note that Boas and Grant shared a number of characteristics. They both believed in the power of science to improve humankind. They both lived in New York City and were associated with the American Museum of Natural History. They both loved the Pacific Northwest. For Boas this was the site of most of his fieldwork on indigenous peoples, and for Grant it was the site where he hunted indigenous big game. However, Grant was born into a very rich, aristocratic American family, whereas Boas, although he came from an upper-middle-class German family, was an immigrant Jew. Spiro (2009, 278–279) characterizes the differences in the views of the two men:

> Boas preached with increasing vigor and confidence against racial preju-
> dice, and consciously and actively worked to thwart the dangerous influ-
> ence of Grant ("that charlatan") and his disciples. . . . He denied that there
> was any correlation between physical characteristics of a population and
> its mental or moral traits. The latter, he asserted, were created by the "cul-
> ture" in which the individual was raised, not by his or her germ plasm. [He]
> opposed every facet of Grant's eugenic program. . . . [To Grant it was clear]
> that the root of Boas' hostility lay in the fact that he was a Jew . . . and that
> Boas "naturally does not take stock in my version of anthropology which
> relegates him and his race to the inferior position that they have occupied
> throughout recorded history."

The two men intellectually battled each other throughout their lives (Grant died in 1937 and Boas in 1942). In his excellent book *Defending the Master Race,* Spiro does an exceptional job of tracing this "cold war" between Boas and Grant. Boas began his lifelong attack against eugenics in earnest with his two 1911 volumes. The earliest and most direct attack was in *Changes in Bodily Form.* Grant countered this volume with a num-ber of nasty letters to influential editors and politicians explaining that Boas's conclusions were absurd and absolutely at variance with scientific anthropology. He wrote to President William Howard Taft explaining that because of the immense antiquity of the races, it was not possible that their physical characteristics could change in one generation due to environmen-tal factors. To a congressman he accounted for Boas's "silly" claims by re-minding him that "Dr. Boas, himself a Jew, in this matter represents a large body of Jewish immigrants, who resent the suggestion that they do not belong to the white race" (quoted in Spiro 2009, 199).

Grant believed that even if Boas's anthropometric measurements were correct, all they proved was evidence of immoral immigrant mothers hav-ing illicit affairs with American-born men. He wrote to one clergyman that Jews "like rats, have formed a race able to survive the gutter conditions which quickly destroy higher types" (quoted in Spiro 2009, 299) and that Boas's work was misleading and should be ignored.

Later in 1911, *The Mind of Primitive Man* appeared. As stated above, in this book, Boas laid out his claims that mental aptitude was not deter-mined by heredity, that any race could achieve civilization given the proper environmental conditions, that there was more variation within races than between them, and that environment accounted for most of the differences that did exist between races. He also introduced the anthropological concept of culture. This was Boas's most popular and widely read book; one of his students later referred to it as "a Magna Carta of race equality" (Spier 1959, 147). In 1933, it was placed on the Nazi's list of books to be burned. Grant realized the danger of this volume to eugenics and, in a 1912 letter to Osborn, he wrote that somebody must publish something "to counteract the evil effects of Boas propaganda" (quoted in Spiro 2009, 300).

That somebody turned out to be Grant himself. Four years later, in 1916, Grant published *The Passing of the Great Race*. Although, this book reiterated polygenecists' history and summarized eugenicists' views of the time, it was also certainly stimulated by Grant's rage over and fear of Boas's antiracist, anti-eugenics writings. Davenport and Laughlin (1917, 10–11) reported in *EN* that *Passing* would finally discredit and silence "certain anthropologists, like Boas," who attempted to deny the inherited, fixed mental differences between the races. But Boas was not silenced. He followed Grant's book with anti-polygenic articles and reviews of the *Passing* in *Scientific Monthly* (1916) and *New Republic* (1917). In the first he wrote that most human personality traits, such as criminality and alcoholism, were determined by environment and not heredity and that instead of being a "panacea that will cure human ills," eugenics was "a dangerous sword." In the *New Republic* review, he criticized the book as being naïve, dogmatic, and dangerous. Grant responded in his fourth edition of *Passing* that this kind of bitter opposition to his book was expected by those of inferior races who refused to believe that their nature was determined by "fixed inherited qualities . . . which cannot be obliterated or greatly modified by a change of environments" (quoted in Spiro 2009, 300).

Grant's book was immensely popular among academics, politicians, policy makers, and the public. It essentially defined, popularized, and advertised the views of much of America about racism and eugenics. It was the conventional wisdom of the day. *Passing* was so popular among academics that Boas was a lone voice against eugenics, Madison Grant, and his colleagues and supporters.

Boas was not deterred. Although many of the leading scientists, including anthropologists, were eugenicists, either actively or passively, at that time, Boas continued to train students and to attract colleagues who were to become devoted to fighting American eugenics and, later, its logical extension, Nazi racism. Boas began to amass a cadre of supporters who were well-trained scholars and who had compiled a massive amount of data that could be used to join the assault against eugenics and the fifteenth-century views of polygenics. These included many anthropologists who earned their PhDs studying under Boas: A. L. Kroeber (PhD 1901), Robert Lowie (1908), Edward Sapir (1909), Alexander Goldenweiser (1910), Paul Radin (1911), Leslie Spier (1920), Ruth Benedict (1923), Melville Herskovits (1923), Margaret Mead (1929), and Ashley Montagu (1937). As Spiro (2009, 302) notes, "By the early 1920s, the members of the first generation of Boas's students were devising the intellectual weapons and amassing the ethnographic data they would need to combat the disciples of Grant." Boas had a significant influence on his students, as a teacher, a theorist, and as a person, especially with regard to his interest in races (Barkan 1992). Colleagues also were beginning to join Boas in his fight against eugenics, including a fellow Columbia University professor, psychologist Otto Klineberg. Physical anthropologist Aleš Hrdlička (1869–1943) also supported Boas in a number of ways, mainly because of his personal vendetta against Grant and not for shared scientific beliefs (see later in this chapter).

Thus, in the early decades of the twentieth century, the 500-year-old debate between the polygenecists and the monogenecists still persisted, with only slight variations on the original themes. The adherents to polygenics still were obsessed with the classification of races and the superiority of one race over others. Monogenecists were convinced that the environment (culture) played an important role in the development of complex human behaviors and that races were socially constructed myths. As Spiro (2009) points out, the former were generally the older generation of amateur physical anthropologists who were mainly aristocratic WASPs with no academic affiliation or were associated with a museum. The new proponents of environment and culture were a younger generation of anthropologists who were professionally trained and held positions in academia. However, as at the beginning of this long-lasting debate, this was still somewhat of a religious battle between native and mainly wealthy Christian Protestants against Jews and other recent immigrants into the United States, who used education as a means to support themselves. After all, with the exception of Kroeber, Benedict, and Mead, all of the Boas students mentioned above were Jewish (Spiro 2009).

Many of these new professionals did most of their work in cultural rather than physical anthropology. I believe this was directly the result of the foothold that eugenics had at that time on all of the biological sciences and especially on physical anthropology (as mentioned above, a similar split had occurred between Haeckel and Virchow earlier in Germany). The big four of eugenics, Davenport, Grant, Osborn, and Laughlin, all considered themselves to be doing real anthropology, following the traditions of the earlier American Anthropological School of the Mortonites, in contrast to the culturally oriented, Boas-influenced group with ties to the American Anthropological Association (AAA), which had been established in 1902 (Spiro 2009). The eugenicists dominated this discipline in the early 1900s. It is in this context that they established the Galton Society in 1918, a racially oriented anthropological organization that they hoped would serve as a rival to the AAA. Eugenicists looked at cultural anthropology as unscientific and trivial. Osborn said of the Boasians in 1908 that "much anthropology is merely opinion, or the gossip of natives. It is many years from being a science" (quoted in Spiro 2009, 303).

To counter this approach, cultural anthropologists focused on professionalizing anthropology, making it a true social science discipline in which practitioners were trained with a specific methodology and a unique paradigm. With this in mind, Boas began to train students in his type of anthropology. He and his followers then focused on turning the AAA into an organization of university-trained, professionally qualified scholars. Before this, the membership was composed mainly of wealthy, untrained hobbyists. Slowly Boas's strategy began to pay off. In 1907, Boas became president of the AAA and his former PhD students soon also attained leadership positions in the association. By 1910, the AAA was a respected society of academic anthropologists and Boasians were in the majority. Graduate programs in anthropology were established at several universities, and by 1912 twenty men had received doctorates in anthropology (Patterson 2001). By 1915, the association's journal, *American Anthropologist (AA),* was in the hands of Pliny Goddard (a relative of Henry Goddard who was an ally of Boas's [Zenderland 1998]) and Robert Lowie (Boas's student and staunch supporter). Eugenics and biological determinism were replaced with the concept of culture within the pages of the journal, and the culture paradigm was becoming dominant within the profession, though, as we have seen, not yet in other scientific disciplines or among the public (Stocking 1968; Degler 1991; Spiro 2009). By 1928, an additional thirty-three men and nine women had been awarded PhD degrees in anthropology; fifteen of these men and seven of the women had been supervised by Boas at Columbia (Patterson 2001; Barker 2010).

A. L. Kroeber (1876–1960), who took up the fight against the eugenicists, further emphasized the importance of the concept of culture in understanding human nature, behavior, and society. He expanded and further defined the concept. Writing mainly in *AA,* he noted the lack of understanding of the concept of culture among eugenicists and biologists. In a letter to Boas he wrote: "Consequently, the sense always crops up in their minds that we are doing something vain and unscientific, and that if only they could have our job they could do our work for us much better" (quoted in Degler 1991, 83). This combination of a failure to understand the concept of culture and the idea that biologists could do a better job than social scientists in understanding human behavior is still common today in some fields of biology. For example, E. O. Wilson (1998, 183–185), the founder of sociobiology and evolutionary psychology, recently stated:

> Advanced social theorists . . . are equally happy with folk psychology. As a rule they ignore the findings of scientific psychology and biology. . . . In short, social scientists have paid little attention to the foundations of human nature, and they have had almost no interest in its deep origins. . . . Believing it a virtue to declare that all cultures are equal but in different ways, Boas and other influential anthropologists nailed their flag of cultural relativism to the mast. . . . This scientific belief lent strength in the United States and other Western Societies to political multiculturalism. . . . It holds that ethnics, women, and homosexuals possess subcultures deserving of equal standing with those of the "majority." . . . So, no biology.

Wilson (1998, 188, 190, 193) continues, "Enough! . . . It is time to call a truce and forge an alliance. . . . If social scientists choose to select rigorous theory as their ultimate goal, as have the natural scientists . . . that means nothing less than aligning their explanations with those of the natural sciences. . . . To summarize . . . [behavior is guided] by innate operations in the sensory system and brain. They are rules of thumb that allow organisms to find rapid solutions to problems encountered in the environment. They predispose individuals to view the world in a particular innate way and automatically make certain choices as opposed to others."

The problem here is that Wilson does not seem to understand the profound influence that culture has on these so-called innate ways and automatic choices. In most cases, it is the particular history of the individual within his/her culture and only very loosely his genetic background or any fixed action pattern (or instinct) that leads to a particular reaction to any stimulus. This misunderstanding leads Wilson and many sociobiologists into thinking that people generally react similarly to similar stimuli or at least that they should do so and that if they don't, they are not acting naturally—not adhering to so-called human universals (see Brown 1991 and Gillette 2011, for example). These human universals are often defined by Western European and American cultural values and mores. Thus, human universals become those that best fit the culture of those that define them. In this sense, culture defines biology rather than the other way around.

For example, one might say, as many sociobiologists and evolutionary psychologists do, that humans are inherently violent, especially men (see Wrangham 1996). However, this observation tells us little about human nature and variation in human behavior. Variation in the expression of violence or peacefulness among and within different cultures and among individual men and women is enormous (Fry 2006, 2013; Pim 2010; Sussman and Marshack 2010; Sussman and Cloninger 2011). Homicide statistics vary greatly in New York, London, and Tokyo, for example, and are not related to innate biological or genetic differences in the populations who live in those places. In order to explain this variation, we must understand cultural and subcultural as well as individual and family differences. Saying that humans are by nature homicidal is uninformative.

Kroeber (1916) turned Weismann's debunking of Lamarckism on its head and used it to defend the concept of culture. He claimed that if accomplishments of one generation could not be passed on to the next generation through inheritance, then the so-called higher races could not claim a biological basis for their so-called superiority. On the other hand, the so-called lower races could not be blamed for creating their own inferiority through lack of effort to improve themselves. Instead, each "race" created its own culture. "The accomplishments of a group, relative to other groups, are little or not influenced by heredity" (Kroeber 1917, 47). Although eugenicists and biologists of the time believed that differences in heredity or race explained why some societies behaved differently from others, Kroeber insisted that "the historian, until such differences are established and defined, must assume their nonexistence. If he does not base his studies on this assumption, his work becomes a vitiated mixture of history and biology" (Kroeber 1915, quoted in Degler 1991, 92).

The new paradigm of Boas's concept of culture quickly began to spread to other social sciences. It gave social scientists a completely new way to interpret the data on human individual and social behavior. The concept of culture did more than provide an alternative theory to Lamarckism as a way to disprove the claims of eugenicists. It also challenged the concept of degeneration. The differences between different races and societies were not inherently better or worse based on their heredity. The differences were not innate but were determined by their social environment and their history. In addition, judgment of these differences was based on facts rather than on values.

In the second decade of the twentieth century, a number of sociologists and psychologists began to be influenced by Boas's concept of culture and views of race. William I. Thomas, a University of Chicago sociologist, was one of the pioneers of research in culture and personality (Stocking 1968). He began his career as a Lamarckian with a nineteenth-century view of race. However, Thomas was in regular correspondence with Boas (Degler 1991), and over time his views of race and culture changed. A year after the publication of *Mind of Primitive Man,* Thomas (1912, 726–727) wrote an article that emphasized that "individual variation is of more importance than racial difference, and that the main factors in social change" were products of cultural environment. Referring to the thinking of Boas, he went on: "Present-day anthropology does not pretend that any of the characteristic mental powers . . . are feeble or lacking in any race." University of Missouri sociologist Charles W. Ellwood had an intellectual transformation similar to that of Thomas's. At first, he accepted Goddard's thesis in his *Kallikak* book that environmental influences could not improve the inferiority of blacks (Stocking 1968). Later, referring to Boas's thought, he stated that "a large and growing school of anthropologists and race-psychologists finds the explanation of the mental and moral differences between the races, not in innate qualities or capabilities, but in differences in their social equipment or machinery" (Ellwood 1906, quoted in Degler 1991, 85).

The nationally known sociologist Carl Kelsey of the University of Pennsylvania was disappointed when Lamarckian theory was disproven and worried about how that development would shape social theory (Degler 1991). However, when Kelsey (1913) reviewed *The Mind of Primitive Man,* he wrote that the book was a genuine service to scholarship and that Boas's conclusion that racial differences are largely superficial was the prevailing belief among social scientists. Degler (1991, 86) notes: "By shifting the explanation of social behavior from heredity to culture, from nature to nurture, Kelsey became even more optimistic than he had been as a

Lamarckian." In 1916, Kelsey published a sociology textbook that was greatly influenced by the writings of Boas. Howard Odum, a sociologist from the University of Georgia, had a very quick reversal of opinion after reading Boas's two 1911 volumes. In his thesis in 1910, Odum opined that not much could be done to make a Negro anything but a Negro. Three years later, Odum quoted Boas on the plasticity of the human mind and went on to say that "it would clearly be impossible for the Negro children to show the same manifestations of mental traits as white children, after having been under the influence of entirely different environments for many generations" (Odum 1913, 205).

Psychologists were also beginning to be influenced by Boas. Columbia University psychologist Robert Woodworth took anthropometric measurements and tested for vision, hearing, and intelligence from simple performance tests on 1,100 people. He concluded that "sensory and motor processes, and the elementary brain activities, though differing from one individual to another, are about the same from one race to another" (Woodworth 1910, 178). He went on to give an alternative cultural explanation for possible "racial" differences. Stocking (1968, 218) points out that "Woodworth had taken his anthropometric and statistical training under Franz Boas, and had gained from him 'some appreciation of the value of anthropology to the psychologist.'" As we shall see below, another prominent psychologist and colleague at Columbia University, Otto Klineberg, was also greatly influenced by Boas.

Grant was aware of these developments and was frustrated by them. In 1912, he wrote to Osborn: "You must have at the head of any anthropological work a member of the North European race, who has no bias in favor of helots and mongrels" (quoted in Spiro 2009, 303). As Kroeber expressed it, there was an eternal chasm between the two sides (Spiro 2009). They each had a completely different way of seeing, interacting, and living in the world—diametrically opposite worldviews. One could say they were operating in essentially different cultures.

During the 1910s through the 1930s, those who followed the views of Grant were mainly his intellectual heirs. "They respected him personally, honored him publicly, and were content to continue following his lead well into the 1930s. They were also exclusively WASPs. In contrast, the professionals taking over the field of anthropology in the 1910s were Grant's intellectual *opponents*. They rejected his theories, denigrated his methods, and publicly castigated his work. They were also overwhelmingly Jews" (Spiro 2009, 304). They also were fighting for their professional lives, and later, as eugenics turned to Nazism, they were literally fighting for the survival of many people the world over.

The Galton Society: Grant's Attempt to Take over Anthropology

Grant, seeing the Boasians' takeover of the AAA and its journal *American Anthropologist,* attempted to respond and take back anthropology from the "Jews." With the aid of Davenport and Osborn, he founded the Galton Society as a racist-oriented organization to rival the "culture-oriented" AAA. To distinguish it from the university-based AAA, it was based at the American Museum of Natural History and was named after the founder of eugenics. The new organization was made up of nine self-perpetuating charter fellows who could elect officers and choose the sixteen, and later thirty-two, members. According to Grant, the membership would consist only of "native Americans, who are anthropologically, socially, and politically sound . . . no Bolsheviki need apply" (Grant to Osborn, 1918, quoted in Spiro 2009, 305). The first meeting was held in April 1918. Thus, the big four of scientific racism and eugenics, Grant, Davenport, Osborn, and Laughlin, tried to monopolize American anthropology in the early twentieth century.

None of the charter fellows of the Galton Society were professionally trained anthropologists. They were biologist E. G. Conklin of Princeton, paleontologist W. K. Gregory of the American Museum of Natural History, anatomist G. S. Huntington of the College of Physicians and Surgeons, zoologist H. H. McGregor of Columbia, paleontologist J. C. Merriam of the University of California, and psychologist E. L. Thorndike of Columbia, plus Charles Davenport, Madison Grant, and Henry Fairfield Osborn. The main purpose of the society was to promote eugenics and racial anthropology (Gregory 1919). The membership soon included the elite of the eugenics movement, such as C. C. Brigham, H. H. Laughlin, P. Popenoe, L. Stoddard, and R. M. Yerkes, to name only a few. Some anthropologists also became members, including R. Pearl, E. A. Hooton, M. Steggerda, T. W. Todd, and C. Wissler (Spiro 2009).

The society met at the American Museum of Natural History from 1918 to the mid-1930s. After eating lunch, members would give talks on racial anthropology followed by discussion, usually one talk per meeting. There were no dues; the "gentlemen's" meetings were funded mainly by Grant, Osborn, Grant's friend M. T. Pyne, and frequent eugenics philanthropist Mary W. Harriman (Engs 2005). At an early meeting, Grant had explained to the members that "from now on, authentic anthropology would be confined to 'the study of man as a *physical* animal and not confused with an ethnologic study of pottery and blankets'" (quoted in Spiro 2009, 306).

According to Spiro, the main goal of the Galton Society was to provide an uncontested, friendly meeting place for the eugenicists to share their points of view and to reinforce and provide encouragement for their ideology. It also provided an excellent environment for discussing new projects and findings and getting feedback from a friendly audience. For example, it was here that Yerkes and Brigham first presented and refined their interpretation of the World War I army intelligence tests that later became *A Study of American Intelligence* (Brigham 1923). Laughlin also first rehearsed his testimony for the 1923–1924 congressional hearings on immigration at a Galton Society meeting. Davenport and Clark Wissler, a devoted Grantian, consulted with members as they prepared *Race Crossing in Jamaica* (1929) and *Man and Culture* (1923), respectively. Thus, as Gregory wrote in a letter to Osborn, the Galton Society inspired and "unified the faithful and bolstered them to venture into the secular world to preach the gospel among the heathen" (Spiro 2009, 307).

In a meeting in 1925, eugenicist Harvard psychologist William McDougall summed up his view of the differences between the Galton Society and the AAA: "On one side of the nature-nurture debate were the sentimental social workers, egalitarian Bolshevists, and intellectual Jews, all of whom were 'biased against racial psychology' and allowed the emotional appeal of humanitarianism to stand 'in place of truth.' On the other side were the 'serious students of race' such as Madison Grant and Lothrop Stoddard, who recognized 'the reality' of inequality and stood for 'the importance of preserving racial distinctions in their purity'" (quoted in Spiro 2009, 306).

The Galton Society was not a peripheral organization in the first third of the twentieth century. One-third of the society's members were also members of the National Academy of Sciences (NAS), half were members of the American Philosophical Society, and over half were members of the AAAS; three served as president of the latter association. Since the Boasians were increasingly infiltrating anthropology within academics and universities, Grant, Davenport, and Osborn used the Galton Society in their attempts to influence anthropology in the federal government and among policy makers and granting agencies. They did this by directing their efforts toward influencing members of Congress and other government officials and by controlling the National Research Council (NRC). They were very successful in these efforts. As stated by Chase (1977, 166), "the Galton Society helped retard the development of modern anthropology in America for many years." In fact, it did much more than that.

The National Research Council and the Early History of American Physical Anthropology

The NRC was a federal agency established in 1916 to coordinate the country's scientific resources in the interests of national security and preparedness in anticipation of the U.S. entering the war in Europe. The council was chaired by a longtime acquaintance of Grant's, astronomer G. E. Hale, and was funded mainly by reliable eugenics funders, the Carnegie Corporation and the Rockefeller Foundation. Of the twenty-seven committees representing the various fields of American science, the eugenicists especially wanted to control two: the psychology and anthropology committees. They were successful in their endeavors. Robert Yerkes was put in charge of psychology and William H. Holmes, past president of the AAA and head curator of the Department of Anthropology at the Smithsonian, was named chair of anthropology. Holmes was a descendent of the Puritans, a confirmed eugenicist, a longtime friend of Grant's, and an opponent of Boas's (Cole 1999). Because Boas and his colleagues then constituted a majority of the AAA and the editorial board of the *AA* but Holmes and other Grant supporters were influential in the NRC, it took months to form the Anthropology Committee of the latter. Grant and his supporters claimed that the Boasian cultural anthropologists, unlike the "physical" anthropologists, were not true scientists (Spiro 2009). Siding with Grant, Hale appointed only *physical* anthropologists (who at that time, of course, were mainly polygenic eugenicists and Grantians) to the Anthropology Committee of the NRC. Thus, when the committee was named in July 1917, all of the members were confirmed eugenicists and "physical anthropologists," including Holmes as chair, Davenport as vice chairman, Grant, and Aleš Hrdlička, who was curator of physical anthropology at the Smithsonian (Spiro 2009; Barker 2010).

The committee then promoted three major projects that they felt would be of immediate value to the war effort and would be of increasing value to America's future. The first was a simple matter: to relax the minimum height standards for the military and allow shorter men to serve. In public, they claimed that this would allow many more fit men to serve in the armed forces. However, the real motivation was to make it possible for more recent immigrants (who were generally shorter than resident Nordics) to serve and be sent to the front in the hope that more of them would be killed (Spiro 2009, 310).

The second project, initiated by Grant, was to use recruits to compile a comprehensive anatomical record of the U.S. population. Since millions of men were being recruited, there was an opportunity to gather a massive amount of data on anthropometry and other measurable attributes in order to ascertain the physical dimensions of the American "type." Grant ultimately hoped to demonstrate that immigration and miscegenation were deteriorating the Nordic race. Davenport was put in charge of the subcommittee on anthropometry. He was the ideal person for this task given his work at Cold Spring Harbor and the fact that he had concluded from an earlier study that success as a naval officer was related to a single inherited recessive trait, "thalassophilia" (love of the sea), that was especially common among Nordics. This was typical of the unit theory of inheritance of complex traits by the eugenicists. Despite this "scientific" examination, the navy, however, continued to use other criteria to select its officers.

Under the auspices of this project, Grant and Davenport began collecting eugenic race data on the soldiers, emphasizing that they were collecting data to do such things as create better uniforms. However, the data collected included such things as eye color, mother's native language, and father's religion—data that was not relevant to helmet size or the length of a soldier's sleeves. After these data were collected, they were transferred to Hollerith punch cards and analyzed according to race (Davenport and Love 1921).

Another major study developed within this initiative was Robert Yerkes's use of the army recruits to legitimate mental testing and again prove that Nordic Americans were superior in intelligence to other "races" within the United States. The data from Davenport and Love's study and their subsequent 1921 volume *Army Anthropology* (in which the Hebrew "race" ranked last and Nordics first in almost all of two dozen anthropometrical measurements) were instrumental in congressional debates about restricting immigration from 1922 to 1924. Yerkes's study and Brigham's 1923 popular volume, *Study of American Intelligence,* which synthesized the data on the army intelligence tests, also were important in these debates.

The third project of the NCR's Anthropology Committee was to establish a journal of physical anthropology (presumably to compete with *AA*). This had been a longtime dream of physical anthropologist Aleš Hrdlička and other eugenicists. However, as I will discuss below, this journal later became a major factor in the downfall of the hold of eugenics over anthropology within the United States.

Aleš Hrdlička (1869–1943) was born in what is now the Czech Republic and his family immigrated to the United States when he was twelve (Ortner 2010). He was trained in medicine and anatomy, and he studied with French anthropologist L. Manouvrier at the Paris Anthropological Institute for four months in 1896. This was his only formal training in anthropology, but it had a major impact on his career. In 1903, Hrdlička was hired by William Holmes (partly at the recommendation of Boas) as curator of physical anthropology at the Smithsonian National Museum of Natural History, where he remained for nearly forty years. Holmes was a mentor and strong supporter of Hrdlička (Brace 2005; Little 2010; Ortner 2010; Szathmáry 2010).

Hrdlička had mixed feelings about the standard eugenics of the day. His general views on race were in accord with the accepted polygenic ideology of the day. For example, he thought Africans were inferior to Europeans and that mixing the two races was deleterious. However, being an Eastern European immigrant himself, he thought that all Europeans were equal and that mixing of recent European immigrants with Americans did not diminish the capacity of Americans (Brace 2005). He thus did not agree with Grant's "Nordicism" or with the immigration restriction laws of the 1920s (Brace 2005; Ortner 2010). As we shall see below, the relationship between Grant and Hrdlička changed as time went on, and this relationship had a great influence on the history of anthropology. Hrdlička did believe in making immigration more selective and in using intelligence tests to limit immigration. He pressed for individual screenings but rejected the notion of placing limits on entire nationalities (Degler 1991). However, Hrdlička's opposition to the conventional, biologically deterministic wisdom of the day was not strong, and he predicted that the "growing science of eugenics will essentially become applied anthropology" (Hrdlička 1919, 25). Hrdlička, after all, had trained in medicine and was a physical anthropologist.

Within the NRC, Hrdlička had the support of Holmes, and he moved closer to Grant and the hereditarians in order to gain support (and financial backing) for the proposed journal of physical anthropology. He became a member of the AES, put Grant on the editorial board of the journal, and solicited articles from members of the Eugenics Research Association (ERA). In his prospectus for the proposed journal he stated that it would assist "in such important coming national movements as . . . the regulation of immigration, eugenics progress, and all other endeavors . . . safeguarding and advancing the physical status of man in this country" (Hrdlička 1918, quoted in Spiro 2009, 313). In 1918, the Anthropology Committee of the NRC approved the creation of the *American Journal of Physical Anthropology* with Hrdlička as editor, a position he held until 1942. With this, the three major goals of the committee were accomplished. However, President Wilson rejected a fourth project, which sought to include Madison Grant in the U.S. delegation to the Paris Peace Conference so that Wilson could use Grant's racial expertise in redrawing the boundaries of Europe (Spiro 2009).

Hrdlička put all of his efforts into the journal and began to neglect the other projects of the Anthropology Committee of the NRC. Grant thought the anthropometry study was more important and more pressing than the journal, which he believed could be launched at any time. Hrdlička was afraid the emphasis on anthropometry would drain resources from his journal and that Grant was attempting to stymie his life's dream. The two became increasingly irritated with one another and their relationship deteriorated quickly.

In Spiro's (2009, 314) analysis of the conflict, "The important point about these otherwise nugatory machinations is that the rift between Grant and Hrdlička had nothing to do with ideology, principal, or scholarship, and everything to do with personal jealousy and ambition." Of course, Grant explained Hrdlička's behavior as "unconsciously biased" because he was "an East European round head" (Grant letter to Osborn, 1912, quoted in Spiro 2009, 314). In 1923, when Hrdlička heard that Grant was being treated for rheumatism, he wrote to John Harvey Kellogg that Grant "ought to be afflicted with everlasting rheumatism of all his writing organs, for he has done a great deal of mischief with his 'Nordicism' " (quoted in Ortner 2010, 97). By mid-1918, Grant and Hrdlička were no longer on speaking terms. The argument between these two men led to major developments in anthropological history. It drove Hrdlička to form an alliance with Grant's main nemesis, Franz Boas.

At the end of World War I, President Wilson turned the NRC into a permanent peacetime institution. He reduced the number of research committees from twenty-seven to thirteen "divisions," merging psychology and anthropology. John C. Merriam, president of the Carnegie Institution, was appointed chairman of the NRC. Merriam, an avid eugenicist, was a charter member of the Galton Society and won the Advisory Council of the AES. When he began his work at the NRC, he and Grant, who were longtime friends, agreed to coordinate their efforts in the interest of science in the future, as they had done in the past (Spiro 2009).

This appointment fueled the fire of the antagonism between Grant and Boas. In 1918, Merriam announced at a meeting of the Galton Society that American anthropologists should stop wasting their time on the cultural practices of American Indians and devote their time to the more relevant problems of the day, such as the racial traits of new immigrants (Gregory 1919). When the organizing meeting for the Division of Anthropology and Psychology was held in 1919, Boas was not invited and cultural anthropology was not represented. Boas and his new ally Hrdlička wrote an open letter to the NRC in the *AA* demanding that cultural anthropology be represented in the division, stating: "The forces which determine the development of human types are to a very large extent cultural forces" (Boas, Hrdlička, and Tozzer 1919). The Grantians responded with letters of their own urging Merriam to exclude Boas and his followers.

The rift between the two anthropological camps widened, with the Grantian biological determinists insisting that the Boasian concept of culture was not true science. Lowie, an immigrant Jewish student of Boas, wrote that anthropology "is divided by a far-reaching difference as to principles and ideals . . . I feel that no scientific bond unites me with our opponents." Like Boas, he despised Grant, labeling him as arrogant and comparing him to a sadistic Prussian nobleman who "gloats over the vision of big blond princes leading little brunet Mediterraneans to be remorselessly butchered" (quoted in Spiro 2009, 316). Given the ideological differences between the two groups and the racism of the eugenicist physical anthropologists, Lowie believed that it was beginning to be a disgrace to be classed as an anthropologist.

However, because the only official professional organization in anthropology was the American Association of Anthropologists, after a year of infighting, the NRC decided to have the AAA elect six representatives to its Anthropology and Psychology Division. This was a major setback for Grant and his followers. The professionalization of the discipline by Boas and his followers had paid off, and five of the six members were, more or less, Boasians, including Boas and Kroeber. Only one, Clark Wissler, was from the aristocratic, self-appointed Galton Society (Barkan 1992).

Six weeks later, Boas (1919), likely not thinking of the professional ramifications, published a short, three-paragraph letter in the magazine *The Nation* claiming that at least four (unnamed) anthropologists had served as spies for the U.S. government during the war under the guise of doing research and that in doing so, they were prostituting science. Boas believed that anthropologists who spied for their country betrayed scientific ethics and damaged the credibility of anthropological research. Furthermore, he believed that when scientists used their research as a cover for political spying, they forfeited their rights as scientists. He worried that other countries would look upon visiting foreign investigators with mistrust and suspect them of sinister motives, and that such action had raised new barriers against the development of international friendly cooperation. These were not frivolous concerns, and the subject of anthropologists acting as spies is a serious ethical issue that is still being debated within the field (Price 2000; Gledhill 2008; Barker 2010).

In his letter, Boas also implied that President Wilson was a liar and a hypocrite and that democracy in America was a fiction. Boas had played into the hands of Grant and his colleagues, putting his influence in the AAA and in the NRC in jeopardy. Ten days after the letter was published, at the annual meeting of the AAA, Boas was expelled from the association's council by a vote of twenty to ten and forced to resign from the NRC's Division of Anthropology and Psychology. An infuriated Lowie viewed this as a degrading spectacle, in which "the foremost representative of our science" was scorned by men "who are anthropologists only by courtesy" (Lowie letter to Wissler, 1920, quoted in Spiro 2009, 317). A few weeks later, Pliny Goddard, a colleague and strong supporter of Boas, was removed as editor of the *AA*. The reign of Boas in these important professional anthropology organizations appeared to be over. Referring to Stocking's (1968) analysis of these incidents, Spiro (2009, 217) states that the censure of Boas "was the result of numerous schisms within American anthropology, but at its heart it reflected years of friction between aristocratic WASPs (mainly physical anthropologists) anxious to stem the rising tide of color and immigrant Jews (mainly cultural anthropologists) seeking to impede the ascent of scientific racism."

A number of Boasians, including Lowie, wanted to leave the AAA and start a new organization, but Boas and Kroeber insisted that they stay and continue their fight from within the association (and we shall see that this strategy paid off in the end). On the other hand, the eugenicists were delighted. Osborn gloated, in a 1920 letter to the head of the Smithsonian, C. D. Walcott, that Boas "now occupies a comparatively obscure and uninfluential position" and that they would never hear from Franz Boas again (quoted in Spiro 2009, 318). But the Grantians were not satisfied and continued to go after Boas personally and professionally. Walcott fired Boas as honorary philologist in the Smithsonian's Bureau of American Ethnology and then proceeded to attempt to get Boas fired from Columbia University. However, Boas had not done anything actionable that would enable Columbia to fire him. Walcott was frustrated but not deterred. He next contacted President Woodrow Wilson to demand that something be done about Boas. After an investigation by the U.S. Justice Department found that Boas had not broken any laws, they were unable to remove him from the Columbia faculty, though funding for the Anthropology Department was limited for many years (Spiro 2009).

With the Boasians out of the way, the Grantians were in control of the NRC and their racist policies could now be pursued without resistance. They proceeded to launch a number of studies in scientific racism. One of the best-funded projects, which had financial support from the Rockefeller and the Russell Sage Foundations, was the Committee on Scientific Problems of Human Migration. The committee operated from 1922 to 1928 and included as directors Robert Yerkes, Clark Wissler, and John Merriam, all members of the Galton Society and the AES. Grant's *Passing of the Great Race* guided much of their research. The committee sponsored sixteen research projects, all of which focused on eugenics questions and were headed by associates of Madison Grant and members of the Galton Society. For example, the committee funded Carl Brigham's publication on the army IQ tests, Henry Fairchild's project on the causes of immigration, and Clark Wissler's project on race intermixture. All of the projects were intended to be used to put science in the service of politics, an effort that Grant had pushed for and an approach that was soon to be epitomized by the Nazi use of "science" to justify their policies. Stocking (1992) argues that the Committee on Scientific Problems of Human Migration simply served as a research arm of the immigration restriction movement.

Thus, in the early 1920s, the eugenics movement, the Galton Society, and the so-called physical anthropologists controlled the influence of the field of anthropology on the U.S. government and its granting agencies. Franz Boas and his type of environmental anthropology were considered to be finished. It is no wonder that Boas and his students did not identify with the physical anthropology of the time.

During the 1920s and early 1930s, while Grant and his eugenics followers were working on creating U.S. policies related to immigration restriction, sterilization laws, anti-miscegenation laws, Boas continued to train and graduate PhD students. These students, who were mainly loyal to Boas, took over essentially all of the major departments of anthropology in the country and comprised the majority of trained professionals in the discipline. They also soon joined the AAA, and within a few years after Boas's letter to *The Nation,* they again controlled the only professional anthropological organization in the country. In 1924, Robert Lowie was appointed editor of *AA,* a position he held until he was replaced in 1933 by Leslie Spier, another student of Boas's. In addition, Hrdlička was still editor of the *American Journal of Physical Anthropology* and was increasingly opposed to Grant and his type of eugenics. Thus, by the mid-1920s, the Boasians and anti-eugenicists had gained control of the only truly professional anthropological association and the two professional journals.

However, the Division of Anthropology and Psychology of the NRC was still in control of funding for anthropology, and it only funded projects on scientific racism and rejected projects related to environmental influences or culture. This at first frustrated Boas, but he soon devised a plan to take advantage of the system. In 1923, the NRC received funding from the Rockefeller Foundation for a graduate fellowship program. Students of Boas began submitting projects in scientific racism that *sounded* like good old hereditarianism. In addition, the representation in the division had changed slightly. Yerkes had left the division to pursue his new interests in primatology (see Sussman 2011). Also, having regained control of the AAA, Kroeber and Lowie were elected to represent that organization at the Division of Anthropology and Psychology of the NRC.

Three of Boas's students were awarded NRC fellowships. In 1923, Melville Herskovitz, who had earlier been denied a fellowship, received funding to conduct an anthropometric study of African Americans (with the support of Wissler and Davenport, who believed that the cultural anthropologists were finally coming around). In 1925, Margaret Mead received funding to study adolescence in Samoa, a project that seemed to be not cultural anthropology but purely orthodox biology. In the same year, Otto Klineberg, a psychologist and colleague of Boas's at Columbia, was awarded a fellowship to study mental differences between the races. These projects were carried out under Boas's direction, were related to his research agenda, and had originally been defined by Boas. Stocking (1968, 300), who has examined Boas's correspondence, believes that he saw "all of these studies as part of a coordinated attack on the problem of the cultural factor in racial difference."

In addition to infiltrating academics and the major government grant-ing agency and proselytizing the concept of culture in this context, Boas began a public campaign against eugenics and Madison Grant and his disciples. Spiro (2009, 324) comments:

> Starting in the mid-1920s, he managed to publish at least one anti-Nordicist book or article a year in the popular press. His basic thesis had not changed since the 1910s, when he had first attacked Madison Grant. But this time [he was] armed with a mass of ethnographic research showing that envi-ronmental factors were at least as important as genes in determining men-tal and social traits. . . . Boas relentlessly denounced the immigration laws as being unscientific, attacked the army intelligence tests for being cultur-ally biased, provided biological evidence to show that mongrelization did not lead to deterioration, and cited the findings of cultural anthropologists to show that culture, and not biology, accounted for the mental (and a good deal of the physical) nature of man.

At the end of the 1920s and early 1930s, the research results of Boas's NRC fellowship recipients began to appear, all acknowledging Boas's guid-ance. Margaret Mead (1901–1978), in *Coming of Age in Samoa* (1928), showed that even such things as biologically based puberty varied greatly from culture to culture and introduced the American public to the concept of cultural determinism. Mead went on to become immensely popular and wrote many scientifically influential and popular books in cultural an-thropology. She was awarded the Presidential Medal of Freedom posthu-mously in 1979 by President Carter. That same year, Melville Herskovits (1895–1963) published *American Negro* (1928), pointing out that "it now becomes apparent that *social* as well as biological factors are of the utmost importance in the consideration of the American Negro type" (66). Throughout his career, Herskovits emphasized that race was a socio-logical and not a biological concept. He also further developed the con-cept of cultural relativism in his writings. Herskovitz later established the first African and African American studies programs at Northwestern University.

Otto Klineberg (1899–1982) investigated the validity of Yerkes and Brigham's IQ testing of army recruits. His results were first published in the *Archives of Psychology* (1928, 1931) and later in *Negro Intelligence and Selective Migration* (1935a) and, in a more popular venue, *Race Dif-ferences* (1935b). Klineberg first tested Grant's hypothesis and Yerkes and

Brigham's supposed findings that IQ scores were higher among Nordic Europeans than among those of Alpine or Mediterranean ancestry. He traveled to Europe and tested 1,000 boys from ten rural and urban communities. He found that urban boys scored higher than rural boys regardless of ancestry and that differences among so-called racial groups were small and unreliable. Klineberg (1935b, 194) concluded that there was "no support to the theory of a definite race hierarchy" among these European categories.

Klineberg then scrutinized the long-held belief that the intelligence of certain races was inferior. He also examined the army test results on African Americans. His research design was similar to that of Boas's study of American immigrants. It had been previously noted that whites from northern portions of America had scored higher than northern blacks and that southern whites scored higher than southern blacks. However, northern blacks scored higher than many southern whites (Boas 1921; Kroeber 1923). Yerkes (1923) had explained these differences as "selective migration," arguing that the more energetic, progressive, and mentally alert blacks had migrated to the North to improve their educational and vocational possibilities.

Klineberg examined the school records of over 500 northern schoolchildren from Tennessee, South Carolina, and Alabama before they had migrated north and found no differences in their scores from the average of the entire population of African Americans in those states. Thus, he found no differences based upon selective migration. Anticipating a rebuttal, he also noted that if the parents of these children (who were the ones that had decided to move North) were more intelligent than those who had stayed in the South, why had their children been only average if intelligence was inherited? He then went further and tested the IQ of southern-born African Americans in New York City in relationship to the amount of time since they had moved from the South. He found that the longer they had lived in the city, the higher their scores were. He concluded that environment alone and not heredity or selective migration accounted for the differences (Klineberg 1928, 1935b). Klineberg urged his fellow psychologists to use the concept of culture (Klineberg 1935b; Degler 1991). Klineberg later became the president of many psychology associations, was the founder of social psychology, and played a major role in advancing the psychology profession generally (Hollander 1993). Based on the data Klineberg and others collected during this period, Boas and other anthropologists insisted that "the I.Q. tests were culture-bound, inaccurate, based on unscientific categories and assumptions and so contaminated by sloppy methods as to be without merit" (Cravens 2009, 162).

Works by other Boasians continued to appear both in scientific journals and in the popular media, including George Dorsey's *Race and Civilization* (1928), Boas's *Anthropology and Modern Life* (1928), Lowie's *Are We Civilized?* (1929), and Benedict's *Patterns of Culture* (1934). Cultural anthropology, the concept of culture, and the legitimacy of environmental over hereditarian explanations of behavior and racial differences were becoming accepted. Boas and Klineberg also were having an effect on other social sciences and psychologists, and many publications emphasizing the culture paradigm were beginning to appear. By the end of the 1920s, many social scientists were also seeing culture as an alternative explanation to race (Degler 1991). The writings of Boas and his students were having an effect on how race was studied and understood, and cultural relativism had freed anthropology from the shackles of biology (Barkan 1992). As the concept of culture began to take over respectable social science, the number of proponents of eugenics, polygenics, and the hereditarian point of view began to dwindle precipitously.

The importance of environmental influences on behavior also was gaining support because of criticisms of the significance or even the existence of instinctual behavior among humans. Many psychologists were beginning to see environment as having a major influence on behavior (see review of this trend in Degler 1991). This culminated in a movement toward behaviorism in American psychology. The extremes of this movement were represented by John B. Watson. In his popular book *Behaviorism* (1924, 103–104) he wrote: "Give me a dozen healthy infants, well-formed, and my own specified world to bring them up in and I'll guarantee to take any one at random and train him to become any type of specialist I might select—doctor, lawyer, artist, merchant-chief and, yes, even beggar-man and thief, regardless of his talents, penchants, tendencies, abilities, vocations, and race of his ancestors. I am going beyond my facts and I admit it, but so have the advocates of the contrary and they have been doing it for many thousands of years."

By the early 1930s, even such advocates of eugenics as Henry Goddard, Carl Brigham, and Lewis Terman were having serious doubts. Brigham (1930, 165) himself repudiated the conclusions that he had reached about the army IQ tests. He stated that given the current evidence, "that study with its entire hypothetical superstructure of racial differences collapses completely. . . . Comparative studies of various national and racial groups may not be made with existing tests. . . . One of the most pretentious of these comparative racial studies—the writer's own—was without foundation." Terman wrote in a letter to Davenport in 1931 (quoted in Degler 1991, 151): "As you know, there are a good many psychologists and anthropologists these days, also sociologists, who are inclined to argue that the intelligence that individuals develop during childhood and adult life is determined largely if not entirely, by his cultural environment and formal training. If that is true, then eugenics has no place as far as intellectual differences in human beings are concerned."

In addition to data from the social sciences, evidence from the biological sciences, especially genetics, began to accumulate that posed challenges to the strict biological determinist approach and eugenics (Allen 1986, 2011). As early as 1913, criticisms of Davenport's ERO work on the inheritance of "feeble-mindedness" as lacking scientific rigor came from the Galton Laboratory under Pearson (Heron 1913; Pearson 1925). One of the major early criticisms focused on the oversimplification of behavioral or personality phenotypes. Terms such as "feeble-mindedness," "criminality," and "intelligence" were seen as so vague and subjective as to be meaningless (Kenrick 1914; Morgan 1925, 1932; see Allen 2011). After 1915, many prominent geneticists, including T. H. Morgan, Herbert Spencer Jennings, Raymond Pearl, H. J. Muller, and Sewell Wright began to privately attack the claims of eugenicists (Allen 1986). Unfortunately, before 1925, much of this early criticism by geneticists and biologists appeared largely in scientific journals. For the most part, the general public was not aware of this literature and thus had the impression that eugenics had the stamp of approval of the scientific community. After all, eugenics was promoted by leading scientists and had wide exposure in the popular media (Allen 2011).

However, Raymond Pearl, an early proponent of eugenics, a biometrist, a geneticist, and a physical anthropologist, began to criticize eugenics in popular media, for example in *American Mercury* magazine in 1927: "The literature of eugenics has largely become a mingled mass of ill-grounded and uncritical sociology, economics, anthropology, and politics . . . full of emotional appeals to class and race prejudices, solemnly put forth as science and unfortunately acknowledged as such by the general public" (quoted in Ludmerer 1972, 84). In 1924, a colleague of Pearl's at John Hopkins University, Herbert S. Jennings, a nationally prominent biologist, wrote that the "unit character" concept, based on the assumption of a one-to-one relationship between a phenotypic character and a single Mendelian gene, had been rejected ten to fifteen years earlier (Allen 2013). In fact, he stated that the concept of the "unit character" as the product of a single gene that was unaffected by the environment was "an illustration of the adage that a little knowledge is a dangerous thing" (quoted in Zenderland 1998, 321). In 1932, he wrote on the subject "Eugenics" in *The Encyclopedia of Social Sciences:* "Both racial arrogance and the desire to justify present social systems find a congenial field in eugenic propaganda" (quoted in Degler 1991, 150). Geneticist A. Weinstein wrote the entry on "Heredity," stating: "Most claims of genetic superiority are inspired not so much by scientific data as by a desire on the part of some classes, nations or races to justify their subjection of others" (quoted in Degler 1991, 205–206). (It is interesting to note that in this encyclopedia the topic of "Race" was covered by Boas, the topic of "Mental Testing" by Klineberg, and the topic of "Instinct" by a psychologist, L. L. Bernard, who rejected the concept.) Degler (1991, 150) emphasizes that "despite the telling and steady flow of objections emanating from anthropologists and sociologists, the death knell for eugenics was not sounded by them. That job was accomplished by geneticists as they learned that their science was being both misunderstood and misused by public advocates of eugenics."

From the time of the split between Grant and Hrdlička in 1918, professional anthropology—that associated with the university and academic training of professionals—was represented by the AAA, the anthropological section of the AAAS, in which Boas had been active since the 1880s (Spencer 1968; Szathmáry 2010), and the publications *AA* and *American Journal of Physical Anthropology*. In 1928, Hrdlička was successful in establishing the American Association of Physical Anthropologists at the annual meeting of the AAAS. The first meeting of the AAPA was held in 1930, thus establishing physical anthropology as a distinct branch of science in the United States. Hrdlička was appointed president. One of the major goals was to establish the discipline at American universities, since almost all of the PhDs in anthropology in the United States were in archeology and cultural anthropology (only 11 percent of doctoral theses between 1900 and 1925 were on topics in physical anthropology). In fact, of the eighty-four charter members of the AAPA, only six had PhDs in physical anthropology; most were anatomists and medical doctors (Szathmáry 2010).

Anthropology in Europe (mainly England, France, and Germany) was basically physical anthropology and eugenics, but in the United States there had still been a division between the eugenicists and the cultural anthropologists. Now, with Hrdlička, who was still diametrically opposed to Grant, physical anthropology was moving away from the amateur Galton Society and the eugenics movement. By 1930, all the truly professional anthropological associations and journals in the United States were in the hands of Boasians, or at least the hands of those who were against the Grantians. Whereas eugenics was on the rise in physical anthropology in Europe, especially in Germany, at this time, it was undergoing a precipitous fall in America. For example, in Germany, the Gesellschaft für Physische Anthropologie was formed in 1926 with Nazi supporter, eugenicist, and longtime friend and disciple of Grant and Davenport, Eugen Fischer, as president. Such staunch Nazis as Otmar von Verschuer, Fritz Lenz, Ernst Rüdin, and Joseph Mengele were prominent members. The Institute of Anthropology, Human Heredity, and Eugenics was added to the Kaiser Wilhelm Institutes in 1928, with Fischer as director (see Chapter 4). In the United States, in contrast, the AAPA was headed by Grant's nemesis Hrdlička, and Boas, Kroeber, Herskovits, and a number of other Boasians were prominent members (Spiro 2009; Szathmáry 2010).

With all of the professional organizations and journals of physical anthropology in the hands of anti-eugenicists and major departments in the hands of Boas's students, the face of physical anthropology in the United States was finally changing. Before this, going back to the American School of Anthropology of the Mortonites in the nineteenth century, physical anthropology in America had been dominated by racism, polygenics, and eugenics, and gentlemen scientists without university degrees. By 1930, although it was still largely focused on static, typological, and mainly descriptive anatomical research, physical anthropology was moving away from eugenics (Little and Kennedy 2010; Sussman 2011).

Thus, in the 1930s, Grant and eugenics had already begun their decline in America. But Grant and the proponents of the eugenics movement were occupied with politics and public policy during the 1920s and 1930s and were not worried about or aware of what was happening in professional physical anthropology. They were successfully getting sterilization, immigration, and anti-miscegenation laws passed. However, while they were winning the battle, they were losing the war. In addition, the Nazi movement in Germany wasn't helping them. The popularity and influence of the eugenics movement was essentially over. Even the influence of the eugenicists at the NRC was over. In 1931, Robert Lowie was named chair of the Division of Anthropology and Psychology. That same year, Boas was elected president of the AAAS. In his presidential address, Boas specifically addressed the issue of race-mixing and pointed out that there was no evidence of any deleterious effects. He stressed the influences of the environment and culture on human behavior (Boas 1931; Farber 2011). In 1932, the *New York Times* proclaimed that Nordicism was a discredited doctrine. In 1936, Franz Boas appeared on the cover of *Time* magazine with a cover story titled "Environmentalism" (Spiro 2009). By 1938, a number of anti-racist declarations had been issued by the AAA, the executive council of the Society for the Psychological Study of Social Issues, and a group of distinguished geneticists at the International Congress of Genetics in Edinburgh, which became known as the Geneticist's Manifesto (Barkan 1992).

In the meantime, the eugenicists were attempting to make a recovery. In 1932, the Third International Eugenics Congress was held at the American Museum of Natural History, sponsored by the ERO, the ERA, and Galton Society. It was just eleven years after the second congress, which also took place at American Museum of Natural History, and twenty years after the first congress in London. It was organized and run by the same old guard as in 1912. Davenport was president and Henry Fairfield Osborn and Leonard Darwin were honorary presidents. The managing committee included Irving Fisher, Harry Laughlin, Frederick Osborn, and Madison Grant. In 1931, *EN* announced proudly that the congress would present the real meaning and content of eugenics. Publication of the proceedings of the congress was financed by the Carnegie Institution of Washington; the volume was dedicated to Mrs. E. H. Harriman (Engs 2005).

Although the theme of the congress was "A Decade of Progress in Eugenics," the papers offered nothing new. They focused on race survival, sterilization, Davenport's anthropometry, intelligence tests, immigration restriction, birth control, and eugenics marriage counseling. The Kallikak family was resurrected to defend mandatory sterilization, "biological disharmony" and polygenic views of racial differences were once again invoked to argue against miscegenation, and army intelligence tests were used to defend immigration restriction laws, although even Brigham had admitted in 1930 that these tests were worthless (Spiro 2009). Essentially, the congress was a reiteration of Grant's *Passing of the Great Race* and all of the polygenic rhetoric that had come before that tome.

The organizers of the congress had sent out over 5,000 invitations, but only seventy-three delegates showed up. Only sixty-five papers were delivered. The second congress, in 1921, had included four times as many delegates and twice as many papers. By 1932, the writing was on the wall when one of the delegates, H. J. Muller, a geneticist (and future Nobel Prize winner) who had worked with T. H. Morgan, "told the delegates that eugenics had become unrealistic, ineffective, and anachronistic pseudoscience. . . . [He] admonished eugenicists for ignoring the obvious influence of environment on IQ, and he suggested that slums were more important than heredity in the creation of criminals" (Spiro 2009, 341). Muller argued that economic inequality masks genetic differences and that the differences between social classes and races are fully accounted for by known environmental effects (Muller 1934; Paul 1995).

For the public that *EN* had hoped to reach in the 1930s, the message seemed to be very clear. In press coverage of eugenics topics, it seemed that eugenicists were blaming the Great Depression on the morals of the unemployed and ignoring the major advances in anthropology, sociology, psychology, and genetics that had been made in the last two decades. Spiro (2009, 341) summarizes the effect: "The Grantians unmasked themselves as a dogmatic group of reactionaries motivated not by science but by bigotry. Eugenics, concluded the *New York Times* at the end of the Congress, 'seems to have become a disguise for race prejudice, ancestor worship, and caste snobbery.' " When Hitler began to implement the very program eugenicists were advocating, the attitude of the American public toward these ancient polygenic ideals grew cold.

A day after the eugenics congress closed, the International Congress of Genetics opened in upstate New York at Cornell University. It was attended by eight times as many delegates as had attended the Third International Eugenics Congress and included more than three times as many papers. The *New York Times* (1932) summarized the difference between the two conferences: "In a single week, we are thus granted the opportunity of comparing the theories and methods of two schools. On the one hand, much loose talk about sterilization of the feeble-minded and restricting marriages of undesirables, on the other earnest experiments [dealing with the mechanism of heredity]. . . . It is easy to choose between the two schools."

The Third International Eugenics Congress was the last eugenics congress. However, Madison Grant made one final attempt to restart the eugenics movement and revitalize polygenics. In 1933, he published *The Conquest of a Continent, or the Expansion of Races in America*. The forward was written by KWIA director Eugen Fischer. The book basically explained that after the great Nordic race had bravely conquered the United States and Canada, it had then made three major mistakes. Lazy southerners had brought in African slaves to do their work for them, greedy northerners had encouraged inferior Mediterranean Europeans to work for low wages, and misguided humanitarians had allowed Jewish victims of persecution to take refuge in America. Grant argued that intermarriage between blacks, Italians, and Jews and even between these races and members of the Nordic race would be the end of civilization.

Grant, however, had four remedies for the situation. These were: (1) prohibit all immigration into the United States from all countries; (2) deport all undesirable aliens and unemployed persons; (3) sterilize all criminals and intellectually unfit individuals; and (4) force birth control on African Americans and other inferior groups (to be defined, we suspect, by Grant and his followers). It was "the duty of all Americans . . . to face the problem boldly and to take all eugenic means to . . . abate drastically the increase of the unfit," he wrote (Grant 1933, 349–351).

The eugenics crowd thought that, just as had *Passing of the Great Race*, *Conquest* would set science and the public straight. Grant and Hitler would be vindicated, and the forces of eugenics would once again reign in the United States, as they were about to do in Germany and Europe. Henry Fairfield Osborn had written in the introduction to the book, that *Conquest* "'explodes the bubble' of the environmentalists and reveals their beliefs to be 'merely myths' by showing 'that moral, intellectual, and spiritual traits are just as distinctive and characteristic of different races as head form'" (quoted in Spiro 2009, 344). Eugenicists welcomed Osborn's introduction as a confirmation of the views of Nazi human geneticists, and the next year Osborn was given an honorary degree from Frankfurt University (Weiss 2010). In 1933, Davenport wrote in a letter to Grant: "Your book will play an important part in destroying these idols of the environmentalists. I can only trust that it may have as much influence on civilization as has your *Passing of the Great Race*" (quoted in Spiro 2009, 343). The publisher, Scribner's Sons, vigorously promoted the book; it sent out massive mailings, it offered discounts, and it offered free copies to educators in an attempt to get the book adopted in high school and college courses. It also sent copies to educational journals and anthropology departments for review. In circulars it claimed that just as Hitler was working out the race problems in Germany, Grant's book offered the solution to those problems in the United States.

The public didn't buy the message or the book. Some reviews of the book were favorable, but the vast majority were extremely negative. Grant and his followers blamed this on Jews and egalitarians. However, by the end of the 1930s, the book was essentially ignored. Only 3,000 copies had been sold. Grantians believed that the book had been the victim of a Jewish conspiracy. Spiro (2009, 346), however, argues that "the eugenic gospel no longer resonated with the public, and not even a work as smoothly argued as *The Conquest of a Continent* was going to prevent the movement's inexorable slide towards oblivion."

And toward oblivion it did slide. The Galton Society quietly dissolved in 1935 (Spiro 2009). By the 1930s, reputable scientists were leaving the ERA in droves. The annual meetings were being attended mainly by eccentrics, amateurs, and Mussolini and Nazi sympathizers. The last president of the association was Charles M. Goethe, a Nazi sympathizer to the end (see Chapter 4). Goethe, a wealthy businessman, spent millions of dollars of his own to publish and distribute racist pamphlets and later became a major supporter of Richard Nixon. During his presidency of the ERA, a number of German eugenicists, including Eugen Fischer, Ernst Rüdin, and Mussolini's science advisor, Corrado Gini, were made members of the ERA. The last meeting of the association was held in 1938 (Spiro 2009).

The Eugenics Record Office was already considered obsolete by 1930. In 1921, the ERO and the Station for Experimental Evolution were consolidated into the Department of Genetics at Cold Spring Harbor, although the CIW still funded it and Davenport continued as director (Engs 2005). In 1923, the highly respected geneticist T. H. Morgan expressed doubts to the CIW about the quality of the research at the ERO (Allen 2004). After the Immigration Restriction Act was passed in 1924, Morgan added a new chapter to the 1925 edition of his text *Evolution and Genetics* in which he argued that almost nothing was known about the causes of mental differences among individuals, much less among nations or races (Allen 1978; Paul 1995). He deplored research that confused nature and nurture, and he wrote that until we know how large a role environment plays in human behavior, "the student of human heredity will do well to recommend more enlightenment on the social causes of deficiencies" instead of adopting the panaceas of eugenicists (Morgan 1925, 201). Danish geneticist Wilhelm Johannsen had expressed concern about similar problems a decade earlier. Johannsen had coined the term *gene* and in 1909 had introduced the concepts of "genotype" (the internal hereditary makeup of an organism) and "phenotype" (its external appearance). He argued that the genotype does not in itself determine the phenotype, since the way genes are expressed are dependent upon their context, which includes the interactions of genes and the environment. Indeed, he was ahead of his times; he argued that the same genotype may be expressed very differently in different environments (Paul 1995).

As did Morgan and Johannsen, many geneticists thought that the eugenicists, especially Grant and Laughlin and those who were working at Cold Spring Harbor, were totally out of touch with the advances being made in the field of genetics (Allen 2004, 2011; see also Farber 2011). In February 1929, in response to these criticisms, John C. Merriam, the president of the CIW, invited a visiting committee to examine the ERO and evaluate its usefulness for future research in genetics. However, most of the committee members, who included Laughlin, Davenport, and Brigham, were strong supporters of eugenics. Although the committee came up with some specific criticisms, it was basically sympathetic to the eugenics agenda (Allen 1986).

The ERO continued to operate as usual for the next five years. Laughlin, the ERO, and *EN* continued their work of propagandizing the Nazi Nuremberg Laws, advising the U.S. Congress on eugenic immigration restrictions for victims of Nazi persecution, advocating Nazi goals and methods of racial cleansing, and praising Hitler for taking away the property and liberty of Jews in Germany and moving into other parts of Europe. Davenport warned Laughlin to be more careful in his public statements and to become less involved in politically inflammatory issues, but Laughlin ignored this advice. In 1934, Davenport retired and the Department of Genetics at Cold Spring Harbor was taken over by geneticist A. F. Blakeslee (Allen 2004). However, Osborn and Grant convinced Merriam to allow Laughlin to continue his role as superintendent of the ERO. In the context of Laughlin's inflammatory statements, Davenport's retirement, and continued criticism of ERO, Merriam convened another committee in 1935. The composition of this committee was very different from the earlier one. Strong proponents of eugenics were not represented. The new committee was made up of scientists, such as L. C. Dunn, A. H. Schultz, H. Redfield, and E. A. Hooton, who did not agree with the simplistic genetic formulations of Davenport and Laughlin (Allen 1986).

After its meeting at Cold Spring Harbor on June 16–17, 1935, the committee concluded that "the Eugenics Record Office was a worthless endeavor from top to bottom, yielding no real data, and that eugenics itself was not science but rather a social propaganda campaign with no discernible value to the science of either genetics or human heredity" (Black 2003, 390). The committee suggested that any future involvement of the Carnegie

Institution with Cold Spring Harbor should be devoid of the word eugenics and should instead use the word genetics. Addressing the problem of mixing research with political activity and propaganda, it recommended that the ERO discontinue its association with *EN* and "cease from engaging in all forms of propaganda and the urging or sponsoring of programs for social reform and race betterment" (quoted in Allen 1986, 252). Some of the committee members suggested closing down Cold Spring Harbor altogether. But the ERO limped on, and Laughlin continued to be an embarrassment to the CIW (Allen 1986).

In late 1938, the CIW severed its connection with *EN*. *EN* then came solely under the aegis of the AES, and Laughlin was not included on the editorial staff in the later reorganized publication (see below). In January 1939, Merriam retired and was replaced as president of Carnegie by Vannevar Bush, who immediately forced Laughlin to retire, giving him one more year of salary. Laughlin was retired on December 31, 1939, and the Eugenics Record Office permanently closed its doors on the same day (Allen 1986; Black 2003; Weiss 2010). Laughlin moved to his home state, Missouri, and after an uneventful few years, he died in January 1943 at the age of sixty-three. In its peak years, as Allen (1986, 226) states, "the ERO became a nerve center for the eugenics movement as a whole. When it closed its doors on 31 December 1939, it was clear that the movement as such no longer existed."

Possibly the only quasi-academic institutional stronghold of eugenics was the American Museum of Natural History, at least as long as Henry Fairfield Osborn served as its president, a position he obtained in 1908. However, by the 1930s, Osborn was losing the respect of serious scientists and even his own staff, and in 1933 he was forced to resign from the museum. The museum was never again a center of eugenics (Spiro 2009).

The AES was also swiftly fading. In 1930, it had over 1,200 members representing every state in the United States, but three years later, it had fewer than 500 members, and six years later it had only 282. Reputable researchers let their memberships lapse. The AES had remained financially solvent because of the financial support of a few very wealthy contributors, such as John D. Rockefeller and George Eastman of Eastman Kodak. However, Rockefeller, seeing the writing on the wall, withdrew his support in 1930. Although Eastman volunteered to support it early in 1932, within a month of his pledge, he committed suicide. The society reduced its activities, shutting down offices in Chicago, Ann Arbor, and New York City, and shutting down its publishing company, leaving only an office in

New Haven. In 1935, Muller wrote what turned out to be the Geneticist's Manifesto against racism among eugenicists in general and Nazi racial policies in particular. The manifesto was published by twenty-three of the most influential reform eugenicists of the time, including Aldous Huxley, Theodosius Dobzhansky, J. B. S. Haldane, Gunnar Dahlberg, and Hermann Muller (Crew et al. 1939). *Eugenical News,* by then simply called *Eugenics,* ceased publication. Although the AES managed to remain in existence to the end of the decade because of sporadic funding from Wickliffe P. Draper's Pioneer Fund (see Chapter 8), for all intents and purposes it was inoperative.

Summary of Boas's Contributions

In his classic book *The Structure of Scientific Revolutions* (1962), Thomas Kuhn examines how scientific paradigms change over time. He states: "Though a generation is sometimes required to effect the change, scientific communities have again and again been converted to new paradigms. . . . Though some scientists . . . may resist indefinitely, most of them can be reached one way or another. Conversions will occur a few at a time until, after the last holdouts have died, the whole profession will again be practicing under a single, but now a different paradigm" (Kuhn 1962, 151). By the mid-1930s, the old guard of strict biological determinism and eugenics was dying off. Osborn died in 1935, two years after his forced retirement. Grant was devastated. He had visited or spoken to Osborn every day since 1895 and had worked with him in creating a number of organizations, including the Bronx Zoo, the New York Aquarium, the Galton Society, and the AES. Grant died two years later, in 1937, at the age of seventy-two, having been debilitated with arthritis for many years. As mentioned previously, in his final days, Grant had been helping Herman Goering organize an international hunting exposition. Goering was commander in chief of the Luftwaffe (Nazi Air Force) and Hitler's designated successor. The exposition did take place and culminated with a few selected delegates accompanying Goering in a hunting party. Grant had been one of these selected guests but did not live to attend the hunt. Laughlin died in 1943 (Black 2003). He remained a staunch eugenicist to the end. Davenport, who had retired from Cold Spring Harbor in 1934, died in 1944 of pneumonia at the age of seventy-eight. He was an active elder statesman of eugenics to the end, protesting the increasing opposition in America to Nazism and to racist policies at home and abroad. None of the big four of scientific racism produced students or academic "offspring." As Spiro (2009, 339) states: "Finally, the simple fact of human mortality played an inescapable role in the demise of scientific racism. . . . The same men who organized the movement were still in charge three decades later, when they were old, reactionary, and tired. And as they began to die off, almost nobody was interested in taking their place."

Thus, from the fifteenth century to the commencement of World War II, Western Europeans and Western European colonists of the United States defined "others" mainly within two fairly unchanging paradigms: polygenism and monogenism. The polygenecists (or pre-Adamites) believed that people who were not Western Europeans were not created by God but were on the earth before Adam and that physical characteristics and complex behaviors were biologically fixed and immutable. No environmental conditions could improve their lot. The monogenecists believed that all humans were created by God but that some had degenerated from the original ideal because they lived in less than ideal environmental conditions (either bad climate and/or uncivilized social conditions). To the monogenecists these poor creatures could eventually be "saved" if they could be reintroduced to Western European civilization. Both paradigms considered Western Europeans to be superior to other peoples or races.

As Spiro (2009, 328–329) states: "Within ten years of 1924, scientific racism was a discredited doctrine in the United States, and the Grantians were being pushed down the path towards irrelevance. One of the reasons, of course, was the prodigious influence of Franz Boas and the cultural anthropologists." Other historians have noted the profound importance of Boas and his anthropological concept of culture in changing attitudes toward biological determinism and human biological and behavioral variation. For example, historian Thomas F. Gossett (1965, 418) wrote: "It is possible that Boas did more to combat race prejudice than any other person in history."

Historian Carl N. Degler (1991, 71, 84) said of Boas that his "exposition of the concept of culture as an alternative to race in accounting for the differences among human groups was a genuine seminal contribution." In his *Human Biodiversity,* anthropological geneticist Jonathan Marks (1995, 21, 63) writes: "Nineteenth-century theories of history failed to divorce racial from cultural history, and were undermined by the development of the culture concept, employed by Franz Boas and his students."

It is interesting that today Boas's contributions to anthropology, to science, and to society in general seem to be more understood and appreciated by historians and practitioners of other scientific disciplines than by anthropologists. His profound significance has been especially ignored by physical anthropologists (Washburn 1984, Little 2010; but see Little and Kennedy 2010). Boas and his students were reluctant to identify with the physical anthropologists of his day. After all, physical anthropology did not divorce itself from typological thinking, or from racism, for that matter, and did not become a modern science until the mid-twentieth century (Little and Kennedy 2010; Sussman 2011; Marks 2012). Boas was ahead of his time.

7

The Downfall of Eugenics

Obviously, it was not Boas and the anthropological concept of culture alone that led to the downfall of eugenics as conventional wisdom in American (and Western European) science, academics, and popular thinking. It took both ideological changes and a number of events to accomplish this. However, as I will argue, the ancient view of polygenism, which finally won out over the alternative explanation of monogenism at the turn of the twentieth century, could not be dethroned from either science or popular thought without an alternative paradigm. Strict biological determinism and eugenics reigned in the United States and Europe from the 1900s to the early 1930s and culminated in the 1940s with its logical climax in Nazism. However, by the 1930s and into the 1940s, both the more than 500-year-old theories of monogenism (degeneration) and polygenism (pre-Adamites) and the newer eugenics theories were no longer tenable. What led to the final downfall of these theories and to the ultimate shift to an entirely new paradigm for at least most scientists and an intelligent populace?

Paradigmatic Changes

Acceptance of the Boasian cultural paradigm as an alternative to the strict biological determinism or eugenic explanation of human nature, behavior, and variation occurred quite rapidly. By 1919, students, colleagues, and supporters of Boas made up the majority of the American Anthropological Association, and most of the anthropological work done in the United States at that time was being done by people who had passed through Boas's department. By 1926, students of Boas's headed every major anthropology department in American universities. As Stocking (1968, 296) has stated, "Indeed, there was at this point no 'other real source of influence' besides Boas. . . . The profession as a whole was united within a single organization of academically oriented anthropologists. By and large, they shared a common understanding of the fundamental significance of the historically conditioned variety of human cultures in the determination of human behavior."

As early as 1912, the biologist Alfred Russel Wallace had expressed his disagreement with negative eugenics: "Segregation of the unfit is a mere excuse for establishing a medical tyranny. And we have enough of this kind of tyranny already. . . . The world does not want the eugenist to set it straight. . . . Eugenics is simply the meddlesome interference of an arrogant scientific priestcraft" (quoted in Paul 1995, 36).

By the mid-1920s, not only were professional anthropologists fighting eugenics but sociologists and psychologists were also either entering the fray or changing their minds about the validity of simple genetic explanations for complex behavior. For example, sociologists such as George A. Lundberg, William F. Ogburn, and Robert E. Park were extolling the importance of complex interactions of nature and nurture in human society and criticizing biological determinism. In 1926, sociologist Frank Hankins, once a staunch follower of the eugenics movement, lambasted many of the eugenics proponents in his *The Racial Basis of Civilization: A Critique of the Nordic Doctrine*, especially Madison Grant. He said of Grant's *Passing* that it was highly doctrinaire, hopelessly confused, and one part truth and nine parts imagination, suspended in a mixture of exaggeration and suggestion. He referred to the book as crudely bald and dangerous (Spiro 2009).

By this time an "increasing number of sociologists [had] abandoned the hereditarians and embraced the culture concept. . . . Along with the sociologists, many psychologists were forsaking the hereditarians in the early 1930s" (Spiro 2009, 332–333; see also Stocking 1968; Degler 1991). In fact, as early as 1920, Kroeber had worked with IQ tester and previous eugenics proponent Lewis Terman to develop a "culture-free" intelligence test. Otto Klineberg led psychologists in this conversion to the acceptance of the culture paradigm. By the 1920s, such psychologists as William Bagley, Edwin Boring, and Kimball Young were arguing that environmental reasons could cause differences in IQ scores and mental differences between racial groups and were discrediting claims of inherent mental differences between those groups. In 1930, Carl Brigham became the major critic of his own work; he refuted the findings of his book *A Study of American*

Intelligence and stated that the entire study "was without foundation" (Brigham 1930). He later stated (in 1933) that the claim that intelligence tests measured innate intelligence, without regard to environment, had been "one of the most serious fallacies in the history of science" (quoted in Spiro 2009, 333). These tests had been accepted as a definitive proof of polygenic theories as they had been spelled out in Grant's *Passing of the Great Race*.

In the 1920s, geneticists joined the social scientists in their criticisms of eugenics and simplistic biological determinism. In 1935, L. C. Dunn was among a growing group of geneticists who were calling for a complete break between eugenics and genetics (Weiss 2010). Geneticists such as Thomas Hunt Morgan and Theodosius Dobzhansky were also highly critical of simplistic concepts of race and of eugenicist views of race mixing (Farber 2009, 2011). In fact, by the late 1930s, a number of natural scientists were developing what was to be called the "modern synthesis" of evolutionary theory. This theory combined Darwinism with information gained from modern experimental geneticists, population geneticists and biologists, mathematicians, and paleontologists. It recognized the population as the unit of study in evolution and led to a better understanding of population variability, the dynamics of population genetics and evolutionary processes, and how species evolved and what constitutes a subspecies or race within biology. Some of the leaders of this integrative evolutionary theory were geneticists and biologists T. Dobzhansky, T. H. Morgan, Julian Huxley (the grandson of Thomas Huxley), R. A. Fisher, S. Wright, L. C. Dunn, E. Mayr, J. B. S. Haldane, and G. L. Stebbins; paleontologist G. G. Simpson; and anthropologists A. Montagu and S. L. Washburn (Milner 2009; Farber 2011). The modern synthesis "reestablished a Darwinian perspective on evolution and constructed a solid foundation in population genetics on which to build" (Farber 2011, 59). In 1950, physical anthropologist S. L. Washburn and geneticist T. Dobzhansky organized a symposium that brought together leading evolutionary biologists, geneticists, and anthropologists to discuss how this new way of thinking related to human evolution. Ironically, the symposium was held at Cold Spring Harbor (Demerec 1951). By the mid-1950s, the modern synthesis had swept through the fields of biology and physical anthropology, and this paradigm and Boas's concept of culture still dominate the life and social sciences today.

Farber (2011, 59) notes that "Boas and his students led the struggle against scientific racism. . . . Less well known are the contributions of life scientists in the new understanding of race and race mixing. . . . To overlook the contributions of the life sciences, however, is to miss an important component of the story. Starting in the late thirties, a set of *theoretical* works on the theory of evolution transformed biology and with that the concept of race." By the 1960s, this new evolutionary theory and the concept of culture were informing how the educated public understood modern biology, evolution, and the concept of race. Stocking (1968, 203–303) describes the impact of the modern synthesis on scientific research: "Over a period of time several psychologists publicly changed their minds on the issue, and by the late 1930s the whole profession had clearly moved a long way toward the acceptance of the cultural critique. . . . These developments in turn must be viewed in the context of changes in scientific disciplines *outside* the social sciences. . . . This process operated in such a way as in the long run to give the anthropological idea of culture a kind of paradigm status for social sciences generally."

Although scientific theories often are thought to be timeless, unbiased, and natural, philosophers of science have shown us that, over time and place, the dominant ideologies of the wider culture influence the theories that scientists develop (see Benson 2011). It is obvious that the theories of monogenism and polygenism were influenced by popular ideologies for centuries. Of course, scientific theories also influence popular thought; there is a feedback loop between scientific and popular ideology. Hacking (1995) refers to this as the "looping effect." He claims that in such fields as anthropology, history, and sociology of science there has been an increasing awareness of the effects of social theory.

Benson (2011, 200) describes how science affects popular thought: "Language used in science affects the empirical world and thus the outcomes of future research in human populations. 'It is about how causal understanding can change character,' writes Hacking, arguing that audiences are influenced by scientific reporting on human nature, such that 'science can change the kind of people that they are.'" Eugenics did this to a vast majority of the population of Western Europe and the United States during the early decades of the twentieth century. I believe that by the 1930s, a feedback loop, or "looping effect," was occurring between the anthropological cultural paradigm in science and the general populace. However, other current events and basic popular attitudes also helped with this shift in paradigmatic thinking.

In the early 1900s, nativism was rampant in America. Between 1890 and 1920, over twenty million immigrants entered the United States. Eugenicists considered that most of these immigrants were genetically undesirable

(Black 2003). They emphasized the differences between earlier immigrants and more recent arrivals and pointed out that this great influx of human-ity had created competition for jobs. This provided the ammunition as well as the "factual" evidence for the eugenics movement. It was easy for the populace to believe and beneficial for certain portions of the popula-tion, especially for the aristocratic perpetrators of scientific racism, poly-genics, and eugenics. However, a number of important changes in the so-cial and cultural context took place over the next few decades.

Changes in Social and Cultural Context

The United States (and Europe, for that matter) was not the same place from 1920 to 1930 as it had been in 1900 or 1910. Social conditions and the demographic landscape had changed quite radically. Unlike most co-lonial Americans and earlier groups of immigrants, the new arrivals to the United States were mainly from Eastern Europe. They were generally poor and Catholic or Jewish, and they did not speak English. They congregated in larger cities. In fact, after World War I, for the first time, the majority of the population lived in urban and not rural areas. This included not only these new immigrants, but many rural people who had moved to cities to find work. Also, many soldiers settled in when they returned from the war. This included African American soldiers who, having served their country, wanted equal treatment. At the same time, the war industry was beginning to shut down and lay off workers. This led to economic, ethnic, and demo-graphic turmoil just after World War I, including massive strikes in 1919.

In the 1920s, things began to settle down and the United States began to prosper again. In addition, the new immigration law had closed the doors to new immigrants from Eastern Europe. As Spiro (2009, 329) notes: "One of the reasons for the decline of scientific racism was that Madison Grant had been too successful for his own good. Immigration policies and the Im-migration Restriction Act of 1924 had done such a thorough job of shutting out the Alpines and Mediterraneans that most Americans assumed that the threat posed by the inferior breeds had been adequately dealt with." In 1914, for example, 1.2 million Europeans immigrated into the United States; immigration numbers then dropped 73 percent to 327,000 as a re-sult of the European war (Degler 1991). After 1924, there was essentially no immigration from Eastern or Southern Europe into the United States.

However, by the 1920s immigrants from those parts of Europe already had settled into the cities, especially in the North. In addition, beginning just before World War I, hundreds of thousands of African Americans began to migrate from the rural South to the urban North in what has been called the Great Migration. This migration intensified after the Immigration Restriction Act cut off the supply of cheap immigrant labor. Degler (1991, 197) believes that this migration, which continued during and just after World War I, "was perhaps the most significant social change among Americans in the first half of the twentieth century," second only to the massive immigration of Eastern and southern Europeans earlier in the century. At first, it caused some racial tension and expressions of racism in the North. At the same time, however, it helped change the perception of African Americans.

African Americans took advantage of education and employment opportunities in the cities. They were able to receive better education, better health care, and better nutrition (and scored higher on IQ tests). Many African Americans also began to enter academe and scientific fields and began to contribute to antiracist science and scholarship. Black Harlem celebrated a new prominence of African Americans in the nation's culture, and in 1928, Chicago elected the first African American member of Congress from a northern state.

Spiro (2009, 330) writes that "many sociologists, psychologists, anthropologists and members of the public realized that they had underestimated the capabilities of African Americans, whom they had heretofore encountered mainly as socially and intellectually stunted sharecroppers." The presence of black scholars altered perceptions of the innate capacities of African Americans. These developments provided further support for those who argued that nurture and environment had an impact on human capabilities. The successes of African Americans in the North led to a new self-awareness among southern blacks as they saw the alternatives for their counterparts in the North. This led to changes in attitudes among both black and white southern scholars and forced the old southern order to confront attitudes and arguments it had rarely had to face in the past.

This new attitude among southern blacks led to a revival of the Ku Klux Klan in the early 1920s. In the meantime, Grant worried about what he called the "Negro invasion of the North." He wrote to one of his friends in 1930, "The enormous Negro population in New York is spreading and ruining real estate values in fairly good sections. In Harlem one constantly sees on the street white and negro boys playing together and white and negro girls walking together" (quoted in Spiro 2009, 330).

Along with the African Americans, immigrants from Eastern and Southern Europe and many of their children were taking advantage of education opportunities. More and more of them were graduating from college, earning higher degrees, and obtaining professional positions—even becoming well-known scientists. As we have seen, Jewish, Catholic, and other European immigrants had begun to directly oppose eugenics. To decry the mental abilities of Jews and other Eastern Europeans when men such as Albert Einstein (1879–1955) were actively involved in science was becoming an absurd and unsupportable proposition. However, major universities, such as Harvard, Yale, Columbia, Princeton, Duke, Cornell, and the University of Virginia, thought it necessary to set quotas limiting the number of Jewish students. In an article in the *Journal of Heredity* in 1919, Grant, echoing the views held during the Spanish Inquisition, wrote that there had been a time when Jews in America justifiably had been denied "access to social circles and to positions of responsibility." And in 1924, he wrote to a friend who was a Harvard graduate that Harvard should never have admitted Jews and that it was now paying the price of its liberalism by having to create "Kykological Tests to save herself from being swamped" (quotes in Spiro 2009, 331). But Grant and the Grantians were dying out, both figuratively and literally.

In 1925, Grant predicted that "those who are alien in race and religion have not amalgamated with the native population. . . . It will take centuries before foreigners now here become Americans" (Grant 1925, 350–351). He was quite wrong, and this may have been Grant's biggest miscalculation. The children of immigrants were anxious to assimilate and become "Americans." Soon Czechs, Italians, Poles, Jews, and Greeks were no longer seen as members of different races, and they became "white"; race became ethnicity.

In Germany, civil society broke down as the fascist government legalized and enforced racial prejudice and eugenics in a process that was legitimized by scientists and doctors. But as Weiss (2010, 307) notes, the United states did not "[adopt] the 'gospel of Galton' as its official ideology nor was it ever a one-party state. Even when some of its citizens were threatened by the work of American human genetics research—as in the case of mandatory sterilization—the United States continued to retain a healthy civil society, even during the troubled years of the Great Depression. The existence of such a society with a plurality of ideologies served as a barrier to the adoption of the sort of radical state policy that was practiced in Nazi Germany."

UNESCO issued a "Statement by Experts on Race Problems" and *The Race Question in Modern Science* in 1950 and 1952, respectively. These summaries of the findings of an international panel of anthropologists, geneticists, sociologists, and psychologists were written primarily by anthropologist Ashley Montagu and geneticist L. C. Dunn. As early as 1940, Montagu, incorporating the ideas of the modern synthesis of evolutionary theory, had stated that human populations are the result of the processes of evolution and are in constant flux. What we call race is just a population with a particular set of gene frequencies that are constantly shifting due to migration, natural selection, social and sexual selection, mutation, random genetic drift, and so forth (Farber 2011). Shortly after completing his thesis under Boas in 1947, Montagu began to put together a series of papers that continued the dynamic view of evolution and populations Boas had developed and combined it with the modern synthesis of evolution that biologists and population geneticists were developing. These essentially extended Montagu's critique of the use of the term *race*. He continued to build his argument as more and more research in population genetics and anthropology accumulated in his popular book *Man's Most Dangerous Myth*. This volume, which was first published in 1942 at 214 pages, had grown to 699 pages by the time of the sixth and final edition in 1997. It has had a major impact on the modern concept of race and is still in print (Farber 2011).

The UNESCO statements asserted that all humans belong to the same species; that "race" is a myth rather than a biological reality and that the term should be replaced by "ethnic group"; that there are no connections between mental and physical characteristics among humans and that different ethnic groups do not differ in innate mental capacity; that intelligence test results are greatly affected by environment; and that there is no evidence that mixture between ethnic groups is in any way deleterious. The UNESCO statements essentially claimed that it was the consensus of scientists throughout the world that all of the major tenets of Madison Grant's and the eugenicists' epistemology were obsolete (Degler 1991; Spiro 2009). The *New York Times* reported on the appearance of the 1950 statement with a story whose headline read "No Scientific Basis for Race Bias Found by World Panel of Experts" (Spiro 2009, 383). Proctor (1988, 174) has written of the UNESCO statements that they were "the triumph of Boasian anthropology on a world-historical scale." The statement reflected the modern synthesis of evolutionary biology, in which populations are the unit of evolution and ideal "types" are not recognized. The findings of the UNESCO panel, of course, are all still upheld by modern science (for example, American Association of Physical Anthropologists 1996; Cartmill 1998; Harrison 1998; Templeton 1998, 2007; Edgar and Hunley 2009; Farber 2009; Tattersall and DeSalle 2011; Smedley and Smedley 2012; and Mukhopadhyay, Henze, and Moses 2014).

Thus, beginning in the 1920s, at least and especially in the urbanized northern United States and in many European cities, there was widespread and increasing interaction between and among peoples of all ethnic groups, and increased and increasing assimilation. This trend has continued. As an example, my parents were both Jewish immigrants to the United States from Russia in 1910 and 1919, and my wife's family immigrated into the United States from Poland and Italy in the late 1890s and early 1900s. My only grandson (to date) is descended from a mixture from the last two generations of Russian Jews, Italians, Poles, Irish, Chileans, Native Americans, and Japanese.

In addition to the rapidly changing population in the United States, two other events occurred during the late 1920s to early 1940s that appear to have had a major influence on the acceptance of the culture paradigm. The first was the Great Depression, which began in 1929. Even many of the upper class became poor overnight, as did much of the middle class. It became difficult to claim that wealth was determined by an individual's genetic endowment when his or her wealth was suddenly gone. On the other hand, it became easy to see that environment could have everything to do with one's socioeconomic standing from one short time period to another. This prompted historian D. J. Pickens (1968, 215) to quip that "Galtonian eugenics was a victim of unemployment." However, the eugenicists claimed that the unemployed were genetically inferior and that the depression proved the claims of Grant's *Passing of the Great Race*. To them the remedy for the Great Depression was massive sterilization and deportation (Spiro 2009).

Of course, the other major event that worked against eugenics and for the triumph of the culture concept was Nazism. The epitome of polygenics, racism, and eugenics became the cause of the downfall of eugenics. Most Americans and other Europeans were repulsed by the atrocities the Nazis and Germany carried out in the name of eugenics and scientific racism. "After the horrors of the Holocaust were revealed, eugenics practices such as sterilization were universally acknowledged to be the summit of a horrifyingly slippery slope. . . . The irony was that by putting Madison Grant's theories into practice, the Nazis discredited those theories forever" (Spiro 2009, 339).

As Stocking (1968, 307) aptly summarizes: "In the long run, it was Boasian anthropology—rather than the racialist writers associated with the eugenics movement—which was able to speak to Americans as the voice of science on all matters of race, culture, and evolution."

Are We There Yet?

We have made major strides against the age-old myths of polygenism, eugenics, and "scientific" racism. However, have we completely shed certain underlying fifteenth-century, Western European, white American views of racist eugenics and the biological basis and fixity of certain complex human behavioral traits? Have we accepted the scientific findings of the 1950s reflected in the UNESCO statements? Has the importance of environmental influences on these factors been completely understood? Have the profound implications of the anthropological concept of culture been misunderstood and indeed trivialized, even within anthropology? I believe that in some fashion, polygenics, eugenics, and anti-environmentalism are still with us today, especially in two venues: in the new scientific racism and in claims by modern biologists and anthropologists that such complex human behavioral and cognitive characteristics as manic depression, schizophrenia, alcoholism, homosexuality, intelligence, and warriorism are still determined by single genes (see Gillette 2011, for example). Each gene is only a single player in an amazingly complex drama involving nonadditive interactions of genes, proteins, hormones, food, and life experiences and affecting a variety of cognitive and behavioral functions. An assumption that a single gene can mediate the development and operation of a human cognitive and behavioral function can lead to unwarranted conclusions, and to an over interpretation of any genuine genetic linkage (Berkowitz 1999; Allen 2001b).

8

The Beginnings of Modern Scientific Racism

The new scientific racism is connected directly and historically to the early 1900s eugenics movement and to Nazism. As I have mentioned above, eugenics and scientific racism peaked in the United States in the mid-1920s. During this period, in 1923, Wickliffe Preston Draper (1891–1972) wrote to Charles Davenport to express his interest in financially supporting "the advancement of eugenics" (quoted in Tucker 2002, 30). This was the beginning of a lifelong relationship, and it was the beginning of the link between the eugenics movement of the early 1900s and the scientific racism of today. This story is well told in William H. Tucker's excellent book, *The Funding of Scientific Racism: Wickliffe Draper and the Pioneer Fund* (2002).

Origins of the Pioneer Fund

Wickliffe Draper came from an aristocratic southern family on his mother's side (Tucker 2002). His maternal grandfather, General William Preston, was a commanding officer in the Confederate Army during the Civil War. His maternal grandmother, Margaret Wickliffe, was the daughter of the wealthiest planter and largest slave owner in Kentucky. The family was listed as "Kentucky aristocracy" in eugenics textbooks. On his father's side, Wickliffe traced his family back to successful Puritan settlers who came to Massachusetts in 1648. His great-grandfather invented weaving machinery and was the largest manufacturer of textile machinery in the world. Wickliffe's grandfather, George Draper, virtually monopolized the production of textile machinery in the United States. The company became George Draper and Sons.

Upon George's death in 1887, William F. Draper, the eldest of three sons, became president of the company, reorganizing it and renaming it the Draper Company. By the late 1800s, the company was worth $8 million (hundreds of millions in current dollars). William, who had fought in the Civil War and risen to brigadier general in the army, was elected to Congress in 1892 and then appointed ambassador to Italy. William's two younger brothers, annoyed by his absentee ownership, maneuvered him out of the business, and the younger brother, Eben Sumner Draper, took over. Eben then also entered politics, first as lieutenant governor and then as governor of Massachusetts for two terms. When he lost his bid for a third term, Eben returned to manage the Draper Company. However, he died suddenly in 1914, leaving the company to George's third son, George Albert Draper, Wickliffe's father, who was treasurer of the family company and a successful businessman in his own right. At the time, the Draper Company was at its peak, worth tens of millions of dollars, including the majority of the property in the town where Wickliffe was born. The company housing stock alone was worth $35.5 million (in 2014 currency) (Tucker 2002).

Wickliffe was raised in an upper-class environment. He attended an exclusive private school, where he had a "dismal academic record"; a single B was his highest grade (Tucker 2002). He then entered Harvard University, which was more interested in pedigree than academic record. He graduated in 1913. His major interests at Harvard were the Aeronautical Society, the Shooting Club, and the Yacht Club. In adulthood, his interests, besides racism and eugenics, were travel, hunting, and exploration.

It appears that two factors might have made a particularly important impression on Wickliffe when he was at Harvard. During his junior year, the International Workers of the World began to organize the Draper Company. In his senior year, the union led a mostly immigrant workforce in a bitter four-month strike against the company that included clashes between strikers and police. The strike focused on demands for a shorter work week and an increase in wages, but it also was directed against Eben Draper, Wickliffe's uncle, who was still head of the company at the time and who as governor had vetoed labor initiatives such as the eight-hour workday. Eben Draper had refused to negotiate with the strikers.

The second important influence on Wickliffe was likely the nation's and Harvard's obsession with eugenics at the time. Harvard, the academic home of Louis Agassiz and Nathanial Shaler and the alma mater of such distinguished graduates as Charles Davenport, was a center of eugenics during the years Wickliffe attended (see Chapter 2). Other events in the eugenics movement that happened while Wickliffe was at Harvard included the publication of Davenport's eugenics textbook *Heredity in Relation to Eugenics* in 1911 and the First International Eugenics Congress in London in 1912. Davenport and Harvard's president, Charles William Eliot, served as vice-presidents of the congress. Many of Harvard's professors were very active in the eugenics movement, and the university offered four separate eugenics courses. The chair of the Psychology Department and the most respected psychologist in the country, William McDougall, greatly influenced by the U.S. army IQ tests (Cravens 2009), was declaring that African Americans and non-Nordic immigrants were a biological threat to American civilization and were incompatible with American democracy. He advocated disenfranchising blacks, restricting intermarriage between races, and complete segregation of blacks from whites (McDougall 1923, 1925). McDougall saw some hope in the South, where a group of "solid, serious minded, pious and patriotic Americans" (the Ku Klux Klan) was opposing the government's attempt to provide equality for blacks (McDougall 1925, 38). These attitudes became central to Wickliffe Draper's life efforts. As we have seen, they were popularized in Madison Grant's best-selling book, *Passing of the Great Race,* in 1916. Grant was a personal acquaintance of Draper's and in many ways a role model for him (Tucker 2002).

Besides the Anglo-Saxon elite and the Ivy League, the officer corps of the U.S. Army was steeped in eugenics and polygenic ideology during the early 1900s (Tucker 2002). In 1914, shortly after graduating from Harvard, Draper enlisted as a lieutenant in the Royal Field Artillery of the British Army, months after the beginning of World War I. He was wounded in 1917 and returned to the United States. After recovering, Wickliffe received a commission in the U.S. Army and spent a year as a training instructor. He was discharged in 1919, but he remained a member of the Cavalry Reserve, where he was promoted as a civilian. During World War II, his commission was reactivated and he served as a "senior observer" in India for the U.S. Army. Wickliffe felt that retired lieutenant colonel was his occupation and liked to be addressed as "Colonel Draper" by his friends and associates.

Aside from these brief periods of military service, Draper did not pursue a professional career and never held a job of any kind. Living on old family wealth, Wickliffe traveled the world and lived in a luxury three-floor apartment in Manhattan. Like many of his fellow leaders of the eugenics movement, he never married or had any children. In 1938, in Harvard's twenty-fifth anniversary report for the class of 1913, Draper described his life as follows (quoted in Tucker 2002, 22): "A dozen years of travel. Shooting jaguar in Matto Grosso and deer in Sonora; elephant in Uganda and chamois in Steiermark: ibex in Baltistan and Antelope in Mongolia. Climbing in Alps and Rockies. Pigsticking in India and fox-hunting in England. Exploring in West Sahara with French Mission."

In 1923, Wickliffe's father, George Draper, died, leaving an estate of over $10 million to Wickliffe and his sister. When his sister died childless in 1933, Wickliffe acquired the whole estate. Upon his inheritance, Wickliffe began a lifelong pursuit, using his wealth in an attempt to keep racism and eugenics alive and flourishing. It was soon after his father's death, in 1923, that the young Draper first wrote to Charles Davenport offering him a "bequest for the advancement of eugenics." Wickliffe probably would have liked to conduct research on eugenics himself, but his biographer, William Tucker, believes that he was too unfocused and lacked discipline. He could not write (he rarely wrote more than a few lines in his letters) and was painfully shy and reclusive. "Draper generally shunned any public attention, preferring to pursue his agenda by putting up the money for research that would support it" (Tucker 2002, 30).

The day after Davenport received Wickliffe's letter, the two men met. Draper outlined a plan for an endowment to the Eugenics Research Association of more than $1 million ($10.5 million in 2014). He asked Davenport to give him an outline of how this money would be spent. Davenport produced an ambitious proposal for Draper. Draper responded that he was leaving for India and would give the proposal careful consideration. More than two years and many attempts later, Draper granted the ERA a meager $1,000 and requested another proposal from Davenport. Davenport immediately proposed another program of research. Draper responded that Davenport's research plan was not in line with what he wanted and that he would prefer to fund research on the effect of miscegenation in such countries as Haiti, Brazil, and the United States. Davenport was delighted, and this led to Draper's funding, in 1926, of Davenport and Steggerda's research, resulting in the 1929 volume *Race Crossing in Jamaica* (see Chapter 2). The ERA and the CIW, which published the study, acknowledged the Draper Fund for financing the project (Tucker 2002).

Davenport wanted continued funding for research on mixed races and offered to appoint Draper as member of the Committee on Race Crossing of the IFEO. Draper again agreed to continue contributions to the ERA but with his own instructions as to how those funds were to be spent. He established prize money (anonymously) for the ERA to organize two essay contests related to Grant's thesis in *Passing* on the worldwide threat to Nordic populations. Grant was delighted and offered to add to the prize money. The contests were first announced in the *New York Times* and in eugenics publications in 1928. The second of these contests drew ninety-six essays, of which eighty-one were from Germany and Austria and only five from the United States. In 1933, the Draper Fund anonymously sponsored another contest through the ERA, this time for original research on the incidence of mental disorders in families.

However, Draper had more overt political goals and wanted to see eugenics developed as an applied science and to see social policy based on eugenic ideals. His funding of ERA projects was about to end. Draper had become close to Harry Laughlin, the most energetic policy-oriented activist among the eugenicists, as we have seen. In fact, Laughlin had been instrumental in getting Draper an invitation to the International Congress for the Scientific Investigation of Population Problems in Berlin in 1935. He was sent as an official delegate of the ERA, and Laughlin gave him a letter of introduction to Eugen Fischer. Laughlin asked Fischer to give Draper a favorable reception and described him as "one of the staunchest supporters of eugenical research and policy in the United States" (quoted in Lombardo 2002, 207).

This brought together a major benefactor of eugenics policy in the United States with leading Nazi race scientists and eugenicists. Draper was ecstatic and immediately thanked Laughlin for paving the way for him with German officials. He informed Laughlin that when he returned, he wanted to talk with him about several of the Germans who might provide him with useful information (Lombardo 2002; Tucker 2002). Clarence Campbell accompanied Draper to the congress. Campbell was then president of the ERA and served on the editorial board of *EN*. At the congress, Campbell's speech included this rhetoric: "The leader of the German nation, Adolf Hitler . . . guided by the nation's anthropologists, eugenists [*sic*], and social philosophers, has been able to construct a comprehensive racial policy of population development and improvement that promises to be epochal in racial history. . . . The difference between the Jew and the Aryan is as unsurmountable [*sic*] as that between black and white. . . . Germany has set a pattern which other nations must follow." At the closing banquet, he made a toast "to that great leader, Adolf Hitler!" (quotes in Tucker 2002, 53–54). Shortly after their return from Germany, in 1936, Draper asked Campbell to join Laughlin and Davenport as judges for his third essay contest (Tucker 2002). The admiration of this group of eugenicists for the Nazis and their policies was unquestionable.

In 1936, Madison Grant introduced Draper to Earnest Sevier Cox and W. A. Plecker, both southern white supremacists and separatists. Cox (1880–1966) had read Grant's *Passing* and immediately devoted his life to promoting the exportation of African Americans to Africa (Jackson and Winston 2009). He was especially struck by Grant's comments, in Chapter 7, that two distinct "species" of humans cannot live side by side without one driving the other out, or forming a population of "race bastards" with the lower type ultimately being preponderant. This stimulated him to write his Gobinesque/Grantian volume *White America* (Cox 1923). In 1924, Laughlin wrote an extremely favorable review of *White America* in *EN*, stating that the removal of Negroes from America would make Cox a greater savior of the country than George Washington (Jackson and Winston 2009). Laughlin had already familiarized Draper with this volume. Cox's familiar thesis was basically that the white race had founded all civilizations, that only the white race could sustain civilization, and that civilization would be lost as the white race became hybridized. This led Cox to propose that it was necessary to put African Americans in concentration camps and then deport them to Africa in order to maintain Nordic purity.

Plecker, as discussed in Chapter 3, was a white supremacist who was one of the authors and main instigators of Virginia's Racial Integrity Act of 1924. Grant introduced Cox and Plecker to his disciples Lothrop Stoddard and Harry Laughlin and they had all become friends and worked together on Virginia's sterilization and racial separation laws. In fact, the three northern scientific racists (Grant, Laughlin, and Stoddard) provided advice and acted as mentors to the southern racists (Cox and Plecker) for years to come (Jackson and Winston 2009; Spiro 2009). Grant also introduced Cox and Plecker to Draper, and these three men also became close friends. Draper funded their projects (mainly anonymously) for the rest of his life. This was the beginning of Draper's long financial support for projects that sought to export African Americans to Africa. In light of the successful racist policies in Virginia, Laughlin suggested that Draper fund a eugenics institute associated with the University of Virginia.

However, Draper established his own independent funding agency, the Pioneer Fund, which he incorporated in March 1937. Draper's close friend Madison Grant, then nearing his death, also provided financial support for this effort (Allen 1986). As I will show, the Pioneer Fund continues to be the major source of funding for research that supports racist ideology and for projects that promote biological determinism and eugenics. The stated goals of the fund at its inception were to help with the education of children of parents who had unusual value as citizens; that is, parents who were descended from white persons who had settled in the original thirteen states. The main reason for this support was to encourage those parents to have more children by defraying the costs of their children's education and thus ultimately improving the character of the people of the United States. The fund's second goal was to study and research the problems of heredity and eugenics in the human race and the problems associated with race betterment, especially in the United States, and to inform the general public concerning the findings related to heredity and eugenics (Kühl 1994; Lombardo 2002; Tucker 2002). However, the Pioneer Fund is basically a propaganda organization for fifteenth-century racism. Its first priority is to continue to attempt to convince the public that ancient, unscientific, eugenic, Nazi views of race are indeed scientifically valid (see Mercer 1994; Brace 2005). As Rosenthal (1995, 44) points out: "The pro-Nazi Pioneer Fund . . . has funded the work of every major advocate of racism, eugenics, and fascism since the late 1930s."

According to its bylaws, the fund was to be managed by a five-member board of directors, three of whom would be officers. The first president was Harry Laughlin. Laughlin was delighted and thought the organization would help the United States achieve the "most patriotic development of racial ideals" through "the conservation of the best racial stocks" and would prevent "increase of certain lower stocks and unassimilable races" (quoted in Tucker 2002, 48). At the time, Laughlin thought Earnest Cox was the nation's savior and saw Nazi Germany as a role model: a perfect choice to lead Draper's funding corporation. The secretary of the fund was Frederick Henry Osborn, the nephew of Henry Fairfield Osborn. At the time, the younger Osborn's ideology was similar to Laughlin's. For example, he advocated that all known carriers of "hereditary defects" be either segregated or sterilized. He estimated that there were up to 5 million carriers of defects at that time, 3 to 4 percent of the United States population. Osborn later considerably modified his views on eugenics (see later in this chapter). The treasurer of the fund was Malcolm Donald, an attorney who had been managing Draper's finances and had executed George Draper's will. The two other members of the board were John Marshall Harlan, a lawyer who performed legal work for the fund (and later became a U.S. Supreme Court justice), and Wickliffe Draper himself. In fact, only Osborn and

Laughlin gave advice on the actual eugenic merits of any project, and all final decisions on any project were made by Draper (Tucker 2002).

From 1937 to 1941, before the United States entered World War II, the Pioneer Fund supported only two projects of any significance. Both were conceived of by Draper and seem to have been influenced by his visit to Nazi Germany. The first was the distribution of the German film *Erbkrank* *(The Hereditary Defective)* with English subtitles. This was a Nazi propaganda film about the costs of hereditary degeneracy such as feeble-mindedness, insanity, crime, hereditary disease, and inborn deformity and how the field of applied negative eugenics was making substantial progress on a long-term plan to prevent such degeneracy. There were a number of anti-Semitic passages in the film. Laughlin gave the film a glowing review in *EN*. The plan was for the Pioneer Fund to advertise the film in 3,000 high schools and eventually loan out 1,000 copies for viewing across the country. This ambitious goal was never achieved, and in the end only twenty-eight schools and a handful of state welfare workers in Connecticut actually viewed the film.

The second project and the only project to ever address the first stated goal of the Pioneer Fund sought to increase the income of families of officers in the United States Army Air Corps, who were considered to meet the standards set out in the original purposes of the fund. That is, the fund considered these families to be of unusual value as citizens and to be descended predominantly from white persons who had settled in the original thirteen states. Financial assistance would be given for the education of children in these families in order to encourage the parents to have more children. The families who were eligible to receive the funds had to have at least three children already. Eleven children from nine families were given these grants of around $33,000 each (in today's dollars). This project was similar to a number of "positive eugenics" projects being carried out in Nazi Germany at the time.

The Pioneer Fund did little during the war, and from 1943 to 1947 there were no meetings of the board of directors and no funds were expended for any project (Tucker 2002). However, Draper continued to support racist initiatives. After meeting Earnest Cox in 1936, Draper was so impressed with his efforts to create an America for whites only that he was willing to aid Cox substantially in his future efforts. Shortly after this meeting, Cox was surprised to receive an order for 1,000 copies of *White America*. Since the book had sold few copies, Draper had had to pay for the work to be reprinted (which he did anonymously). The funds were for

Cox to strategically distribute the book for the maximum political effect. Cox sent copies to members of Congress and to state legislators in North Carolina and Mississippi.

Cox soon reported to Draper that he had caught a willing fish. After reading his copy of *White America*, Mississippi senator Theodore Gilmore Bilbo wrote to Cox that he had decided "to specialize on the repatriation of the negro" (Tucker 2002, 35) and to make Cox's dream a reality (Jackson and Winston 2009). Bilbo has been described as "one of the most ardent racists ever to serve in the upper house" (Tucker 2002, 34)—or more vividly as "the colorful, corrupt, race-baiting, anti-Semitic, Dixie demagogue who had memorized entire passages of *The Passing of the Great Race*" (Spiro 2009, 264). Bilbo had served in prison in 1923 after two terms as governor of Mississippi and after leaving the state in a huge financial hole. After hearing from Cox of Senator Bilbo's interest in taking the cause of repatriation to Congress, Draper replied in a 1938 letter to Cox, "if I can be of help, pray let me know" (quoted in Tucker 2002, 36). In 1939, Bilbo introduced the Greater Liberia Act, which would have mandated the removal of every African American from the United States within a generation or so.

From this time on, Draper helped Cox and Bilbo, both financially and with strategic planning, in their decades-long effort to get Congress to pass a repatriation bill. However, thanks to the growing civil rights movement, the efforts for repatriation legislation finally died in Congress in 1959 (Tucker 2002; Spiro 2009). In 1954, Draper finally withdrew financial (though not moral) support for Cox's effort at repatriation after the Supreme Court had issued its decision on the *Brown v. Board of Education* case, although Cox continued to work on returning blacks to Africa until his death in 1965 (Jackson and Winston 2009). The *Brown* case rejected the opinion that separate education was equal education, paving the way for desegregation in the United States. Although President Truman had ordered desegregation of the armed forces in 1949, segregation was still the law of the land, underwritten and guaranteed by the U.S. Constitution (Human Rights Library 2000, Cravens 2009). Around this time, Draper shifted the funding strategies of the Pioneer Fund towards other more powerful racist, eugenic projects and anti-integration initiatives. After all, Cravens has stated (2009, 171), "The United States was still a deeply segregated society with daunting, widespread racism literally in the bones of its institutions."

When Cox died in 1966, Draper wanted to honor him; he considered him "a zealous defender of our race and nation" (quoted in Tucker 2002,

39). Draper's way of doing this was to reprint a pamphlet Cox had written on repatriation. The republication was done by Willis Carto, considered by some to be the most active and influential neo-Nazi and anti-Semite in the United States at the time (Jackson and Winston 2009). Carto and Cox had become friends after Carto read Cox's review of *Race and Mankind* and his *White America*. In fact, Carto believed that Cox was "the dean of American racists" (Jackson and Winston 2009, 75). In letters to Cox written in 1954 and 1955, Carto expressed the belief that repatriation would be the only way to combat the "niggerfication of America" and would also be the strongest blow against the power of organized Jewry (Tucker 2002, 40). Cox had made his views on Jews clear in 1947 when he reviewed Ruth Benedict and Gene Wetfish's *Races of Mankind;* he wrote that it was a part of the "Boas Conspiracy" of the Jewish plan for world domination (Jackson and Winston 2009). At the time he was approached by Draper, Carto was in the process of republishing *White America*. The funding for this project was done, as usual, anonymously, funneled through an intermediary, the Virginia Education Fund (Tucker 2002).

Frederick Henry Osborn, who had assumed presidency of the Pioneer Fund when Laughlin died, began to part ways with Draper in 1954. By that time he had modified his views on eugenics and refuted outright racism. Even in the late 1930s, Osborn had believed that there was little evidence of innate differences among groups and had warned against assuming the superiority of any class or race over another (Paul 1995). Believing that Draper's views had no basis in science, he rejected Draper's attempt to continue his funding of the AES (of which Osborn was also the head [Tucker 2002, 57]), and in 1958 he ended his association with the Pioneer Fund. After that, Draper reorganized the fund to mirror his own personality and viewpoints more closely. He was eager to take concrete actions to achieve ethnic homogeneity in the United States (Tucker 2002, 58).

Draper chose Harry Frederick Weyher Jr. (1921–2001) to replace Osborn. Weyher and Draper were kindred souls, and he eventually made Draper's causes his own. He was a tax attorney from North Carolina who had been educated at the University of North Carolina and then Harvard Law School. Weyher was an unadulterated racist, he was a friend of white separatist senator Jesse Helms's, and he was sympathetic to the struggle of southern whites to maintain racial supremacy. Draper named John B. Trevor Jr., who was head of the American Coalition of Patriotic Societies, as treasurer.

Trevor Jr.'s history was linked to that of his father. Trevor Sr. had been a close friend of Madison Grant's and, like Grant, a member of the social elite who traced his family roots to colonial times. He was a rabid anti-Semite and pro-Nazi and was one of the principal architects, with Grant and Laughlin, of the Immigration Restriction Act of 1924 (Beirich 2008). Trevor Sr. had once developed a plan to suppress potential mass uprisings of Jewish "subversives" in New York City and had amassed 6,000 rifles and a machine-gun battalion for deployment in Jewish neighborhoods. He also had worked on a strategy (with Grant, Laughlin, and Draper) to stop a last-ditch attempt to give asylum in the United States to a few victims of Nazi oppression. In 1929, Trevor Sr. founded the American Coalition of Patriotic Societies, which became an umbrella organization for a number of far-right political groups and which later helped distribute Nazi propaganda. Trevor Sr. was named in an indictment by the U.S. Justice Department for pro-Nazi activities. When his father died in 1956, Trevor Jr. took charge of the coalition.

The coalition had a long list of enemies and expressed its opposition to any attempt to "undermine" the immigration laws by admitting more "aliens"; membership of the United States in the United Nations and the presence of the United Nations in the United States; diplomatic relations with the USSR; NATO's attempts to promote political and economic co-operation; fluoridation of public water supplies; and the National Council of Churches and the World Federation for Mental Health; all of these, according to Trevor Jr., sought to destroy the U.S. government and the American way of life. On the other hand, the coalition supported South African apartheid for its "well-reasoned" racial policies and amnesty for Nazi war criminals. One of the official spokespersons of the coalition included outspoken anti-Semite General Pedro del Valle, who was a staunch supporter of George Lincoln Rockwell, the "commander" of the American Nazi Party.

After 1960, the Supreme Court began to take legal steps to dismantle segregation, and in 1964, Congress passed the Civil Rights Act banning discrimination in employment, places of public accommodation, and federally funded institutions. At that time, the coalition added the Supreme Court to its list of enemies. During this period, and likely much earlier, the majority of the coalition's activities were probably being funded mainly or even exclusively by Wickliffe Draper (Tucker 2002). Tucker (2002, 63) characterizes the coalition in the 1960s: "With Draper, Weyher, and Trevor Jr. on board, Pioneer was firmly controlled by virulent opponents of civil rights and Nazi sympathizers." An organization that was dedicated to keeping the fifteenth-century myth of polygenics and its racist agenda alive and active was, indeed, alive, well, and well funded.

For the rest of his life, Draper used his own funds (mainly anonymously) and the Pioneer Fund to support two major endeavors. They first focused on the "study and research into the problems of heredity and eugenics in the human race . . . and . . . into the problems of race betterment" (Pioneer Fund Certificate of Incorporation, 1937 [ISAR 1998]); this was essentially research centered on "scientific" racism. The second supported "the advancement of knowledge and the dissemination of information with respect to any studies so made or in general with respect to heredity and eugenics." This was basically an attempt to re-educate policy makers and the public by publishing literature and other propaganda about supposedly scientific findings, especially concerning race differences. The fund also supported direct political and clandestine attempts to overturn and undermine the civil rights agenda.

In this first cause, Draper wanted to recruit scientific authorities with academic credentials and scholarly records who believed in the necessity of racial purity and that integration posed a threat to civilization. This eventually led a dedicated group of activists to work with Pioneer Fund board members Weyher and Trevor in the campaign to use science to oppose civil rights. Basically, Draper wanted to provide current scientific justification to support his belief that blacks (and Jews) were inferior to "whites" and were a danger to civilization as we know it. Furthermore, he wanted to get this information to leaders (e.g., politicians, CEOs, etc.) and to the public. To Draper, the public had forgotten and needed to relearn the anthropology of Morton and the anthropological eugenics of Charles Davenport, Madison Grant, and Eugen Fischer.

An earlier recipient of Draper money in this regard was University of North Carolina anatomy professor Wesley Critz George. Two of his pamphlets, *Race, Heredity, and Civilization* and *The Race Problem from the Standpoint of One Who Is Concerned with the Evils of Miscegenation*, were distributed with Draper funds in the early 1960s. They claimed that blacks were genetically unacceptable and that the mixing of races would destroy "our" race and civilization. This was precisely what Draper and the Pioneer Fund wanted—the negative response of a recognized scientist to the civil rights movement. After financing the distribution of these pamphlets, Weyher asked George if he could supply "the names of any professors or other personnel at any university" who could provide similar assistance to the cause (Tucker 2002, 69). This began a pattern of inviting scientists to join the cause of racist eugenics and then helping them publish their materials, often by distributing them widely at no cost, a practice that continues today.

In order to fulfill his second major agenda, from the time of the civil rights movement until his death, Draper, with Weyher's assistance, provided substantial funding for a number of projects opposing the notion that African Americans had equal rights. Some of these projects were supported directly by the Pioneer Fund and others were supported clandestinely by Draper himself. Draper realized that any association with an overtly political agenda would diminish his claim that he was purely interested in subsidizing scientific progress and would jeopardize the contention that the Pioneer Fund was a "scientific" and academic organization (Tucker 2002).

The International Association for the Advancement of Ethnology and Eugenics (IAAEE)

The first project supported by Draper and his fund was the establishment of the International Association for the Advancement of Ethnology and Eugenics (IAAEE). This association published and distributed reprints, monographs, and an edited collection on race and science. It also published the journal *Mankind Quarterly (MQ)*. This association was incorporated in 1959 with the goals of providing scientific arguments against the civil rights movement and resuscitating the Nazi ideology of racial hygiene. It was part of an international white supremacy movement to keep African Americans separate and powerless in the southern United States, South Africa, and Zimbabwe (then Rhodesia). The Executive Committee reflected the views and goals of the association and was a group that remained very close to Draper and Weyher for years to come.

Four of these committee members were particularly influential. They give us a good picture of what the association stood for. The most important (and Draper's favorite) was Henry Garrett (1894–1973). He had been president of a number of professional psychological associations, including the American Psychological Association, and was a fellow of the AAAS. He chaired the Psychology Department at Columbia University for sixteen years and in 1956 had returned to his home state to be a visiting professor at the University of Virginia. Garrett was among the most eminent of the scientific segregationists of his time. After the *Brown* decision, for which he was a witness for the segregationists (Brace 2005), he began to spout his "scientific" racism in letters to scientific journals and (especially) in the IAAEE's *MQ*. Garrett and Bonner were still using the example of the Kallikaks in their 1961 psychology textbook (Zenderland 1998). He made such "scientific" observations as "no matter how low . . . an American white may be, his ancestors built the civilizations of Europe, and no matter how high . . . a Negro may be, his ancestors (and kinsmen still are) savages in an African jungle" (quoted in Tucker 2002, 154). He also wrote that a normal black person resembled a white European only after the latter had had a frontal lobotomy (Tucker 2002). Dr. Garrett soon became a contributing editor to neo-Nazi journals (*Western Destiny* and *American Mercury*) and a contributor to *Citizen*, the monthly journal of the racist Council of Conservative Citizens (CofCC). In 1972, he became president of the Pioneer Fund (Brace 2005).

A second influential member of IAAEE's Executive Committee was Robert E. Kuttner (1927–1987), the founding director and president of the association. Kuttner was a lifelong neo-Nazi. In the mid-1950s, when he was in his twenties, Kuttner was the associate editor and frequent contributor to *Truth Seeker*, a virulent anti-Semitic publication. Kuttner was a zoologist with a PhD from the University of Connecticut. He believed that race mixing was unnatural and an obstacle to the upward evolution of the human race and that this kind of abnormal behavior is found only among zoo animals. He explained that the treatment of Jews in Nazi Germany was an instinctual result of the natural repulsion of different racial groups when they are compelled to interact. Thus, by being too pushy, Jews had provoked their own fate in the Third Reich. He also argued that if blacks continued to demand equality, they risked justifiable violent retaliation (Tucker 2002). Both Garrett and Kuttner regularly received ample financial support from Draper.

A third important member of the committee was Dr. A. James Gregor (1929—) a prominent political scientist at the University of California, Berkeley, and the Hoover Institute. He professed that Nazi policy was based on the scientific recognition of differences in racial "archetypes." People favored others in their own race; in fact, all social animals had a natural preference for their own kind and a natural aversion and inborn tendency to discriminate against individuals who were physically different. Thus, racial integration naturally led to inevitable and irresolvable tensions and disharmonies. Racial prejudice was a biological, normal social behavior. Gregor believed that segregation actually protected African

American children from psychic tension and from serious feelings of inferiority as a result of contact with whites and that they should be protected by insulating them from the latter.

The fourth insider and important committee member of IAAEE was the association's treasurer, Donald A. Swan (1935–1981). He was the most active administrator of the group and served much like an executive director. He began graduate school at Columbia University in economics but was expelled for stealing books. Swan was a self-proclaimed Nazi sympathizer. At the age of nineteen, he wrote to an obscure magazine, *Exposé*, in response to an article ("I Am an American Fascist") by his Nazi friend, H. Keith Thompson: "Despite our relatively small numbers, we, the Nordic people, have been responsible for almost all the scientific, literary, artistic, commercial, industrial, military, and cultural achievements of the world. The other races can merely imitate, without any contribution of their own. The world we live in today is a product of Nordic inventiveness, and genius. . . . I too am an American fascist" (Swan 1954, 4). In the same piece, Swan proclaimed that Yiddish stooges in Moscow were intent on preventing the Nordic peoples from realizing their rightful aspirations. In 1966, Swan was indicted on twelve counts of mail fraud related to crooked business deals and was sentenced to three years in prison. When U.S. marshals arrested him, they found in his home an arsenal of weapons, photographs of Swan with members of the American Nazi party, a cache of Nazi paraphernalia, and hundreds of anti-Semitic, anti-black, and anti-Catholic tracts (Tucker 2002). Although he temporarily lost affiliation with IAAEE while in prison, after his release Swan remained close to Draper and Weyher and continued to contribute to IAAEE publications and to receive Pioneer Fund grants.

This small group and a few other similar-minded neoracists were constantly interlinked, often in association with other anti-integration and pro-segregation causes. As Tucker (2002, 88) has stated: "As a group, the original American directors of the IAAEE represented probably the most significant coterie of fascist intellectuals in the postwar United States and perhaps the entire history of the country." One of the major strategies of IAAEE, which appears to have been more enduring than briefly organized attempts to defeat equal rights initiatives, was to continuously present the age-old "scientific" evidence of white supremacy and the dangers of miscegenation. "Using the Colonel's preferred weapon, the printer, the IAAEE produced a steady stream of publications for years, all arguing the impossibility of blacks and whites existing on equal terms in the same society" (Tucker 2002, 90).

The most important venue for these publications was *Mankind Quarterly,* which was established in 1961 to give a small group the opportunity to be published, since their views were no longer tolerated in most other scientific outlets. The editorial board included the members of the IAAEE (Garrett, Kuttner, Gregor, and Swan) plus a few other noted Nazi sympathizers such as Robert Gayre, a disciple of Hans Günther who thought the only real problem in Europe during World War II was the lack of racial homogeneity, and R. Ruggles Gates, who believed that humans were divided into five separate species. He was considered, by this group, to be the authority on the biological problems caused by intermarriage between Jews and Germans in the 1930s. The deadly Nazi doctor Otmar von Verschuer was listed as a member of the honorary advisory board of *MQ* from 1966 to 1978, long after his death in 1969 (Kühl 1994; Winston 1996). In 1979, *MQ* was taken over by Roger Pearson, who had bragged of helping hide Mengele after World War II (Anderson and Anderson 1986). As with Gayre, Pearson's racial attitudes were inspired by the Nazi "anthropologist" Hans Günther.

The book review section of *MQ* served as a bulletin board for new eugenically oriented publications, usually lavishly praising new publications that were neo-Nazi and anti-Semitic, anti-black diatribes. The IAAEE also published a number of its own pamphlets, mainly reprints from *MQ*, and a smaller series of monographs for nonprofessional audiences. One of its last projects was to counter the modern scientific research and views of race that UNESCO had published in the 1950s. In 1961, UNESCO updated its research and reprinted its statements on race in an edited volume entitled *The Race Question in Modern Science: Race and Science.* This volume included eleven articles by mainstream, highly respected anthropologists, geneticists, psychologists, and sociologists. Later UNESCO statements on race that made the same points were issued in 1964 and 1967 (Farber 2011). The UNESCO statements, which were based on the latest scientific evidence and information available and were authored by Boas's student Ashley Montagu, summarized the findings of a panel of international scientists. They were distributed internationally to senior scientists in the fields of anthropology, biology, genetics, psychology, sociology, and economics for their input (Brace 2005). Some of the main scientific points were as follows:

Statements that human hybrids frequently show undesirable traits, both physically and mentally, physical disharmonies and mental degeneracies, are not supported by the facts. There is therefore, no biological justification for prohibiting intermarriage between persons of different ethnic groups. . . . Intelligence tests do not enable us to differentiate safely between what is due to innate capacity and what is the result of environmental influences, training and education. . . . There is no proof that the groups of mankind differ in their innate mental characteristics, whether in respect to intelligence or temperament. . . . For all practical purposes "race" is not so much a biological phenomenon as a social myth. (Montagu 1951, 14–17)

Here, Montagu has substituted the term ethnic group for race, given the fact that races among humans, in a true zoological/biological sense, were no longer considered to exist by most social and biological scientists. These statements have stood the test of time, and scientific evidence has only strengthened each one of them (see Templeton 2003, 2013; Brace 2005; Lewontin 2005; Tattersall and DeSalle 2011; Fish 2013; and Mukhopadhyay, Henze, and Moses 2014 for examples).

In 1967, in an attempt to counter the 1951 UNESCO statement, the IAAEE published *Race and Modern Science*, edited by Kuttner. Publication of the book was delayed until 1967 because no reputable publisher was interested in it, even with a subsidy to cover the costs. A private publisher, Social Science Press, was set up for its publication with funding mainly from the Pioneer Fund and an IAAEE-subsidized Human Genetics Fund. Thirteen of the authors were IAAEE members and three others were former Nazi scientists. In the introduction, Kuttner stated that the purpose of the book was to provide a fuller understanding of the evolutionary value of race prejudice as an isolating mechanism favoring group survival. Because the book was directly opposed to the UNESCO statements, it was intended to balance the scales and be a major resource for anti-egalitarians, a defense of racism as innate, inevitable, and of biological value (Tucker 2002).

The IAAEE's major goal with *MQ,* its monographs, its edited volumes, and its reprints was to use tactical maneuvers in order to convert conservatives, who were already troubled by the civil rights movement and other liberal policies, into full-blown bigots who were strongly opposed to civil rights for African Americans and any similar attempts to obtain equal rights for minorities. The journal also considered itself to be an antidote to issues not politically correct. It was openly written *by* racists and *for* racists. A constant theme was that the "real science" of race was being thwarted mainly because of political issues and that *MQ* was providing an outlet for this true science of race.

For forty years, *MQ* published the same polygenic and eugenic theories that racists had used for centuries. This is an interesting phenomenon because it directly relates to the anthropological concept of culture, the scientific paradigm that ultimately was able to refute these theories. It is a paradox. People in particular cultures (or subcultures within a larger culture) develop a worldview through their history and their myths, seen through filters of their ideology, and will often hold onto this worldview regardless of alternative views and direct empirical evidence to the contrary. Of course, there also can be different subcultures within the larger culture with different worldviews about various aspects of their perceived reality. People in Western European culture can share many aspects of their culture and yet differ in certain important aspects of their worldviews. Polygenicists saw the world of human variation differently from monogenicists. Strict biological determinists see it differently from environmentalists. These culturally developed worldviews can be extremely conservative and very difficult to change, even when they are no longer tenable in light of all the empirical facts. Some cultures or subcultures are more conservative than others and less able to accept any change, whereas others might be much more open to change.

It is interesting to note that even scientists often hold onto certain paradigms very conservatively and find it difficult to accept new evidence, even when the old paradigm is no longer tenable (Kuhn 1962). This is worthy of note because one of the goals of empirical science is to constantly test all theories in an attempt to disprove them. Scientific hypotheses should be developed in such a way that they can be disproven, and the goal of the scientist is to constantly attempt to find ways to disprove them—*not* just to look for evidence that fits a particular hypothesis and ignore or reject any evidence that does not fit that hypothesis. This approach to science and anthropology was emphasized by Boas. The biologically deterministic, racist worldview appears to be extremely conservative and has been tested and disproven consistently and yet its proponents have remained resistant to all empirical scientific evidence for more than 500 years, especially during the past 100 years. That is why it is now, and has for some time been, rejected by the vast majority of scientists who actually are trained and work in the science of human variation, mainly theoretical and human geneticists and most biological anthropologists.

Being antiracist is not simply political correctness, it is proven science. Human races cannot be unequal when biological races do not exist among humans. It is impossible, scientifically and otherwise! It is the concept of culture that enables us to disprove the biologically deterministic, anti-environmental view of racial inequality, and yet it is the reality of culture that continues to keep this same racist view of humanity alive.

Carlton Putnam and the National Putnam Letters Committee

Carlton Putnam (1901–1998), a wealthy New Englander who came from a family who had been in the United States for many generations, was upset by the desegregation cases. In 1958, he wrote a letter to President Eisenhower to enlighten him about the reasonableness of segregation, hoping the president would then enlighten the nation. After earning a law degree, Putnam founded and became CEO of the airline that later became Delta. However, by 1958, he had retired to become an amateur historian. The president ignored the letter. However, Putnam also sent a copy to a friend who was the editor of a Virginia newspaper. His friend published the letter and sent copies to every editor in the United States. It made a big splash in the South, generating thousands of positive responses from the general public, students, schoolteachers and administrators, lawyers, judges, and members of Congress. However, only one small newspaper in the North published the letter, with an unfavorable comment. This stimulated a small group of "distinguished public figures" to form the National Putnam Letters Committee with the goal of publishing paid advertisements in northern newspapers.

The main ideas in Putnam's letter and in this ad campaign were that desegregation violated "the white man's right to freedom of association" and that anyone could see that blacks lacked "that combination of character and intelligence" necessary for the development of civilization (quoted in Tucker 2002, 102). In the letter, Putnam also asserted that anthropology, the public, and the Supreme Court had been deceived into adopting a social policy by an equalitarian "pseudoscientific hoax."

Tucker (2002, 103) wrote that "the assertion that support for civil rights had been consciously predicated solely on the conclusion of modern science would become the basis for a campaign, spearheaded by Putnam, to shift public discourse from the ethical to the empirical." This approach was music to Draper's ears and the Pioneer Fund, and Putnam soon found that his writings were supported by Draper's money. Putnam subsequently published many books and articles for the cause, but the most important one of these, published in 1961, was *Race and Reason,* originally titled *Warning to the North: A Yankee View.* Draper financed the publication and distribution of this short book. It was quite a successful publication for the movement, selling more than 60,000 copies within the first six months, but most of these were purchased by "the financial backer" for free distribution. This became a strategy of the Pioneer Fund: massive free distribution of racist materials to prepared lists of recipients in an attempt to propagandize an unsuspecting audience.

In *Race and Reason*, Putnam identified Boas and his disciples (all members of the same racial minority) as the mastermind of the "pseudoscientific," subversive conspiracy of equalitarian thought. After centuries of failure, this group had migrated into the United States, mainly in northern cities, and was attempting to prove their own worth by claiming that all races were equal. They were deliberately promoting theories that would weaken the white race as a whole. Putnam also warned of the dangers of miscegenation. People of mixed race were a disharmonic mixture of white ambition and black inadequacy. The genetic inferiority of the black race, according to Putnam, precluded not only integration but also any measure of social or political equality. The book's introduction, written by four IAAEE authorities, Gates, Garrett, Gayre, and George, claimed that it provided "logic and common sense." The book was sent out with a massive publicity campaign by the National Putnam Letters Committee (composed of Weyher, Putnam, Trevor, Garrett, Swan, and a few others), which had moved to New York. In 1961, a full-page paid advertisement for the book appeared in the *New York Times,* mass mailings were sent to scientists and educators, and 10,000 copies were sent to members of Congress, members of the judiciary, governors, and clergy, and to scientists on a list compiled by Swan that included 14,000 members of professional associations in sociology, anthropology, psychology, and biology. All of this was supported by the Pioneer Fund.

Race and Reason was a great success for the new race science movement and for keeping modern polygenics alive, especially in the South. Putnam was given the key to the city in Birmingham and New Orleans, the governor of Mississippi proclaimed a "Race and Reason" day for the state, and the state of Louisiana purchased 10,000 copies for required reading for education officials and for high school and college students. Putnam gave speeches to many southern audiences. After this success, the IAAEE and the Pioneer Fund funded another book, this time by W. C. George, with the same lavish resources. Commissioned by the governor of Alabama for the purpose of racial litigation, *The Biology of the Race Problem* (George 1962) criticized integration, egalitarianism, and other "false" ideas based on the "Boasian conspiracy" that were prevalent in universities. George worried that many universities had mandatory courses for wholesale indoctrination of unsuspecting students. In his own University of North Carolina, an integration course was required of all first-year students that was taught by Otto Klineberg. George argued that it lacked scholarly merit and existed for the sole purpose of indoctrinating naïve young people. Hoping that *Biology of the Race Problem* would have the same effect that *Race and Reason* did, the IAAEE and the Pioneer Fund published and reprinted the book, and it was distributed by the National Putnam Letters Committee. Thirty thousand free copies were distributed. It appears that Draper had a clear role in decisions about what would appear under the guise of the National Putnam Letters Committee. The committee also published a number of anonymous open letters as advertisements in major newspapers, more than a dozen pamphlets (many by Putnam), and reprints of *MQ* articles. In 1964, it became the Patrick Henry Group, which published another five anti-integration pamphlets, which Garrett wrote and Draper and his cronies paid for and distributed. These were distributed in the millions, mainly to teachers. Their major claims were that race mixing would lead to educational disaster by spreading "Negro culture" and causing the eventual decline of intellectual and cultural assets. These pamphlets presented integration as the primary goal of intermarriage. Garrett claimed that the main strategy of the civil rights movement was for African Americans to obtain equality with superior whites by making them "Negroid" by mixing descendants of Shakespeare and Newton with a race that was not "capable of rising above the mud-hut stage" (quoted in Tucker 2002, 111).

Court Cases to Repeal Integration

Draper, the IAAEE, and the Pioneer Fund gang also attempted to overturn the *Brown* decision. They were heavily involved in two court cases, one in Savannah, Georgia, and the other in Jackson Mississippi. They believed that the *Brown* case had been decided based not on the law and constitutional principle but rather on "factually erroneous claims of equalitarian scientists." Since they implicitly assumed that African Americans were genetically inferior, the segregationists' strategy in the *Brown* case had been to ignore scientific testimony, assuming that it was irrelevant to the central issues of states' rights and judicial precedents. In the new cases, the Draper segregationist scientists believed that if they were able to present the age-old science of Morton/Davenport/Grant, of good old American eugenics, the "truth" would prevail.

In both cases the Draper people used to test the Supreme Court ruling, the main goal was to find judges who would be sympathetic to the segregationist "scientific fact-based" interpretation of the court's ruling. They were successful in this endeavor. Their strategy was to claim that white children would suffer educational harm if they were forced to associate with other ethnic groups; they believed that this strategy would force the NAACP to defend the "scientific" evidence that was the basis of the *Brown* decision. In Georgia, Weyher and Draper's eugenic "scientists" were left to develop the substance of the case. The motion the segregationists presented was circulated to Putnam, Garrett, George, Gregor, and Kuttner for comment, and Swan wrote most of the brief. It was based mainly on a study of aptitude and achievement test scores for students in Savannah-Chatham schools that had been conducted by a University of Georgia psychologist, R. Travis Osborne (see Osborne 1962). Osborne was a member of IAAEE Executive Committee. He had received both Pioneer Fund money and regular cash gifts directly from Draper to support his work.

The expenses for the case were essentially covered by Draper. In 1963, in *Stell v. Savannah-Chatham County Board of Education*, L. Scott Stell and thirty-five other African American parents sued to compel integration in the public school system. The defendants presented a line of expert witnesses: Osborne presented his test scores from the segregated Savannah schools; Garrett testified that the differences in educability of blacks and whites were inherent and so great that no environmental changes could alter these differences; George stated that these differences were related to the size, proportion, and structure of the brain and cited the research of University of Pennsylvania anthropologist Carlton Coon that claimed that blacks had been the last of the races of humans to reach the threshold of *Homo sapiens* and thus had not had time to advance psychologically (Coon 1962). Although the neopolygenicists eagerly embraced Coon's theories as proof of their views, scientists from many fields had already presented evidence that Coon's theories were theoretically impossible and could actually be disproven by available empirical evidence (see Shipman (1994) 2002; Brace 2005). Other evidence was presented that purported to show that integration caused all students to develop a "collective neurosis" because it confused their identification with their own group and that it only caused pathological disturbances in the few capable African American students and intensified feelings of rejection in the others.

A sympathetic judge ruled in the segregationists' favor, stating that integration would seriously injure both white and black students. However, the Fifth Circuit Court of Appeals overruled this decision and the Supreme Court declined to hear an appeal (Tucker 2002).

The second court case, *Evers v. Jackson,* took place in 1964, in Jackson Mississippi. This case, which was much like the *Stell* case, was presented in much the same way and, although the judge was convinced by the evidence presented by the segregationists and wanted to rule in their favor, he realized that the results of the case in Georgia made such a ruling impossible.

Last-Ditch Attempts in Racist Mississippi

With the failure of the attempts of these two cases to overturn *Brown v. Board of Education* and with the passage of the Civil Rights Act in 1964, the segregationists accepted the fact that "the conquest of the public schools by the Negro revolutionaries had been accomplished" (William J. Simmons quoted in Tucker 2002, 126). They then adopted another strategy, focusing mainly in Mississippi, which had been the most radical racist state for generations. One approach was to create an "island of segregation" by creating a private school system for whites only. This was done through the School Committee of Jackson's Citizens' Council and a Council School Foundation, which was chartered in 1964. This foundation was made up of many of the people who had been involved in the Stell and Evers cases. The council implemented a curriculum that basically taught students that segregation was Christian, that God did not want races to mix, that white men could make the rules because they had built the United States, that famous scientists supported God's plan, and that integration was wrong because the white man was civilized while blacks in Africa still lived like savages (Tucker 2002).

The first school in the system, which opened in 1964, had twenty-two students in six elementary grades; the next year there were 110 students in all twelve grades. By 1969, there were two twelve-grade segregated systems with 3,000 students. By 1972, three new complexes were being built and the two existing systems were adding extensive and expensive facilities. These schools were funded by the state, by local governments, and by private donations. Draper was one of the major contributors to these segregated school systems prior to his death and, through his will, after he was buried (Tucker 2002).

Mississippi was also consistently the most Nazi-like state in its treatment of African Americans from before the Civil War to after the Civil

Rights Act. It continued to be the most pernicious in attempting to over-turn or simply ignore integration legislation. The state set up a number of government and private organizations that enabled it to operate much like a police state (Silver 1984; Oshinsky 1996; Tucker 2002). Two of these intertwined groups, the Mississippi Citizens' Council, a private organization, and the Mississippi State Sovereignty Commission, which was established by the state legislature, were amply, though anonymously, funded by Draper. In fact, without this funding these committees would probably not have been able to exist (Tucker 2002). The sovereignty commission kept files on 87,000 citizens, much like those the ERO and the Nazis kept. These files were used to help keep track of individuals "whose utterances or actions indicate they should be watched with suspicion on future racial attitudes" (quoted in Tucker 2002, 120). The main responsibility of the Citizens' Council was the enforcement of segregation practices. This involved social and economic pressure, the loss of jobs or businesses, intimidation, false arrests, and, though possibly unofficially, outright murder. Medgar Evers was assassinated by one of the most active members of the Citizens' Council. A fund for the assassin's legal expenses was established by other members (Tucker 2002). Draper made the greatest contribution of his resources to the Mississippi State Sovereignty Commission and the Citizens' Council of Mississippi. This funding continued until Draper's death in 1972.

The Satterfield Plan

During the mid-1960s, with the promise of Draper funding, a new, third organization was being planned in Mississippi. This would be a more permanent organization with a broader scope. It would be designed to influence policy by convincing the public of the "completely different nature of Negro citizens and white citizens" and showing them that the inferiority of African Americans was not due to environmental factors, mistreatment, or discrimination but to inherent biological shortcomings (Blackman 1999). This would be a large-scale, unified plan to combine "scientific" research with public relations, including a legal component. It planned to provide scientific documentation of racial differences and award grants to educational institutions and individuals for research projects. The results would be disseminated to newspapers, periodicals, wire services, and radio and television stations. The organization would subsidize its own periodical and fronts would be created to distribute more subtle propaganda. A legal arm would take care of legal matters and legislation.

Tucker (2002) has dubbed this the "Satterfield plan," after Mississippi attorney John Satterfield (1904–1981). Satterfield had drafted legislation for the Citizens' Council and was legal counsel to the Mississippi State Sovereignty Commission. He presented his plan to the governor of Mississippi. Draper, through Weyher, offered to fund the new organization and sent funds for its implementation to the Mississippi sovereignty commission. However, in the wake of the murder of three civil rights workers in 1964, worried that Mississippi would become a pariah state, the governor returned the money to Weyher. Thus, the project was never implemented in the state of Mississippi. Though Draper and his followers were unsuccessful in opposing *Brown v. Board of Education* in 1954 or the Civil Rights Act in 1964, "one can see how such well-financed opposition slowed the pace of social justice in the South to such a reluctant crawl" (Brace 2005, 243).

However, when Draper died in 1972, Weyher essentially converted the Pioneer Fund into the organization Satterfield had proposed, not only in Mississippi but also at the national level. From that time on, the fund closely followed the Satterfield plan. Tucker (2002, 126, 130) notes that it provided "financial support for scientific evidence of racial differences, numerous front groups that would distribute the results, journals published by other organizations themselves subsidized by Pioneer, and groups engaged in legislation and litigation on race. . . . In the post-Draper era the fund became the instrument for carrying out the Colonel's agenda. His troops had lost every battle so far, but . . . they looked to a new organization to 'win the war.' That organization was the Pioneer Fund." As we shall see, the fund has continued to carry out the agenda of the Satterfield plan to this day.

9

The Pioneer Fund, 1970s–1990s

My first direct experience with the Pioneer Fund's efforts was in 1970, when I returned to the United States after conducting my PhD field research on lemurs in Madagascar. Arthur Jensen (1923–2012) had recently published an article on race in the *Harvard Educational Review* (1969). Jensen was a professor of educational psychology at University of California, Berkeley. In his article, he argued that general intelligence was a simple genetic trait, an absolute measurable quantity referred to as the *g* factor; that this biological quantity could be measured by IQ tests; that this factor was heritable; that race was a biological reality in humans; that there were real, genetic differences in intelligence among presumed human races; that the intelligence of blacks was congenitally inferior to that of whites; and that because of these factors, it was futile to attempt to use compensatory and enrichment programs in any effort to improve the congenitally inferior intelligence of blacks (Jensen 1969).

Arthur Jensen, William Shockley, and the Foundation for Human Understanding

Jensen received his PhD in psychology from Teachers College at Columbia in 1956 (Brace 2005). He had once believed that environmental factors were more important in explaining differences in intelligence than genetic factors. However, he spent a year (1966–1967) at the Center for Advanced Study in the Behavioral Sciences at Stanford University, where he met the physicist and Nobel laureate William Shockley (1910–1989). The two men became very close friends and Shockley converted Jensen to his own beliefs about race.

In 1965, Shockley was invited to deliver an address at the first annual Nobel Conference, a conference on Genetics and the Future of Man that was held in the United States but was authorized by the Nobel Foundation. At that conference, Shockley revealed his racist ideology. He claimed that social policies were allowing genetic defectives to proliferate. Reverting to the old Humeian, Kantian, Mortonite, and eugenics views, he assumed that the first European inhabitants of the original thirteen states were the most competent peoples and that African Americans were the least capable but were producing the most offspring. Furthermore, he said, environmental intervention could not improve the lot of African Americans. Attempts to provide better education or better health care were also hopeless. To Shockley's mind, only by systematic reduction of the African American population by sterilization and other methods of birth control could we improve our society. This would lead to survival of the fittest, and the fittest were the original European settlers into America. Racial discrimination was not prejudice, he claimed, but was justified based on statistics: "Nature has color-coded groups of individuals so that statistically reliable predictions of their adaptability to intellectually rewarding and effective lives can easily be made and profitably be used by the pragmatic man in the street" (Shockley 1972, 307).

Although the Pioneer Fund had not known of Shockley before this address, he was precisely what they had been looking for—a highly respected scientist who was also a true, unadulterated racist. Even though his area of scientific expertise had nothing to do with human variation, human evolution, or anthropology, an uneducated public would surely be impressed by a Nobel Prize–winning scientist no matter what he talked about. An alliance was quickly formed between Shockley and the Pioneer Fund. Indeed, they were made for each other (Tucker 2002). Shockley was a godsend after the fund's general failure to get the attention of the media and the public. The intelligence of African Americans again became a central issue on the front page of the popular media. Shockley became the authority on and the spokesperson for opposing racial equality in abilities and in treatment. As Tucker (2002, 143) has summarized: "The fact that Shockley had done no research on genetic differences was of no consequence; he was a Nobel laureate, saying what the Draper clique wanted desperately to hear and eager to proselytize, and they quickly set out to exploit the opportunity presented by Shockley's prestige through a public relations campaign in which the physicist was not only an enthusiastic participant but often the chief strategist."

The Pioneer Fund provided substantial grants to the physicist. He also became a close colleague and confidant of the neo-Nazi Carleton Putnam, who even recommended that some of his own Pioneer Fund money go to Shockley in his war against both blacks and Jews. Funds were given to Shockley personally and to Stanford University to further his research. Shockley organized a nonprofit publicity foundation, the Foundation for Research and Education on Eugenics and Dysgenics, to launder these funds "to further public understanding" (Tucker 2002, 145). From 1968 to 1976, the Pioneer Fund gave the equivalent of approximately $1.4 million (in 2014 dollars) to Shockley (Tucker 2002).

Shockley never conducted any research, although he had proposed some highly questionable studies and projects, each of which was more outrageous than the last. For example, he proposed a "Voluntary Sterilization Bonus Plan" that would have given volunteers for sterilization $1,000 for each point their IQ fell below 100, with a bonus award for anyone who persuaded qualified individuals to participate. Instead of doing research, Shockley organized a complete propaganda blitz for eugenics views. Shockley referred to this as "my campaign," but one journalist called it a "traveling carnival of racism" (King 1973, 36).

Shockley used Pioneer Funds that had been granted to his Stanford University lab to hire neo-Nazi colleague Robert E. Kuttner (see last chapter) and to distribute decade-old articles from neo-Nazi publications (Tucker 2002). He basically used Pioneer Fund money, with the fund's complete support and blessing, to attempt to sway public opinion. Using the old Pioneer tactic, Shockley developed specialized mailing lists, depending upon the particular racist issue, to send out tracts free of charge to his chosen public. Free racist propaganda was sent out to presidential candidates, members of Congress, state legislators, members of the NAS, the wire services, major newspapers, smaller southern papers that supported segregation, sympathetic journalists, business executives selected from *Who's Who in America,* and other notable figures. An enormous amount of Pioneer Fund money was used to keep the myth of racism alive.

Perhaps Shockley's most expensive, ambitious, and successful endeavor was his work to change Arthur Jensen's views on race and then bring him into the fold of the Pioneer Fund. He then designed an expensive mailing campaign to market Jensen's work to other scientists and to the public. Among the members of the Pioneer Fund Executive Committee, this soon became known as the "Shockley-Jensen team" (Tucker 2002). Jensen's 1969 *Harvard Educational Review* article, the longest article the *Review* had ever published, became the centerpiece of the Shockley campaign.

In his article, Jensen wrote that it was not possible to improve the educational performance of minority school children because of their inferior genetics and that spending money on this endeavor was a waste of public funds. He also argued that it was a disservice to attempt any improvement because these children should not have high aspirations but should instead be happy with their inferior lot. Thus, minority children (essentially African American children) should have an education that was more suitable to their inferior intelligence. In addition, eugenic measures should be undertaken to reduce their numbers (Jensen 1969). Shockley sent Jensen's article to hundreds of scientists and politicians in his campaign to inform the public about the shortcomings of African Americans. Shockley's and Jensen's main worries were that race mixture was causing dysgenics, or racial reverse evolution. One of their major goals was to force the NAS and members of Congress to study this question in an attempt to solve our "national Negro illness." In this endeavor, Jensen sent out over 1,500 photocopies of his article, which Shockley paid for with Draper money, to members of the NAS and other scientists, politicians, members of Congress, and journalists.

With Shockley's massive publicity campaign, "Jensenism" (as a *New York Times Magazine* science writer dubbed it) became a cause célèbre. In 1969, headlines read "Born Dumb" and "Can Negroes Learn the Way Whites Do?" A highly complimentary article on Jensenism was published in *U.S. News & World Report*. This followed another massive mailing of Jensen's article. Was this massive media blitz successful? When I first became aware of the Jensen article upon my return to Duke University in North Carolina from my field work in Madagascar in 1970, I was quite surprised by this resurgence of scientific, eugenics-like racism among many of the students at Duke, especially those from the South. Many of them had regained a "southern pride." The past racist views of southerners and even slavery were now vindicated by Jensen's and Shockley's claims. It was not unusual to find large Confederate flags decorating their apartments, something I had not seen earlier. The valid criticisms by Boas and his colleagues and by modern geneticists and anthropologists were easily ignored and quickly replaced by the centuries-old cultural myths of polygenics and eugenics, even among many, especially southern-born, anthropology students. Needless to say, I was appalled.

The Pioneer Fund was delighted by this publicity. Indeed, it was this kind of public media, and not scientific research, that the fund was actually paying for. After all, even though Shockley and Jensen were writing about "genetics" and "anthropology," neither was a geneticist or an anthropologist, and they understood exceedingly little about legitimate research in these scientific fields. During the late 1960s and early 1970s, the Pioneer Fund tried, mostly in vain, to get Shockley's views of Jensen even more widely distributed. During all of this, the fund made sure that Shockley had ample funds for this propaganda campaign and Shockley made sure that the Pioneer money kept coming both as public funds (to Stanford University) and private (tax-free) funds to the Foundation for Research and Education on Eugenics and Dysgenics.

The Pioneer Fund people still wanted to use these two scientists in their attempt to overturn the civil rights laws. That was their major strategy. Shockley, Jensen, and the members of the Pioneer Fund believed that desegregation would be harmful to both races because African American students would be treated like normal, average white students when they believed they should be treated as retarded. Shockley and Jensen were especially useful to the fund, and the assumption was that they would be taken seriously by the NAS and other scientists. However, by the mid-1970s, the Pioneer Fund leaders realized that Shockley was becoming detrimental to their cause. Shockley was showing symptoms of extreme paranoia. He was proposing to use polygraphs on fellow faculty to assess their actual views, he was suing those who called his views biased, he was taping his telephone calls, and he was saving all of his papers, including even his laundry lists (Shurkin 2006). The Pioneer Fund was giving Shockley money to support his ability to publicize the fund's agenda, and he was becoming so nettlesome that he could no longer influence people. The Pioneer Fund members recognized this and ended his funding (Tucker 2002).

Jensen, on the other hand, began receiving Pioneer Fund money in 1973, and the fund continued to support him until 1999. This support financed a special institute, the Institute for the Study of Educational Differences, which was set up to escape the University of California, Berkeley's oversight. Jensen was president and his wife was vice-president. It received over $2 million in today's currency. Jensen was never very outspoken about his racist views but he did participate in a few activities that illustrated his racism. For example, he contributed to and served on the scientific board of *Neue Anthropologie,* a German journal dedicated to restoring *Rassenhygiene* and preserving the racial philosophy of Nazi theorist Hans Günther.

This journal was a clone of *MQ*; it featured similar contributors and even on occasions the same articles. Throughout his career, racial differences in intelligence remained an obsession, and recently he teamed up with the recently deceased president of the Pioneer Fund, J. Philippe Rushton (see Chapter 10) (Jensen 1998, 2006; see Rushton and Jensen 2005, 2006, 2010).

Although his statements and publications had no real scientific validity, Jensen gave "scientific" validity to the claims of current racists and, of course, thus, did exactly what the Pioneer Fund wanted. When a person with some scientific legitimacy repeats this cant, it is easier to convince an uneducated public of the old racist myths. Jensen's myth (coined as "Jensenism") was basically that IQ is largely hereditary and that different races have inherited different and racially stratified IQs. He failed to understand the devastating criticisms of the IQ argument Otto Klineberg provided in the 1920s and 1930s and those that other scientists have provided ever since (see articles in Fish 2002, 2013; and Cravens 2009 for examples). He failed to understand the devastating criticisms of the concept of race put forward by anthropologists ever since Boas, by the UNESCO statements (1950 and later), and by anthropologists and geneticists ever since. There can be no genetic association of IQ with race if IQ is not a simple unitary genetic characteristic and if biological races do not exist among humans. However, again, real science is not at question here. The Pioneer Fund is simply interested in continuing the Western European myth of racism, regardless of science. They believe in their cultural myth just as many literal Christians believe that the Earth was created in 4004 B.C. Science cannot trump mythology or culture if the cultural practitioners do not understand the science and have no desire to believe it. This, of course, is the underlying core of prejudice.

Most people have accepted scientific proof that the world is not flat, even though it certainly looks flat to the naked eye. However, even though there is much scientific proof that IQ is not a simple hereditary trait, that it is not easily measured, that all measures of what we assume to be intelligence have some cultural bias, and that biological races do not exist within our species, these modern racists still insist that a simple measured score for IQ is inherited and that there is some relationship between something called race and that measure. Even Jensen admitted that "race" is a social and not a biological construct (Jensen 1995, 42). As Brace (2005, 245) has noted, before Jensen's ill-fated 1969 *Harvard Educational Review* article, "Oddly enough . . . Jensen had written that lower tested achievements of blacks and Hispanics 'cannot be interpreted as evidence of poor genetic potential.' . . . 'Because barriers to social mobility' existed, such socioeconomic and cultural disadvantages suggested the 'reasonable hypothesis that their low-average IQ is due to environmental rather than to genetic factors.'" At one point, then, Jensen seemed to understand the science and agree with the most current research on this topic. Later, he essentially admitted that his views were based on his own prejudice (see Jensen 1998).

Besides Shockley and Jensen, the next largest recipient of money from the Pioneer Fund was a nonprofit organization devoted to publishing and distributing segregationist and racist literature, the Foundation for Human Understanding. It received over $750,000 to publish and disseminate books and articles that no other press would accept. It also served as a repository for racist literature, including such tomes as *Race Crossing in Jamaica,* the study by Davenport and Steggerda that Draper had funded (see Chapter 2). For example, over $50,000 was spent on the distribution of Jensen's *Straight Talk about Mental Tests* (1981) to college presidents and administrators in an attempt to influence admission policies of universities.

The most notorious of the Foundation for Human Understanding publications was Stanley Burnham's, *America's Bimodal Crisis: Black Intelligence in White Society* (1985). This racist diatribe claimed that African Americans were an expense to their employers because they stole, they were incompetent, and having them as employees generated legal fees. Burnham (a pseudonym) also claimed that he knew a number of African American PhD candidates who had such a poor grasp of language that their theses had to be rewritten by trained secretaries and editors. In addition, he doubted that "they has [*sic*] researched and written their own papers." Harking back to the Mortonites, Burnham argued that black people have smaller brains than Europeans and that Africa's problems are the result of the "ignorance and irresponsibility of the African mind." Black people's problems in other parts of the world were the result of "genetic deficiencies that they cannot remedy." Burnham finally claimed that because of the bimodal distribution of intelligence between blacks and whites, "black students whose intelligence and academic performance persistently fall below acceptable standards should be steered as quickly and effectively as possible into job situations in which performance standards are less stringent . . . such as farm labor, dish washing, garbage collecting and the like" (Burnham 1993, 116). Burnham believed that his "monogram" [*sic*] was so important that it "will be read and appreciated a century from now" (Tucker 2002, 158). The Pioneer Fund, through the

Foundation for Human Understanding, published three editions of this "monogram" between 1985 and 1994. In the 1980s and 1990s, the Pioneer Fund continued to fund the Foundation for Human Understanding and a few other distributors until it began to shift its funding more exclusively to the publishing efforts of Roger Pearson, a British economist (Tucker 2002).

Roger Pearson

Roger Pearson was born in Britain in 1927. In 1958, he founded the Northern League to continue Nazi racial theory after World War II and encourage Nordics around the world to think in terms of ethnic and racial identity instead of nationality (Kühl 1994). The Northern League was a eugenics organization that reflected Pearson's views and continued the neo-Nazi ideology that promoted hatred of blacks, Jews, and immigrants. The league attracted a number of past members of the Third Reich, including Franz Altheim, a former assistant to Heinrich Himmler, and Dr. Wilhelm Kesserow, a former SS officer. Ernest Cox of the Ku Klux Klan was an honorary member (see Chapter 8) (Tucker 2002; Jackson and Winston 2009; Jonkers 2008). Another prominent Nazi and co-founder of the league was "anthropologist" Hans Günther. In fact, Pearson was a close friend and protégé of Günther and praised him as "one of the world's greatest names in the field of raciology" (quoted in Tucker 2002, 159–169). The monthly newspaper of the league, *Northlander,* became a major voice for eugenics propaganda, spreading the word of Nordic superiority, the Aryan underpinning of civilization, and the dangers of miscegenation. Soon every British neo-Nazi was a member of the league. Pearson, eager to spread his polygenic views, spent five weeks in the United States in 1959.

In the mid-1960s, Pearson moved permanently to the United States and began working with prominent racists and neo-Nazis in this country. With Willis Carto (see Chapter 8), who had established a Northern League chapter in California, Pearson (using the pseudonym Edward Langford) edited the journal *Western Destiny* as a continuation of *Northlander.* This was a blatantly pro-Nazi publication. In 1966, under the same pseudonym, Pearson edited another journal, *New Patriot,* which addressed itself mainly to the "Jewish question" (Langford 1966; Tucker 2002). In this journal's skewed view, Jews were the oppressors of the Germans, who had been the victims during World War II, and it was the goal of Jews to dominate the world. According to Pearson and the *New Patriot,* they were attempting to do this by suppressing the Nordic culture through planned mongrelization (Langford 1966). Pearson often supported his anti-Semitic views by quoting from well-known Nazis such as Alfred Rosenberg, who had been hanged at Nuremberg for his part in the mass killings of Jews (Tucker 2002).

Also in the 1960s, the Pioneer Fund provided financial support for an international and intellectual group of academics led by Pearson, the IAAEE. It was formed to help keep scientific racism alive and to stop race crossing. This loosely organized community included some "mainstream" academics such as psychologist Henry Garrett and geneticist R. Ruggles Gates, as well as Earnest Cox and Willis Carto (Jackson and Winston 2009).

In 1967, Pearson went on an extensive visit to South Africa, Rhodesia, and Mozambique and allowed his journals to be absorbed by *Thunderbolt,* a journal published by his ally Edward R. Fields. Fields, along with J. B. Stoner, an Atlanta attorney, had founded the Christian Anti-Jewish Party, which had evolved into the National States' Rights Party in 1968. *Thunderbolt* was the official journal of the party. The constitution of this organization began with the statement that "Jew-Devils have no place in a white Christian nation." Stoner believed that being a Jew was a crime punishable by death. Fields was more tolerant; he believed that Jews in positions of power or authority should be removed from those positions. However, if that did not work then a final solution would have to be devised. Besides continuing the *New Patriot's* focus on anti-Semitism, *Thunderbolt,* added the familiar U.S. racists' anti-black theme and a new focus on the desirability of deporting African Americans.

Pearson had two degrees from the University of London, a master's degree in economics and sociology in 1954, and a PhD in economics in 1969 (Pearson 1969). Although he is often referred to as an anthropologist, he is not listed among the PhD recipients in anthropology at that university (Leslie Aiello, personal communication, 2012). In the United States in 1968, Pearson attempted to pass more as a respectable academic than as an active neo-Nazi and began to search for a university position. He became a houseguest of Elmore Greaves in Jackson, Mississippi, a favorite spot of Draper's and the Pioneer Fund.

Greaves was an outspoken segregationist who had received campaign funding from Draper for an unsuccessful run for political office. He recommended Pearson to William D. McCain, the president of the University of Southern Mississippi (USM), for a professorship. McCain, a known segregationist, hired Pearson to chair a new Department of Anthropology and Comparative Religious Studies in 1968. Pearson soon was instrumental in getting the anti-segregation chair of the philosophy department fired and became chair of an expanded Department of Comparative Religion, Anthropology, and Philosophy (referred to on campus as CRAP). He soon succeeded in replacing most of the qualified untenured professors with racists such as Donald Swan and Robert E. Kuttner. Kuttner was hired at USM a few weeks after he began working in Shockley's lab. Both were Pioneer Fund people and were well-known neo-Nazis (Tucker 2002; see also Chapter 8). In 1974, Pearson accepted a position as professor and dean of academic affairs and director of research at Montana Tech (Lichtenstein 1977).

In 1975, Pearson moved to Washington, DC, where he attempted to create a new image as a respectable, mainstream conservative. Conservative DC organizations and think tanks began to embrace him. He served on the editorial boards of a number of these, including the Heritage Foundation, the Foreign Policy Research Institute, and the American Security Council Foundation. A number of conservative Republican members of Congress wrote articles for his *Journal of American Affairs* and other publications (Tucker 2002; Michael 2008). He also sat on the board of trustees for the American Foreign Policy Institute, founded the Council on American Affairs, and served as director of the Council for Social and Economic Studies and president of the University Professors for Academic Order (Kühl 1994; Sautman 1995; Jonkers 2008). Around the same time, he began his own institute, the Institute for the Study of Man (ISM). This was categorized as a book publishing company, but its purpose was to study "the origins and nature of man in order that contemporary Western society and its pressing problems might be more clearly perceived" (quoted in Tucker 2002, 170).

In the mid-1970s, Pearson made an attempt to form a new international Nazi organization through the newly formed U.S. chapter of the World Anti-Communist League. The old chapter had renounced its membership because Pearson as president of the organization had filled it with neo-Nazi, anti-Semitic, and fascist groups and with actual former German army Nazis, including former S.S. officers (Tucker 2002). By the end of the 1970s, the World Anti-Communist League, under Pearson's leadership, "represented one of the greatest fascist blocs in postwar Europe" (Anderson and Anderson 1986, 100). In 1979, Pearson was listed as a member of the Comité de patronage (the advisory board) of the French neo-Nazi journal *Nouvelle École* (Jonkers 2008). Because of his radical ideology,

Pearson was forced out as head of the league, though he managed to ensure that his segregationist landlord from Mississippi, Elmore Greaves, succeeded him.

Unsuccessful at creating a new international Nazi organization, Pearson turned his energy toward his own institution, the ISM. In this effort, he was generously supported by the Pioneer Fund. The fund may have given some money to Pearson before his first academic appointment at USM, and it was certainly aware of him before his move to Washington. Probably as early as 1960, Pearson had a connection with *MQ*, and he featured many of the Pioneer Fund individuals in his early publications; he was likewise featured in many of theirs (Tucker 2002). However, Pioneer Fund support of Pearson officially began when he joined the USM faculty in 1968.

This was a match made in heaven, or, we might say, in hell. Pearson shared many of Pioneer founder Draper's beliefs; in fact, Tucker (2002) refers to Pearson as Draper reincarnated. Both shared the views of earlier U.S. and Nazi eugenicists such as Davenport, Laughlin, Grant, Fischer, and Günther. Like them, they believed that other "races" (such as blacks and Jews) should not be allowed in "Nordic" countries and that Jews were the main enemies of social cohesion. They saw the science of the Third Reich as their model; they viewed Günther and Hitler as their mentors; they both saw "Nordics" as beautiful, both physically and as warriors; and they both had what might be considered an odd fascination for blue-eyed, "Nordic" pilots.

As usual, the Pioneer Fund supported Pearson not because of his "anthropological" research (of which there was none) but because of his publications and his promulgation of neo-Nazi, anti-black, anti-Semitic ideals. The first official Pioneer gift, in 1973, was Draper's private library of rare books and other materials on racial hygiene that he had acquired from the Third Reich. This was to be given to either Pearson or Swan at USM, to add to Swan's similar collection. In 1981, when Swan died, Pioneer gave Pearson the funds (over $50,000 in 2014 dollars) to purchase the whole collection and an additional amount (close to $200,000) for a building to house it. From 1973 to 2000, the Pioneer Fund gave Pearson close to $2 million (in 2014 dollars) most of which was to administer the ISM (Tucker 2002; Jonkers 2008).

Thus, Pearson concentrated on publishing books and journals that promoted the fund's agenda. In 1978, in fact, he acquired *MQ*, the Pioneer Fund's journal. The name of the publisher of many of the ISM's early books was given as Cliveden Press. This was a reference to Pearson's affiliation to the Cliveden Set, the name given to a group of British aristocrats who had supported Hitler (Tucker 2002). The articles and books produced by Pearson's ISM had the same old eugenicist, racist themes. As Tucker (2002, 176) states, "Pearson's journals at the institute . . . provided a pseudoscholarly outlet for fellow Pioneer grantees, and other authors with similar thinking on race, whose work would have little chance of acceptance elsewhere." In many of these publications, Pearson himself was the most prolific author, either in his own name or under a number of pseudonyms. He would often write extremely similar articles or even the same article under different names or he would praise articles written in one pseudonym in another written with a different pseudonym or in his own name. Authorship, ethics, truth, and reality seemed to have little meaning to Pearson and the Pioneer Fund.

The themes of these publications were reminiscent of those of the eugenicists and Nazis of the first third of the twentieth century (Pearson 1996). Aryan or Nordic "noble" families were genetically distinct from others. Nordic societies were of greater innate value and thus had the right to command their inferiors. Some of the threats to Nordic civilization included the Christian belief that all humans were created equal before God and the Catholic belief in the universality of all of humanity. Another threat was the fact that Nordics were being killed off disproportionately in wartime. This was attributable to the propensity of brave, warrior-like Nordics to go to battle for their beliefs, whereas the inferior races were cowards. In addition, modern medicine was keeping more members of the inferior races alive than would the pruning knife of natural selection, and the natural fitness of the Nordics was being diminished demographically. Pearson (1966, 26) stated: "If a nation with a more advanced, more specialized or in any way superior set of genes mingles with, instead of exterminating, an inferior tribe, then it commits racial suicide and destroys the work of thousands of years of biological isolation and natural selection."

To Pearson, the members of the Pioneer Fund, and their neo-Nazi and racist colleagues, the intelligent, scientific response to these threats was the eugenics and Nazi movements of the first third of the century. Their heroes were men such as Davenport, Grant, Laughlin, McDougall, Günther, Hitler, and Cox. These modern racists saw the earlier part of the twentieth century as a happier and more optimistic time, when there was still a possibility that whites would dominate the world and when many states had sterilization and anti-miscegenation laws. Even more important, a fascist dictatorship had been completely free to carry out the most extreme eugenics policies imaginable, and for Pearson and his colleagues, the Nazis and their policies provided a model for all of the Western world.

What went wrong, what stalled this ultimate domination by the Nordic "race"? According to Pearson, it was "a cabal of scheming egalitarian scientists, almost all of them Jewish and disciples of Boas, who saw eugenic programs in conflict with their desire 'to demolish the unity and coherence of national units' by insisting on racial equality" (Tucker 2002, 174; Kühl 1994). Pearson's anti-Semitism seemed to stem from his opposition to what he saw as the Jewish, Boasian attempt to obtain world power and rule over the vast subnormal mass of subhumanity (Pearson 1995, 1996). Thus, neoracists and modern eugenicists still point to Boas as the main reason for their failure to succeed in promoting polygenic racism and for the fact that eugenics is not widely accepted in the world today. They seem to have understood or at least feared the concept of culture that Boas and his followers developed.

To Pearson and his followers, Christian universalists and biological egalitarians were creating disastrous consequences because they were going against the sociobiological goal of perpetuating white, Nordic superior genes by being unnaturally altruistic with inferior races (they weren't following the guidelines of their "selfish" genes). Racial prejudice was an evolutionary necessity—it was a natural human tendency, and it perpetuated and maintained the integrity of the gene pool. They saw mating of different races (species?) in zoos as unnatural and a perversion of natural instincts. Thus, over the years, the main purpose of ISM publications and *MQ*, the journal of the Pioneer Fund, was to encourage racial conflict (often reprinting Pearson's earlier Nazi Northern League pamphlets). The Pioneer Fund basically has been publishing and dispersing the same, unscientific propaganda for five decades and continues to do so. As Tucker (2002, 179) states: "What Pioneer received for its almost $2 million investment, not in Pearson the anthropologist but in Pearson the activist, was largely a campaign for the United States to emulate the Nuremberg Laws, denying to nonwhites the benefits of citizenship." Pearson provided warmed-over Nazi propaganda and a regurgitation of fifteenth-century polygenics and eugenics.

In the early 1980s, Pearson served as an advisor to Senator Jesse Helms, and, in 1982, he received a letter of praise from President Ronald Reagan for his "substantial contributions to promoting and upholding those ideals and principles that we value at home and abroad" (quoted in Kühl 1994, 4). In 1986, Pearson was discovered to be closely associated with a number of former leaders of U.S. intelligence agencies and American Security Council members (Covert Action 1986). Pearson served as editor of *MQ* from 1979 to 1996 (Winston 2013). Currently he serves as director of the Council for Social and Economic Studies, which owns Scott-Townsend Publishers, the publisher of most of his recent books, and he continues to publish several journals in Washington, DC, including *MQ*, *The Journal of Social, Political and Economic Studies*, and *The Journal of Indo-European Studies*. He remains on the board of trustees of the American Foreign Policy Institute. Chip Berlet (2003), a senior analyst at Political Research Associates, a think tank near Boston, considers Pearson "one of the most virulent race scientists operating today."

10

The Pioneer Fund in the Twenty-First Century

Besides financing Pearson, during the past thirty years, the Pioneer Fund has continued to fund a major twin study, seven new scientific racists, and those who oppose immigration into the United States (Tucker 2002). By 2000, the Fund had provided more than $10 million (in 2014 dollars) to support these enterprises. I will discuss the first two of these topics in this chapter and the Pioneer Fund's spearheading of the modern anti-immigration agenda in the next.

The Minnesota Twin Family Study

In one of the many horror stories of the Third Reich, von Verschuer and Mengele conducted torturous research on twins in concentration camps (such as the twin camp at Auschwitz) as part of the Nazis' attempts to determine the genetic basis of human physiology, health, behavior, intelligence, and "race" (Black 2003). This interest in twin studies had been a major focus of eugenicists since Galton's time (Galton 1875) and had continued with Charles Davenport (1911, 1920; Davenport and Steggerda 1929) and American and German eugenics (Black 2003). In the 1940s through the 1960s, Sir Cyril Burt (1883–1971) perpetrated a scientific scandal in his research on IQ scores and identical twins.

Burt was a Kantian who believed that "primitive races" were unable to acquire civilization and that slum dwellers and the Irish were innately inferior. His work on the correlation between IQ scores of identical twins raised apart and those raised together was proven to be patently fraudulent (Kamin 1974; Brace 2005). Known cases of identical twins separated at birth and raised apart are extremely rare. There had been only three published studies of this kind before Burt's. Burt began with fifteen sets of twins in 1943 and expanded his sample over twenty years; by 1966, his study included fifty-three twin pairs. His data showed an extremely high hereditability of 80 percent for IQ performance. He also reported correlation coefficients for his data to three decimal places. However, even though the sample size increased over the years, the coefficients remained exactly the same, a statistical impossibility (Kamin 1974). There were a number of other inconsistencies and ambiguities with his data as well, and he was accused of fraud. Eventually, Burt's friend and official biographer, Leslie Hearnshaw (1979), who had started out with the goal of disproving the allegations against Burt, was forced to accept that the charges of fraud were justified. Later, Tucker (1997) examined the numbers of research subjects who had been used in all of the twin studies conducted between 1922 and 1990 and found that no other study came close to having fifty-three sets of twins that would have satisfied the conditions in Burt's study (Plucker 2012). There were not enough twins to study.

The Pioneer Fund attempted to continue the tradition of twin studies with its support of Thomas Bouchard's Minnesota Twin Family Study (MTFS), a longitudinal study that began in 1979. Arthur Jensen had introduced Bouchard to the Pioneer Fund (Sedgwick 1995). The fund dispersed the equivalent of around $2.6 million (in 2014 dollars) from 1979 to 1999, and during this period the twin study was its largest grantee (Tucker 2002). The MTFS ended in 2000.

At Bouchard's Minnesota Center for Twin and Family Research, twins raised apart were studied to determine how much of their behavior is determined by heredity. In the spirit of the Pioneer Fund, Bouchard and his researchers used study data to argue for the preponderance of genetics over environment as the main influence over behavior. Although race was not a focus of the MTFS research, it is important to remember that modern scientific racism is based on the argument that most human variation in such things as physiology, behavior, and intelligence is biologically based and is not determined or influenced in any important way by environment.

The eugenicists thought they had won this battle. However, Boas's concept of culture and a more sophisticated understanding of genetics meant that environment once again had become a problem for modern neo-eugenicists and race scientists. Thus, Bouchard's reporting that tendencies toward such things as religiosity, political radicalism, tolerance of sexual minorities, preferences and capacities for certain careers, and intelligence and learning capacities are mainly inherited was used as "important proof that genetic factors set the potential limits of human behavior, while the influence of environmental circumstances is determined by heredity" (Kühl 1994, 9).

The initiative for Bouchard's studies came from a report of a highly publicized case of a pair of identical twins who had already been the subject of sensational stories in the popular press (Joseph 2001). Jim Lewis and Jim Springer (commonly called the "Jim Twins") had been separated at birth and were reunited in Ohio at age 39. It had been reported that they shared a number of uncanny similarities: the names of their wives and children, career choices, and preferences for particular brands of beer and cigarettes. They became the first pair in Bouchard's studies of separated twins and were given a battery of personality and mental ability tests, interest and value inventories, psychomotor tests, information-processing measures, and several life history interviews (Bouchard 1984). In a similar case used in Bouchard's studies, identical twins Oskar and Jack shared a number of idiosyncrasies: they liked spicy foods and sweet liqueurs, were absentminded, had a habit of falling asleep in front of the television, thought it funny to sneeze in a crowd of strangers, flushed the toilet before using it, and stored rubber bands on their wrists (Holden 1980, 1324). When they were brought together to participate in the study, both were wearing wire-rimmed glasses and had mustaches, and both sported two-pocket shirts with epaulets. What was not mentioned is the fact that the twins had met previously and had been in contact with each other for more than twenty-five years and that after their cases were reported in the press, they sold their life stories to a Los Angeles film producer (Horgan 1993).

On the basis of his research data, Bouchard (1994, 1701) claimed that "the similarity we see in personality between biological relatives is almost entirely genetic in origin." However, as Farber (1981) and Joseph (2001) have pointed out, Bouchard's and most other conclusions in the literature about "separated" identical twins suffer from a number of problems. In some cases, the twins were separated after being raised together for several years, in others they were raised by different members of the same family, in others they were placed into families of similar socioeconomic status, in others they were aware of each other's existence and had frequent contact during their lives, in others they were brought to the attention of researchers on the basis of their similarity or knowledge of each other. In some cases the material used to evaluate the similarities was collected by the same researchers or reported similarities were not assessed by blinded raters. In addition, all of the twins shared a common prenatal environment. In fact, any two people who are the same sex, were born on the same day, and were brought up in the same culture are likely to have far more in common than two randomly selected people. Thus, generally, from the environmental perspective, monozygotic twins should be far more similar than two randomly selected members of the population.

In case studies of separated twins, it must be proved that the similarity of traits is not correlated with similarity in environments of the twin pair. It cannot simply be assumed that the twins were raised in unequal environments. Bouchard's research results are also questionable because he did not share case histories or his research data. This makes it extremely difficult for independent researchers to analyze the data and to offer alternative interpretations (Joseph 2001). As psychologist Jay Joseph (2001, 25–26) summarizes the results Bouchard and his colleagues have presented: "MISTRA [Minnesota Study of Twins Reared Apart] researchers have not demonstrated that MZAs [monozygotic twins reared apart] were reared in uncorrelated environments. . . . A far better way to determine similarities in the rearing environment would have been a blind evaluation of the never-published information gathered in the extensive life history interviews that were given to all participants in the study. . . . The Minnesota separated twin studies, like the studies that preceded them, are sufficiently flawed that no conclusions about the role of genetic influences on human behavioral and personality differences can be drawn from them."

Sedgwick (1995, 147) succinctly sums up the importance of the Bouchard twin studies for the Pioneer Fund and the new scientific racism: "Bouchard's research may well have paved the way for the current resurgence in hereditarian lines of inquiry that has, for example, led to the enormous public receptivity to news of the 'discoveries' of the genes for alcoholism, homosexuality, and schizophrenia, reports that were later either retracted or mired in qualifications. . . . [His work] marshaled support for the current Human Genome Project."

These twin studies may be one of the most influential factors in leading many biologists, social scientists, and members of the public to think about genetics and heredity in much the same false and simple-minded way that they were thought about by the eugenicists and racists at the turn of the twentieth century. The studies used new, poor science to "prove" that biology was more important than environment in all aspects of human variation. After all, Bouchard's work was published in *Science,* one of the world's most prestigious scientific journals.

In recent, more carefully administered studies, it has been discovered that the similarities among identical twins (monozygotic [MZ] twins) are not surprising. As discussed above, there are many reasons for this (Sternberg 2007; Joseph 2004, 2010; Richardson and Joseph 2011; Miller 2012). Such pairs are scarce; normally separation of twins is incomplete (that is, it usually happens well after birth and the twins are raised with different branches of the same family); the twins remain in regular contact; and they shared their intrauterine environment before birth. In addition, many similarities of behavior are strongly influenced by nongenetic cohort effects that are related to certain characteristics of the historical period and cultural milieu. As Farber (1981, 77) observed, monozygotic twins are "not so much similar to each other as they are similar to people of their eras and SES [socioeconomic statuses]." To really know the importance of reared-apart monozygotic twin resemblance it would be necessary to control data on similarities found in pairs of age-matched strangers (Rose 1982). As Richardson and Joseph (2011) have stated: "Far from experiencing different environments, as twin researchers often claim, in most cases reared-apart MZ pairs share at least seven different cultural influences at the same time: national, regional, ethnic, religious, economic class, birth cohort, and gender cohort."

Epigenetic effects also are now known to be extremely important (Meaney 2010; Miller 2012). Even though twins are born with the same DNA, they can be surprisingly different in behavior, health status, and even in physical appearance. Research on epigenetics has shown that even the same genes can react in different ways given different internal and external environmental influences. In an exceptional analogy, Miller (2012, 58, 64) describes the situation as follows: "Identical twins are born with the same DNA but can become surprisingly different as they grow older. A booming field of epigenetics is revealing how factors like stress and nutrition can cause this divergence by changing how individual genes behave. . . . If you think of DNA as an immense piano keyboard and our genes as keys— each key symbolizing a segment of DNA responsible for a particular note, or trait, and all the keys combining to make us who we are—then epigenetic processes determine when and how each key can be struck, changing the tune being played."

Even though each piano is constructed in much the same way, an infinite number of tunes can be generated. Thus, environment not only impacts our individual histories, it can also directly impact our genes and how they operate (Cloud 2010). As Richardson and Joseph (2011) state:

"Claims about the presence of additive, independently deterministic, genes producing direct genotype-phenotype correlations, arising from twin studies, has inspired genome-wide association (GWA) searches for such genes at the molecular level. Decades of searches, however, have failed to discover them. . . . This provides additional evidence that standard interpretations of twin studies . . . are incorrect, that such simple genes do not exist, and that we need to consider the roles of non-genetic and 'systemic' factors in individual differences."

The New Bigot Brigade

In addition to the twin studies, during the 1980s and 1990s, the Pioneer Fund granted more than $5 million (in 2014 dollars) to seven scientists in an attempt to continue propaganda about their cause. Loring Brace (2005) aptly dubbed the eugenicists of the early part of the twentieth century as the "bigot brigade" and it seems appropriate that this new group of Pioneer-funded scientists be anointed as the "new bigot brigade." It consists of a team of racists who have attempted to use basic polygenic ideology to promote the Pioneer Fund's views of racial classification and white superiority.

Hans J. Eysenck (1916–1997) was a British psychologist from the Institute of Psychiatry at King's College, London. He was a student of Cyril Burt's and an "honorary advisor" for *MQ*. Eysenck was a pure biological determinist who believed that all of the major differences between people and races were the consequence of their genetic worth and that, for example, any compensatory education for African American children or other solutions for bettering their education was useless. Eysenck also was funded by R. J. Reynolds and the Tobacco Research Council, both of which wanted to demonstrate that illnesses caused by smoking were genetic and were not primarily due to the effects of tobacco (Brace 2005). He maintained that there was a decline in IQ scores from Mongoloid through northern Europeans to southern European Caucasoids to Indians to Malays to Negroid groups, in that order (Eysenck 1971, 1985). Like his mentor, Eysenck believed that the Irish have low IQs (Eysenck 1971; Benson 1995). He was the mentor of Arthur Jensen and J. Philippe Rushton (Sautman 1995; see later in this chapter). He also was a strong supporter of Jensenism, the idea that a person's IQ is attributable to heredity, including his or her racial heritage.

Robert A. Gordon (1932–) is a retired sociologist who taught at John Hopkins University from 1963 to 2005. Gordon's ex-wife, Linda S. Gottfredson (1947–), who for many years was the only woman to receive funds from the Pioneer Fund, is a professor of educational psychology at the University of Delaware and co-director, with Gordon, of the Delaware-Johns Hopkins Project for the Study of Intelligence and Society. Gordon believes there are direct correlations between race, IQ, poverty, juvenile delinquency, criminality, and even involvement in conspiracy rumors, all of which he argues are genetically based characteristics (Miller 1995; Tucker 2002). His major theme is that genetically based IQ differences are more accurate determinants of black-white differences in crime statistics than are income, education, or occupation. In 1999, when asked by a reporter if he thought people with low IQs should be paid to be sterilized, as Shockley had suggested, Gordon flippantly answered: "I think that's a rather shocking proposal! I would prefer not paying people. As far as black people are concerned. . . . I think it's right to give them the proper information in order for them to do that" (quoted in Papavasiliou 1999). He has defended the Pioneer Fund as one of "the last sources of private support that courageously operates at all in this intellectually taboo arena" (quoted in Kühl 1994, 192).

Gordon and many of his colleagues have attempted to turn the tables on their critics, accusing them of fascist tactics in their criticism of neoracism, the Pioneer Fund's Nazi connections, and the fund's past and current agenda (Kühl 1994, 8). This is a tactic often used by conservatives, such as the right-wing radio talk show host Rush Limbaugh. They often refer to liberals as Nazis in their attempt to deflect criticism of their own fascist, racist, and Nazi leanings, while defending racist ideology and policies. These policies lead to continuing and increasing racism and to massive separation of the rich and poor in the United States and other industrial nations.

Gottfredson is mainly interested in the relationship between IQ scores (which she believes measure a simple but biologically real unitary intelligence that is only slightly, if at all, influenced by environment), race (which she believes is a true biological entity), and occupational qualification (she believes that some races, especially African Americans, are unqualified for more skilled jobs). The following is one of her typical statements (Gottfredson 1997, 124–125):

Much social policy has much been based on the false presumption that there exist no stubborn or consequential differences in mental capability. . . . Educators routinely overpromise and schools, accordingly, consistently disappoint. Welfare reformers do not take seriously the possibility that today's labor market cannot or will not utilize all low-IQ individuals, no matter how motivated they may be. Civil rights advocates resolutely ignore the possibility that a distressingly high proportion of poor Black youth may be more disadvantaged today by low IQ than by racial discrimination, and thus they will realize few if any benefits (unlike their more able brethren) from ever-more aggressive affirmative action.

Gottfredson believes that socioeconomic inequality between races is the expected outcome of the lower intelligence of African Americans and that much current liberal social policy is based on the fraudulent claims of scientists who refuse to acknowledge that intellectual inferiority (Tucker 2002, 180). The statements quoted above from Gordon and Gottfredson were not written in the early part of this century but in the late 1990s. It is too bad that, like many of her colleagues, Gottfredson does not understand that there is no simple, unitary measure of intelligence; that measures of intelligence are greatly influenced by education and culture; and that almost all competent biologists, anthropologists, and geneticists now agree that biological races do not exist among humans.

Gottfredson wrote a letter that was published in the *Wall Street Journal* and signed by fifty-two scientists in defense of the volume *The Bell Curve* (Herrnstein and Murray 1994), which summarized the rhetoric of the new scientific racism. The letter was written after the book was criticized by a large group of competent scientists (see Jacoby and Glauberman 1995, for example). In this letter, the members of the new bigot brigade claimed that since as many as fifty-two scientists signed this letter, the contents of the book and of the letter must be true. Using this logic, since the vast majority of anthropologists and other social scientists and of geneticists do not agree with the conclusions of this volume, does that mean it must be false? Obviously not. However, it is the actual *science* that began with Boas and his colleagues and that has continued to this date that makes these authors' beliefs and approach untenable at any level.

Gottfredson also is a staunch apologist and supporter of the Pioneer Fund. She claims that this organization did not have connections to the Nazis in the past and that they are not supporting pure racists today. These claims are simply wrong.

Glayde Whitney (1939–2002), another Pioneer Fund grantee, was a geneticist at Florida State University who mainly conducted genetic research on taste sensitivity in mice. However, the Pioneer Fund became interested in him because of his blatant racism. Whitney essentially reasserted the views of Madison Grant. He claimed that just as "Pit Bulls raised by Cocker Spaniels grow up to be Pit Bulls," so "blacks will be blacks." No matter what their environmental circumstances, he believed, they display "evidence of maladjustment." No attempt to improve the cognitive skills or morals of African Americans would succeed. Completely ignorant of and naïve about any environmental or cultural factors, Whitney (1995) made simple statements such as: "We can do a pretty good job of predicting differential murder rates, simply by considering racial composition of the population. . . . The simple correlation between murder rate and percent of the population that is black, is $r = +0.77$." For Whitney and many other members of the new bigot brigade, correlation is equated with causation.

In 1995, Whitney used his keynote address as president of the Behavior Genetics Association to present the "data" that African Americans are genetically predisposed to criminality and that the association should focus its energy on demonstrating the genetic "roots" of the purported inferiority of African Americans. Many of the members walked out, and there was a call to expel Whitney from the association. However, in the end it was argued that this would be a violation of the principle of intellectual freedom. A number of members, including the incoming president, resigned in protest (Panofsky 2005). Whitney believed that the disinformation campaign of the "egalitarian priesthood" was attempting to keep the public ignorant of the facts about race and teach tolerance among races. Repeating nineteenth- and early twentieth-century dogma, Whitney claimed that black and white mixture leads to hybrid incompatibilities, disharmonious combinations, and a wide range of health problems. He related success in long-distance running by members of the Kalenjin tribe of Kenya to the fact that they had been naturally selected through their propensity for stealing cattle. The best cattle robbers in this polygamous tribe, he said, could afford to buy more wives and have more "little runners" (Tucker 2002, 181–182).

Whitney epitomized the Pioneer Fund's view of Jews. He believed that egalitarian conspiracies about equality between the races were generated by Jews. He thought that theories of racial equality were fakes that were similar to (what he believed was) the Jewish invention of the Holocaust. Whitney, a member of the Institute for Historical Review, a Holocaust denial organization, made the case for its views (Whitney 2002). By "leveling blacks and whites," Whitney claimed, Jews believed they could contribute to their own ascendency, "because if the whites could be convinced to accept blacks as equals, they could be convinced to accept anyone" (Whitney 1998). This latter quote comes from the forward to the autobiography of David Duke (Duke 1998), the former grand wizard of the Ku Klux Klan, which glorified Klan members and neo-Nazi activists. In his foreword, Whitney wrote that the book was "a painstakingly documented, academically excellent work of sociobiological-political history." Finally, in true fifteenth-century polygenic fashion, Whitney regurgitated the view that different races did not belong to the same species (Whitney 1998).

Michael Levin (1943–) is the fifth member of the new bigot brigade. He is a professor of philosophy at the City University of New York. He seems to me to be just an old-fashioned, isolated, prejudiced person who fears African Americans. His personal experiences have colored his intolerant and narrow worldview. In an interview in *Rolling Stone* in 1994, Levin stated that he had been "mugged so many times." (Always by blacks? He doesn't say.) "Blacks just have fewer inhibitions, a greater readiness to express anger, an impulsiveness.... What do they do that for? Because the alternative—to work and save—is not psychologically available" (quoted in Miller 1995, 166).

He believes "blackness is a sign of danger" and that given the rate of crime among African Americans, whites should fear and avoid blacks (Miller 1995, 162). Like his Pioneer Fund colleagues, Levin believes that intelligence is a unitary hereditary factor, that blacks are less intelligent than whites, and that this should drive policy recommendations on affirmative action, school segregation, housing policy, welfare reform, and criminal justice. He advocates such things as store owners refusing blacks entry into their stores, police searches of blacks in circumstances that are prohibited for whites, treatment of black underage offenders as adults, and making blacks ride in separate subway cars patrolled by police. He also believes that the standards in universities have been reduced by the presence of intellectually inferior black students. However, Levin's main focus is on black criminality, which he claims is related to the genetic incapacity of blacks to abide by white norms, and he proposes that even "free will" is correlated with race. Such traits as honesty, self-restraint, and cooperativeness "did not have the same evolutionary value in Africa that they did in Eurasia" (quoted in Tucker 2002, 187). Blacks just aren't as moral as whites, genetically, Levin argues. He holds that blacks have two unalterable characteristics: less intelligence and greater proneness to violence (Miller 1995). Thus, segregation is needed in housing, employment, and schools. Ultimately, Levin would have us eliminate affirmative action, welfare, and the civil rights laws.

Levin does no research in these areas and his views, like those of the Pioneer Fund members who support him, are based solely on accepted "truths" and pure prejudice, not on scientific facts. It is amazing to me that such people as Gordon and Levin could be so prejudiced, intolerant, and hateful. Both are of Jewish descent. Levin's grandparents fled Russia during the czar's pogroms against Jews, and members of his family were killed in the Holocaust. Gordon and Levin obviously do not understand how Jews fit into the history of polygenics, eugenics, Nazism, scientific racism, and the Pioneer Fund and how their new friends would treat them if they ever obtained political power. Being a descendent of Russian Jews myself, I cannot understand how anyone whose ancestors have been subjected to the horrors of prejudice could themselves be stricken by such ignorance and hateful intolerance. I consider these two neo-eugenicist Pioneer Fund recipients to be intolerant, hateful individuals who have missed the train of modern science.

The remaining two Pioneer Fund major grant awardees are J. Philippe Rushton and Richard Lynn. In fact, these are the all-stars of the new bigot brigade team. Rushton (1943–2012) was a professor of psychology at Western Ontario University. He was a longtime recipient of grants from the Pioneer Fund. Just as Jensen had done, in 1989 Rushton created his own nonprofit corporation, the Charles Darwin Research Institute, so that he could receive Pioneer Funds. In 2002, when Harry Weyher died, Rushton was appointed president of the Pioneer Fund, a position he held until his death in October 2012.

Rushton based all of his racist ideology on three major premises: that a genetically determined, simple IQ is real; that there are true biologically based races (in fact, three) among humans; and that certain races have higher IQs than others. However, Rushton is not a geneticist. As the world-renowned geneticist David Suzuki (1995, 281) explains: "To geneticists, classification based on skin color gives us groupings that are biologically meaningless. Besides, so long as society imposes such totally different social conditions and pressures on the basis of skin color, the cause basis for differences in IQ scores of blacks and whites can never be answered scientifically."

What does Rushton use as the scientific basis for his claim that three biological races exist? He relies on what he imagines a team of "extraterrestrial scientists" would identify if they were called upon to immediately subdivide the human population—presumably a classification based purely on skin color (Rushton 1995). Another justification for his racial classification is "common usage" (Rushton 1988, 1009). However, we now know that there is more genetic variation within than between races. For example, the genetic distance between Pygmies and Ethiopians is greater than that between Ethiopians and southern Europeans. Geneticists, biologists, and anthropologists now agree that biological races do not exist in the human population (UNESCO 1950; Montagu 1951, 1964, 1997; Livingstone 1962; Molnar 2010; Littlefield, Lieberman, and Reynolds 1982; Brace 1995, 2005; Jacoby and Glauberman 1995; American Association of Physical Anthropologists 1996; Harrison 1998; Templeton 1998, 2002, 2003, 2013; Graves 2001, 2002a, 2002b; Fish 2002, 2013; Marks 2002; Jablonski 2006; Edgar and Hunley 2009; Harrison 2010; Fuentes 2012; Tattersall and DeSalle 2011; Mukhopadhyay, Henze, and Moses 2014). As Graves (2002a, 132) states: "In reality [Rushton] is a spider spinning a pseudoscientific web of incorrectly stated hypotheses supported with dubious evidence. His ideas have not graduated beyond those of the anthropologists of the early 20th century."

Rushton's view of IQ is just as antiquated as his view of human races. Most modern scientists' view of IQ is similar to that reached by Otto Klineberg, who in the late 1920s concluded that environment alone and not heredity or selective migration accounted for the differences in IQ (Klineberg 1928, 1935a). Klineberg urged his fellow psychologists to use the concept of culture (Klineberg 1935b; Degler 1991). In 1944, Klineberg edited and contributed to a volume on the intellectual ability of African Americans. He concluded that the notion of ranking races on the basis of their mental or psychological characteristics lacked any scientific evidence and that competent researchers in the field generally agreed with that conclusion (Klineberg 1944; Farber 2011). More recently, Jefferson Fish (1999) has tried to convince his psychologist colleagues of much the same thing in his provocative article "Why Psychologists Should Learn Some Anthropology." In response to the Jensenism of the 1960s and 1970s, Brace, Gamble, and Bond (1971, 2) wrote that one

> gets the feeling that anthropology has been caught napping to a certain extent because it may have felt that the basic issues were resolved a generation and more ago. The insights of Tylor and Boas from before the turn of the century and their application to combat the racism of the Immigration Restriction

League, the eugenics movement, and the grisly rise of Hitler's "Aryan" theories have now retreated so far into the past that they have become practically a dead issue. [The anthropological view of culture] . . . has become so patently self-evident to anthropologists that one feels a little foolish at being put into the position of having to repeat it before an intelligent audience.

In 1974, psychologist Leon Kamin made a detailed examination of the empirical evidence on IQ test scores. He concluded: "There exists no data which should lead a prudent man to accept the hypothesis that IQ test scores are in any degree heritable" (1). More recently, in his preface to an edited volume summarizing current research on race and intelligence, Fish (2002, xii) summarizes the conclusions of the authors: "Access to schools, school quality, modes of instruction, attitudes toward formal education, and educational values vary cross-culturally. Biological-sounding concepts, especially inheritability, have been misused to imply a genetic basis for group differences in IQ scores. . . . A wide variety of data . . . imply that group differences in IQ are social in origin and can change as the result of changing social circumstances or social interventions."

However, Rushton ignores such scientific evidence and builds his imaginative theories on his false premises of biological races and biologically based IQ. One of his theories is based on a misuse and misunderstanding of an outdated model based on life history that seeks to explain radically differing reproductive and life history strategies among animal species: the r- and K-selection model. Rushton (1985, 441) summarizes this model in this way: "Oysters, producing 500 million eggs a year exemplify r-strategy, while the great apes, producing one infant every 5 years exemplify the K-strategy." In 1950, geneticist T. Dobzhansky (1950) proposed that certain aspects of habitat, lifestyle, and life history might be correlated and might be caused by differences in the evolutionary history of certain species. Biologists Robert MacArthur and E. O. Wilson (1967) developed this theory further. They predicted that r-selected species would develop in unstable environments and would maximize reproduction by producing a massive number of offspring with little parental care; their strategy was to flood the environment and "hope" that some survived. In contrast, K-selected species would be found in stable environments and produce fewer offspring and expend their energy in caring for them, enabling a much higher percentage of infants to survive. Biologist E. R. Pianka (1970) proposed a list of traits that presumably could be associated with the extreme poles of the r-to-K continuum.

However, upon further investigation, these predictions were shown to be wrong. Since the 1970s, biologists who study life history theory have proven the *r*- and *K*-selection theory to be false. Many biologists have shown that the premises of the theory do not hold, and the general theory is now considered virtually useless, even for studying vastly different species (Stearns 1976, 1977, 1983; Begon and Mortimer 1981; Boyce 1984; Weizmann et al. [1990] 1999). In fact, 50 percent of the between-species tests of the relationship between the characters proposed in the *r*-/*K*-selection theory fail where there is enough data for comparison. Applying this theory within species is completely indefensible (Stearns 1977, 1992, 206; Templeton 1983; Graves 2002a).

Rushton does the indefensible. He uses old racist stereotypes to develop a theory that posits a continuum of life history strategies *within* humans, in which "blacks" are considered to be *r*-selected, Orientals are considered to be the most *K*-selected, and "whites" are somewhere in the middle. He explains that *r*-selected individuals (blacks and poor people) have a stronger sex drive and are sexually more active, have less sexual restraint, have lower marital stability, have more multiple births, have higher rates of infant mortality, have more sexual partners and thus a higher risk of AIDS, have more children that are not well cared for, have a weaker work ethic, use less forward planning and personal restraint, mature sexually at an earlier age, have a shorter life span, have smaller brains and lower intelligence, are less law abiding, are more aggressive, have different attitudes toward the environment, are less altruistic, have a lower degree of social organization, use different economic and business practices, have poorer family cohesion, and have lower cultural achievements than whites. He believes the ordering of these characteristics are also mirrored in social and economic classes, with lower classes resembling blacks behaviorally and upper classes resembling whites and Orientals (Rushton and Bogaert 1988; Rushton 1999; but see Weizmann et al. [1990] 1999).

Rushton asserts that *r*-selected traits are related to the easy life of the tropics and that *K*-selected ones are found in the more difficult climates of the north, a completely ridiculous assertion that underscores his lack of understanding of evolutionary processes and data and ignores the paleontological literature (see Dobzhansky 1950; Pianka 1970; Stearns 1976, 1977, 1992; MacArthur and Wilson 1967; Weizmann et al. [1990] 1999; Graves 2002a, 2002b; Brace 2005). (It also is interesting that Rushton's theory reverses the relationships between habitat and life history traits that were proposed in the original MacArthur and Wilson theory.) Rushton's belief that "Negroids" are *r*-selected leads him to make such claims as parents in Africa "do not expect to be the major providers for their children" (Rushton 1995) and that Africans did not have stable political systems before colonization. These claims are based on a complete lack of knowledge and understanding of the massive anthropological and ethnographic literature and are "a classic example of racially motivated and ethnocentric ignorance" (Brace 2005, 259). It is a reversion to the mid-1700 "inductive" naturalistic philosophy theory of David Hume, though Hume did not have over 100 years of ethnographic and anthropological literature to ignore.

To believe that these traits are interrelated and based on an outmoded life history model based on *r*- and *K*-selection is nonsense. There is no scientific justification for Rushton's proposed interrelationships and no rational justification is ever given. As summarized by Weizmann et al. ([1990] 1999, 206): "In the end . . . there is neither any justification from the biological literature nor any strong theoretical justification for Rushton's ascription of r-selected or K-selected status to traits like altruism, criminality, and so on. What Rushton has done is to employ the terminology of r/K theory to justify a number of stereotypic beliefs, all the while ignoring not only the limitations of the r/K model, but frequently the model itself."

Brace (2005, 257–258) summarizes Rushton's documentation of *r*- and *K*-selected traits in humans as follows:

> In many instances, what he reports are simply assertions in the absence of data. "Facts" are cadged from a wide and disparate variety of sources with little concern for their documentation. Many of those sources are blatantly racist and unreliable. Some are misrepresented or selectively used without telling the reader that the authors in question have documented exactly the opposite of what Rushton uses them to support. The results of various different efforts are combined by "aggregation" in a manner that is statistically indefensible and masks within-group variation.

Geneticist Joseph Graves Jr. (2002a, 2002b) has done extensive work on life history theory and has summarized the flaws in Rushton's understanding of *r*- and *K*-selection theory. Further, he points out that over the years, reference to this *r/K* theory has essentially disappeared in legitimate biology journals: from 1977 to 1982 there were an average of forty-two references per year to the theory in the BIOSIS literature search service, but in the period 1984 to 1989, the average dropped to sixteen per year

(Stearns 1992). In 2001, there was only one reference to the theory (Graves 2002a). In 1992, major summaries of the criticisms of this theory were published by Stearns (1992) and Roff (1992), both pointing out that the theory of r-selection and K-selection no longer served any purpose in life history theory.

In 1997, Graves personally presented these and other rebuttals to this theory to Rushton. Yet in his 1999 abridged version of *Race, Evolution, and Behavior*, Rushton did not mention the fact that biologists rejected the theory nor did he discuss any of the criticisms of it by life history biologists (of which he is not one). He offers no justification for his use of this outmoded theory and makes no attempt to counter any of the specific criticisms. In response to this, Graves (2002a, 139) writes that "the absence of such a response only supports my view that Rushton does not understand life history theory. Thus he employs it incorrectly and through this error his work serves racist ideological agendas."

I agree and add that this is exactly what he and the Pioneer Fund want to accomplish: pure propaganda for race and racism and nothing to do with actual scientific reality. They present a scientific veneer for those who don't read or understand legitimate science and who want justification for their own racial prejudice. Graves (2002a, 147) concludes: "J. P. Rushton's view of human evolution suffers from the use of antiquated and simplistic theoretical models concerning life history evolution. In addition his methods of data analysis, results and data sources call into question the legitimacy of his research." As we have seen with many of his new colleagues, Rushton's theories follow in a direct historical line from past racist eugenicists such as Galton, Gobineau, Chamberlain, Shaler, Davenport, Laughlin, Fisher, Günther, and Grant. History does indeed repeat itself.

Another fantastic theory of Rushton's is his idea that there is a relationship between race, brain size, and penis size. As Rushton asserted in an interview: "You know it's a trade-off: more brain or more penis. You can't have everything" (quoted in Miller 1995, 170). Any theory that claims there is a relationship between intelligence and brain size, skull size, penis size, genetics, and a vast number of other sexual and intellectual behavioral characteristics is pure fantasy.

Rushton's information about penis size (length and thickness), angles of erection, and hardness is based on a paper written by a French "surgeon" who used the pseudonym Dr. Jacobus X (A French Army Surgeon 1896). Dr. X does not disclose how he obtained his information, and there are contradictions throughout his paper (see Weizmann et al. [1990] 1999).

This paper is not a scientific journal article but what can be regarded as a prolonged exercise in "ethnopornography" (Weizmann et al. [1990] 1999). However, Rushton (1988) describes the work as part of the ethnographic record and uses it as a major source for most of his comparative data on male and female genital size and shape. In fact, Weizmann et al. ([1990] 1999) point out that the anonymous author, Dr. X, gives his readers descriptions of a variety of sexual perversions along with his descriptions of human physiological traits. He gives details of "genitalia of varying size, shape, texture and color, and the strange sexual customs of a large number of 'semi-civilized' peoples" (Weizmann et al. [1990] 1999, 209). Dr. X even gives a recipe that includes eggplant and hot peppers that he claimed could be used to enlarge penis size.

This is a good example of the "scientific" literature Rushton uses. Rushton's assertions about the relationship between penis size and sexual appetite also is not supported by actual scientific data (Masters and Johnson 1966; Weizmann et al. [1990] 1999). Rushton also claims that white women have larger birth canals than black women, which enables them to have babies with larger brains, another statement that has no scientific justification. Within human populations, among normally healthy individuals, there is no correlation whatsoever between head size, brain size, cranial capacity, and intelligence (Herskovits 1930; Tobias 1970; Brace 1999, 2005; Beals, Smith, and Dodd 1984; Zuckerman and Brody 1988; Cain and Vanderwolf 1989; Gould 1996; Graves 2002a, 2002b; Weizmann et al. [1990] 1999).

Part of Rushton's suite of racist ideas is his belief (which Pearson shares), that racism has a biological basis. He claims that "xenophobia [is] . . . an innate trait in human beings" that enables them "to preserve 'purity' of the gene pool" (quoted in Tucker 2002, 179). Also, he believes that racial differences allow one to naturally distinguish "friend from foe" (Miller 1995, 171). He says such things as the statement that "the Nazi army was effective in battle in World War II because it was racially homogenous, while the U.S. army was ineffective in Vietnam because it was racially mixed" (Sautman 1995).

To his credit, and this is directly related to his lack of understanding of the power of culture, learned worldviews, and environment, Rushton states: "I guess my upbringing led me to believe there really were genetically based class, ethnic and racial differences" (Miller 1995, 170). Rushton was born and raised through the fourth grade under apartheid in South Africa. As Brace (2005, 256) points out: "Rushton's entire career has been devoted to providing proof in favor of his childhood prejudice."

In 1999, Rushton used Pioneer Fund resources given to his Charles Darwin Research Institute to print and distribute a paperback pamphlet "for the public" (Tucker 2002, 210). Rushton sent out tens of thousands of copies of an abridged edition of his book *Race, Evolution, and Behavior* to unsuspecting social scientists. The original book had been published in 1995. Of course most of the recipients did not want any part of this racist diatribe. Since Rushton used purchased mailing lists of professional organizations, many academics received more than one copy of this propaganda pamphlet; I received at least three copies. Some of the pernicious prose in the preface of this hate literature conveys the following views: from the eighth century to European colonialism and onward, visitors to Africa agreed that blacks were like "wild animals" naked, dirty, impoverished; the children often did not even know their fathers, though they had a natural sense of rhythm and oversized sex organs; and the IQs of blacks living in Africa are the lowest ever recorded and neglect and decay are seen everywhere in that nation. This could have come right out of the mouths of fifteenth- and sixteenth-century colonial racists or nineteenth-century proponents of slavery such as Josiah Nott or Louis Agassiz.

Rushton was able to get a full back-cover advertisement for this abridged version of his book in a major anthropological journal, *Evolutionary Anthropology*. He did this not with the editor's approval but by dealing directly with the vice-president of the publishing company, John Wiley & Sons Inc., who at first rejected the advertisement. The vice-president, however, was convinced by Rushton and by Irving Horowitz (the owner of Transaction Press) that the "racial pornographic" (as one pamphlet recipient called it) contents of Rushton's book were worthy of publicity in a respectable scientific journal (Horowitz 1995). Although Horowitz disagrees with much of what Rushton writes, he believed that it was correct for advertisements to be published in scientific journals because of the right of everyone to freedom of the press (Horowitz 1995, 197).

I disagree with Wiley & Sons and with Horowitz. Would a respectable scientific journal have published an advertisement for a book promulgating neo-Nazism or white supremacy or one that claimed the world was flat? Rushton's book is not science and should not be treated as such by respectable publishers. Other authors agree that Rushton's writing is not science. Attorney Steven Greene (1994, 75) wrote: "Nothing could be further from the truth. . . . The bulk of his controversial claims are published in journals that do not adhere to the standard procedures of review and criticism by scientific colleagues." Anthropologist C. Loring Brace said:

"*Race, Evolution, and Behavior* is an amalgamation of bad biology and inexcusable anthropology. It is not science but advocacy, and advocacy for the promotion of 'racialism.' . . . Quite evidently it is a manifestation of blatant bigotry" (Brace 1996, 176–177). Biologist David Barash (1995, 113) wrote: "Bad science and virulent racial prejudice drip like pus from nearly every page of this despicable book." Professor of psychology Richard M. Lerner (1992, 238) wrote, "Rushton's thinking, so redolent of Nazi-era political and scientific pronouncements . . . is nothing more than the most recent instance of genetic determinism promoted as science. . . . [It is] poor science and represents a fatally flawed basis for prescribing social policy."

This was the kind of thinking that fueled slavery, the American eugenics and immigration laws of the early twentieth century, and the deadly medicine and other horrific policies of the Third Reich and continues to fuel racial prejudice and hatred today. To my mind, it is simply propaganda for racism and racial hatred with no scientific validity or value whatsoever. It is an outdated worldview. Rushton did not distance himself from groups on the far right in the United States; he was a regular contributor to the newsletters of the group American Renaissance (see Chapter 11) and spoke at many of its biennial conferences.

At the time of the publication and mass distribution of Rushton's little booklet, I was editor of the flagship journal of the American Anthropological Association, *American Anthropologist*. I suspected that Rushton might use the same strategy with the *AA* as he had with *Evolutionary Anthropology*, attempting to publish an advertisement by going to the administration and not the editor or the professionals in the organization. In fact, he did just that, and I and the president of the organization, archeologist Jane Hill, had to threaten to resign our positions at the journal before the ill-informed executive director (who was not a professional anthropologist) was forced to withdraw the advertisement. Rushton complained that we were obstructing his freedom of speech. I replied in the *Anthropology Newsletter* of the AAA (Sussman 1998): "This is an insidious attempt to legitimize Rushton's racist propaganda and is tantamount to publishing ads for white supremacy and the neo-nazi party. If you have any question about the validity of the 'science' of Rushton's trash you should read any one of his articles and the many rebuttals by ashamed scientists."

Rushton's propaganda pamphlet was yet another example of the strategy and agenda the Pioneer Fund uses to promulgate their view of the world. There is no science here. To the members of the Pioneer Fund, the world is still flat and they want everyone to know it. Although the Pioneer Fund has supported some bad science, its main purpose since it was initiated by Wickliffe Draper has been essentially to serve as a lobby for eugenics, Nazism, and racist ideology. "In both its official and unofficial activities, Pioneer has been as much a pamphleteer as a source of support for scientists," Tucker (2002, 201) writes. The fund has been and continues to be driven by a simple-minded, ancient, and deep fear and hatred of human variety and multiculturalism. It invigorates and panders to a small but dangerous, ignorant, mostly lily-white audience. As long as the Pioneer Fund and its band of bigots can keep the "race" issue in the public press and as long as the press assumes that it is a real issue, the fund has succeeded in its goal. As Horowitz (1995, 180–181) states: "The media drives the data as much as the data drives the media. . . . Such foundations measure success as much by media coverage as by scientific results." As long as the media keeps these ideas alive and treats racial science as real science, the new bigot brigade remains very dangerous. As Reed (1995, 268) has written: "What makes this international vipers' nest so dangerous is that many of its members have maintained academic respectability." (I would put quotation marks around "respectability.")

Richard Lynn (1930–) is a professor emeritus of psychology at the University of Ulster, Ireland. He served on the board of directors of the Pioneer Fund and on the editorial board of *MQ* even before it was taken over by Pearson. When Rushton died in 2012, Lynn became co-director of the Pioneer Fund with Michelle Weyher, the widow of H. F. Weyher Jr., who served as Pioneer Fund president for forty years (1958–2002) (see Chapter 8). Lynn is a good old-fashioned eugenicist and is the scientific mentor of Arthur Jensen and J. Philippe Rushton (Sautman 1995).

In their book on race and IQ, *The Bell Curve*, Herrnstein and Murray rely on Lynn's work as the scientific proof for their claims; they describe him as "a leader and scholar of racial and ethnic differences" (Herrnstein and Murray 1994, xxv). Herrnstein and Murray accept the old eugenicists' assumptions that (1) there is a single, general measure of mental ability; and (2) IQ tests that measure this ability are not culturally biased. They then go on to relate IQ to race and class and draw "sweeping political conclusions about the meaning of class and race based on IQ scores" (Zenderland 1998, 350). However, they never discuss the underlying scientific basis for these claims (Jacoby and Glauberman 1995; Gould 1995; Kamin 1995). Furthermore, they continuously confuse correlation with cause. For example, they conclude that children of the working poor, like their parents before them, are born with poor genes. They conclude that most African Americans are children of poor parents and that they also have poor genes. They also claim that there is also a correlation between poverty and IQ and this is also related to poor genes. However, there is another possible explanation. The majority of poor laboring families and poor African Americans have few opportunities to be in environments that nurture the development of the skills needed to obtain high IQ scores, and they have even fewer opportunities to succeed economically. Herrnstein and Murray never consider this possible explanation because of their eugenic and biological deterministic bias and their complete lack of understanding of the influence that culture has on lifestyle (Kamin 1995).

Lynn is cited twenty-four times in this volume. The other source Herrnstein and Murray rely on heavily is the Pioneer Fund's *MQ* (Lane 1995; Linklater 1995). American psychologist Leon Kamin, who demonstrated that the methodology of Cyril Burt's work on twins was faulty, calls this book a "shame and disgrace" (Kamin 1995). After a detailed and careful review of the literature and research methods Lynn used, Kamin concludes that his "distortions and misrepresentations of the data constitute a truly venomous racism, combined with scandalous disregard for scientific objectivity. But to anyone familiar with Lynn's work and background, this comes as no surprise" (Kamin 1995, 86).

I am not going to go into Kamin's examination of Lynn's bad science here, but I would suggest that anyone who might question the quality of Lynn's research read Kamin's paper (see also Lane 1995; Benson 1995). Many of the authors cited in *The Bell Curve* were recipients of Pioneer Fund grants (Lane 1995). Kamin and other scientists have examined Herrnstein and Murray's highly popular book in *The Bell Curve Debate: History, Documents, Opinions* (Jacoby and Glauberman 1995; see also Gould 1994). I highly recommend this volume for readers who have any questions about the scientific validity of the new bigot brigade. Lynn's scholarship was most influenced by the works of Cyril Burt, Hans J. Eysenck, and Ray Cattell; this should be a clue to his regard for truth in science. The latter was an IQ tester of the same ilk as Goddard and Yerkes who claimed that "intelligence tests point to significant differences among the races" and that the black race "has contributed practically nothing to social progress and culture (except in rhythm)" (quoted in Kamin 1995, 99, 104). In referring to both Jensen's *Harvard Educational Review* article and *The Bell Curve*, Zenderland (1998, 350) states: "The strong parallels between political arguments produced by some hereditarian theorists in the latter half of the twentieth century and those used in its opening decade were hardly lost on modern critics."

Lynn reverts to nineteenth- and early twentieth-century eugenics and Nazi ideology, maintaining that the Nordic "race" is intellectually superior to other Caucasians and that they are more inherently suited for constitutional government than people from Alpine and Mediterranean countries (except for the Irish, who Lynn contends also have low IQs [Lynn 1978]). In the bigger picture, using simple-minded and outmoded racial categories and the similarly simple-minded idea that IQ is a unitary, easily measured, inherited trait, he believes that the IQ of Mongoloid people is slightly higher than that of Caucasoids, and that both are far superior to Negroids. "The Caucasoids and the Mongoloids [are] the two most intelligent races and the only races that have made any significant contribution to the development of civilization" (Lynn 1991a, 117). Lynn bases his views of racial intelligence on a theory that claims that, through evolution, the tropics presented an easier environment for early humans than the colder northern climates did and that this led to higher intelligence (and bigger heads) in Mongoloids and Caucasoids than in Negroids (Lynn 1987). The fact that he ranked Mongoloids slightly above his colleagues' beloved Nordics caused Lynn some problems with them, for example with Pearson. Of course, as with his new colleagues, Lynn assumes that environmental influences have no effect on intelligence and that the differences between blacks and whites are entirely genetic (Lynn 1994a). In fact, he claims, IQ is directly related to the degree of whiteness, because of admixture among the races. As for the intelligence of Jews, Lynn explains that this is a happy by-product of Darwinian evolution: during their "intermittent persecutions" the "more intelligent may have been able to foresee and escape" (Lynn 1991b, 123). Since head size is related to IQ in Lynn's fantasy, he also claims that males are more intelligent than females.

Returning to another old eugenic theme, Lynn claims that intelligence can be related to class differences (Lynn and Vanhanen 2002). Thus, he says, there is a correlation between wealth and IQ, both within countries and between different countries of the world and of course with race. He sees IQ differences among different nations as directly related to economic wealth and growth and to the degree of whiteness or brownness within the nations being compared. Just as in *The Bell Curve,* the "science" Lynn used to support his economic claims is completely bogus (see Brace 2005, 264–267, for an excellent review and rebuttal of Lynn and Vanhanen's racist tome) and, as Brace points out, Lynn's approach is "one of advocacy and not of science or scholarship" (263).

Lynn (2011) argued that the eugenicists of the early twentieth century correctly predicted the deterioration of Western civilization as a result of medical technology and charitable assistance to the poor. This allowed an underclass of genetically less intelligent and less moral people to persist. Thus Lynn (1972) has concluded: "What is called for here is not genocide, the killing off of the population of incompetent cultures. But we do need to think realistically in terms of the 'phasing out' of such peoples. . . . Evolutionary progress means the extinction of the less competent. To think otherwise is mere sentimentality." He has repeated this sentiment many times since (Lynn 2011). For example, adding to this quote, Lynn stated in 2004, "Survival of the fittest, extinction of the unfit. This is the way to a better world."

Lynn urges that immigration into England be completely eliminated but believes that this will not be possible as long as England remains a democratic country, which he thinks is a bad idea (Lynn 2001). In fact, he regards "individual rights and political freedom as 'dysgenic'" (Brace 2005, 253). In a recent online interview, one can see that Lynn's racist view of the world and of the "dangers" of immigration mirror those of Madison Grant's and are triggered by similar events, albeit decades apart. Just as Grant was shocked by the changes in Manhattan due to immigration in the early nineteenth century, Lynn responded to changes in the Britain he knew as a child in the 1940s (Kurtagic 2011):

> Up to around 1950, Britain was a very law abiding country. Crime rates were about 10 per cent of what they are today . . . and again up to around 1950, Britain was an all-white society. I do not remember ever seeing a non-European before this time. This began to change as a result of two developments. The first was the British Nationality Act of 1948, which conferred citizenship and the right to live in Britain on all members of the Commonwealth and Empire. . . . This act meant that huge numbers of non-Europeans—some 800 million—had the right to live and work in Britain. . . . The second development responsible for the transformation of Britain into a multiracial society was signing up to the 1951 Geneva Convention Relating to the Status of Refugees, which allowed entry of a large number of asylum seekers. The result of these two developments was that the number of non-Europeans (of all races) living in Britain recorded in the census of 1951 was 138,000. By 1971, that number had increased to 751,000, and in

the 2001 census it had increased 3,450,000. These non-European immigrants are almost entirely in cities, in a number of which they today comprise approaching half the inhabitants. . . . I am deeply pessimistic about the future of the European peoples because mass immigration of third world peoples will lead to these becoming majorities in the United States and western Europe during the present century. I think this will mean the destruction of European civilization in these countries.

Just as Grant and many earlier eugenicists were, Lynn is xenophobic and fears human variety, change, and multiculturalism; this cultural heritage has colored his worldview for the past forty years. Lynn is the perfect Pioneer Fund grant recipient and leader. He does very little science and the "science" he does is extremely poor. However, he is extremely prolific and thus spreads racist themes backed by a false and ill-begotten reputation of academic respectability. As his admiring interviewer commented in 2011, during the past fifteen years, Lynn has written numerous research papers and no less than seven "scientific books" and still has found time to do television and press interviews and speak at conferences overseas (Kurtagic 2011). You can witness some of Lynn's recent lectures on YouTube (NBC Nightly News 2011). Unfortunately and frighteningly, he has an audience of intolerant, hate-filled racists. Lynn's propaganda, backed by the Pioneer Fund, helps enable modern white supremacists and neo-Nazis to continue to perpetuate the racial hatred and intolerance that has persisted for over 500 years. As Brace (2005, 267) puts it: "It is evident that white supremacy is alive and well and anxious to extend its noxious influence into the new millennium."

11

Modern Racism and Anti-Immigration Policies

In the last couple of chapters, I stressed that the Pioneer Fund was implementing an agenda, essentially mirroring the Satterfield plan that it had unsuccessfully attempted to establish in Mississippi in the 1960s. The fund was determined to influence policy throughout the United States and in other Western countries. One major goal was to convince the public of the different nature of black (now often referred to more vaguely as brown) and white citizens and to convince them that brown inferiority is due mainly to inheritance and not to environmental factors.

Funding of Anti-Immigration Organizations and Right Wing Politics

Thus, beginning in the 1970s, the Pioneer Fund was quite successful in instigating and, to a great degree, underwriting a relatively unified plan to combine "scientific" research with public relations and then build on that to launch campaigns to propose laws that will advance its agenda. The major goals of this plan were (1) to provide "scientific" documentation of racial differences and award grants to educational institutions and individuals for research projects that were in line with that goal; (2) to widely disseminate results to the media; (3) to organize and subsidize its own periodical and other fronts in order to widely distribute more subtle propaganda; and (4) to support legislation and legal matters related to their racially motivated agenda. Though it had been unsuccessful in its attempts to reverse the Civil Rights Act and other civil rights laws in the 1960s, the Pioneer Fund continued its efforts to influence legal and policy matters. In recent years, as the eugenics movement had done in the 1920s, the fund has focused much of its energies and funding on immigration policies and legislation.

In the 1980s and 1990s, the Pioneer Fund began supporting organizations that would enable it to continue its racist propaganda campaigns. The main groups that have received this support are the New Century Foundation (mainly to support the foundation's major project, the American Renaissance Foundation), the Federation for American Immigration Reform (FAIR), and the American Immigration Control Foundation (AICF). These organizations work diligently to return the United States, Canada, and other Western European nations to past racist policies, such as laws that constrict the civil rights of certain individuals and groups, the restoration of segregationist and eugenics policies, and policies that would restrict the immigration of people who are not from Western Europe. They have become major lobbying groups in the United States and seem to have made some major inroads into the right wing of the Republican Party. The Pioneer Fund has had close relationships with several racist politicians, including U.S. senators Theodore Bilbo and Jesse Helms.

In the 1980s, for example, the Pioneer Fund served as a small part of "a multimillion dollar political empire of corporations, foundations, political action committees and ad hoc groups" (Edsall and Vise 1985, 1) developed by Tom Ellis, Harry Weyher, Marion Parrott, Carter Wrenn, and Jesse Helms. The fund was part of an interlocking set of associates that linked the Pioneer Fund to Jesse Helms's political machine. Ellis, for example, simultaneously served as chairman of the National Congressional Club and the Coalition for Freedom, co-founder of Fairness in Media, board member of the Educational Support Foundation, and director of the Pioneer Fund. We shall see that similar interlocking associations appear to be alive and well in today's political arena.

The American Renaissance Foundation

The American Renaissance Foundation is an extremely conservative right-wing organization that also publishes a monthly magazine of the same name, *American Renaissance (AR)*. The magazine's first issue appeared in November 1990 (American Renaissance 2014b). The foundation was established by Jared Taylor (1952–) who serves as president of the New Century Foundation and as editor of *AR*. Taylor has ties to a variety of domestic and international racists and extremists. He is on the editorial advisory board of *Citizens Informer*, the newspaper of the Council of Conservative Citizens, a virulently racist group whose website has referred to blacks as "a retrograde species of humanity" (Southern Poverty Law Center 2012a).

He has contributed writings to *The Occidental Quarterly,* a racist journal. He also has been a member of the board of directors of the National Policy Institute, a self-styled racist think tank, and has received funding from this institute (Anti-Defamation League 2012).

Taylor has close ties with members of various neo-Nazi groups and with Gordon Baum, the CEO of Council of Conservative Citizens. He is a frequent radio guest of Don Black's, operator of Stormfront, a white supremacist online forum that also advertises American Renaissance conferences. He also has ties to Mark Weber, head of the Institute for Historical Review. European racists are among his close associates, including members of the British National Party, a racist, far-right political party in England, and the National Front, a racist, far-right political party in France. Nick Griffin, the head of the British National Party, has been a speaker at two American Renaissance conferences. Frédéric Legrand, a member of the National Front, is a frequent contributor to *American Renaissance* (Anti-Defamation League 2012).

Jared Taylor has written and edited a number of books with racist themes (Taylor 1983, 1992a, 1998, 2011). One of *AR*'s and Taylor's favorite and oft-used quotes, recalling fifteenth-century polygenecist thought, is that of zoologist Raymond Hall from an early issue of *Mankind Quarterly* (Hall 1960, 118): "Two subspecies of the same species do not occur in the same geographic area." Taylor believes that slavery may have been wrong but the alternative was "Negro pandemonium." He complains that civil rights laws prohibiting racial discrimination have turned "common sense" into a "crime" (see Tucker 2002, 184–185). In 2005, he wrote in *AR:* "When blacks are left entirely to their own devices, Western civilization—any kind of civilization—disappears" (quoted in Southern Poverty Law Center 2012a). He argues that racial diversity is a negative for society and that racial integration has failed. Taylor promotes the main Pioneer Fund themes of a genetic basis for differences in intelligence between the races, the benefits of racial homogeneity, a propensity of African Americans to commit crimes at higher rates than whites, and the need to reverse an alleged reconquest of the American Southwest by Mexicans (Anti-Defamation League 2012).

Both the American Renaissance Foundation and its magazine are dedicated to the ideal that the United States is a white European nation (Tucker 2002). Its publications and biennial conferences state that the foundation's goal is to demonstrate the purported superiority of the white race and the threat nonwhite minorities pose to American society. "The stated purpose of the journal, from the outset, was to create 'a literate, unde-ceived journal of race, immigration and the decline of civility.' It held that 'for a nation to be a nation—and not just a crowd—it must consist of people that share the same culture, language, history and aspirations'" (Anti-Defamation League 2011). To the foundation, race is an essential ingredient for citizenship. "Blacks and Third World immigrants did not really belong in the United States and certainly could not be 'real' Ameri-cans" (Taylor 1992a, quoted in Tucker 2002, 182).

AR's list of some of its subscribers provides insight into the historical underpinnings and views of the American Renaissance Foundation. In-cluded on its list of "Americans Who Have Advanced White Interests" are Jared Taylor; David Duke; Robert E. Lee; Arthur Jensen; William Shock-ley; Wilmot Robertson, who viewed Hitler as defender of the white race; Revilo P. Oliver, who argued that Hitler should be recognized as "a semi-divine figure;" William Pierce, founder of the Nazi group National Alliance; George Wallace, past governor of Alabama; Madison Grant; and Theodore Bilbo. In a survey of subscribers, Adolf Hitler ranked first (by a large mar-gin) among "Foreigners Who Have Advanced White Interests." Hitler also ranked as first among "Foreigners Who Have Damaged White Interests," probably because his policies were a public relations disaster for the racist and anti-Semitic causes (Tucker 2002). Among the list of "Americans Who Have Damaged White Interests" is every U.S. president since 1932 except Gerald Ford and Ronald Reagan. The first four people on this list are Lyn-don Johnson, Franklin Roosevelt, William Clinton, and Abraham Lincoln (the list was compiled before Obama's presidency).

A major debate that has been extensively covered in the pages of *AR* is the question of what of two strategies whites should follow in order to preserve "racial survival" in the face of the "brown" threat (Tucker 2002). The first strategy is racial segregation and immigration restriction and, if that is not effective, to enter into negotiations with black and Hispanic nationalists to establish racially based nations within the United States (McCulloch 1995; Schiller 1995; Taylor 1998). The second alternative, as articulated by Samuel Francis, a regular contributor to *AR,* is to exercise the white man's "instinctual . . . proclivity to expand and conquer" and thus achieve a "reconquest of the United States." To accomplish this goal, Francis believes that the "phony" rights of nonwhites should be revoked. These include voting, holding office, attending schools with whites, serv-ing on juries, marrying across racial lines, serving in the armed forces, buying homes near whites, and eating at lunch counters with, riding on buses with, holding jobs with, or even associating with "superior" whites. Further, he proposes that we should seal the nation's borders and impose fertility controls on nonwhites. Francis also points out that the first solution of partitioning nonwhites would pose problems related to the definition of Eastern and southern Europeans, Jews, and even the Irish (see Francis 1995, 1999; Taylor 1999).

For Francis, the statement in the Declaration of Independence that "all men are created equal" was "one of the most dangerous sentences ever written, one of the major blunders of American History" (quoted in Tucker 2002, 184). So we see in the pages of the *AR* a reflection of the history of polygenics and racism, mirroring ideas that have persisted through the ages, beginning with the Spanish Inquisition, going through colonization, slavery, the eugenics movement, Nazism, and the fight against the civil rights movement.

The American Renaissance conferences, which began in 1994 (American Renaissance 2014a), have become a gathering place for white supremacists, white nationalists, white separatists, neo-Nazis, Ku Klux Klan members, Holocaust deniers, and eugenicists (Roddy 2005). It has become a place where many of the usual Pioneer Fund proponents can voice their vitriolic racial hatred to a mostly friendly audience. The conferences are open to the public and usually attended by 200–300 people. Frequent speakers at the conferences have included Richard Lynn, J. Philippe Rushton, Jared Taylor, and many Pioneer Fund grantees. Many of these amazingly racist lectures can be viewed on YouTube.

Many of the papers in *AR* and lectures at American Renaissance conferences are given by the same old Pioneer Fund people. Here are just a few examples, as pointed out by Tucker (2002). Arthur Jensen in an *AR* "conversation" stated that the country's attempt to build a multiracial nation "is doomed to failure." He also claimed that at least one-quarter of all blacks are "mentally retarded" and "not really educatable" (Taylor 1992b). Glayde Whitney, as a contributing editor of *AR*, wrote a regular column in which he suggested that different races did not belong in the same species. Lynn wrote in *AR* that environment has no effect on IQ scores and that differences between blacks and whites are entirely genetic (Lynn 1994c). In a lecture at an *AR* conference, Rushton said that race differences were nature's way of allowing people to distinguish friend from foe (Tucker 2002, 186). Michael Levin (1995) declared at a conference that whites dominate mankind in all of the important criteria for evaluating accomplishments in society, and in *AR* he wrote that the average black is "not as good a person as the average white." Levin also believed that whites would realize this and once again discriminate against blacks in housing, employment, and schools. Taylor (1997, 6) approved of Levin's predictions and speculated that his inspired vision would lead to "policies strikingly similar to those of the pre-civil rights American South." At the 2000 AR conference, Taylor was greeted with a burst of applause when he speculated that whites may have lost so much ground in the last century because of Jews (Tucker 2002, 187).

In 2010, the American Renaissance conference was scheduled to be held in Washington, DC, as usual. However, the conference had to be canceled because of protests by a number of groups over the racist content of the upcoming conference. Late in October 2010, American Renaissance was planning to hold the conference in February 2011 at an undisclosed location in Charlotte, NC. This was the first time in over a decade that the conference was not held in the Washington, DC, area, and the first time it was to be held in an "off" year for the biennial conference (American Renaissance 2014a). After it was discovered that the conference was to be held at the Airport Sheraton and the underlying theme of the conference became known, the hotel canceled the group's reservations. Other hotels in the area followed suit, also shutting their doors to the conference. Taylor was eventually forced to cancel the event. Instead, he held a session in another hotel where the planned speakers and a few spectators gathered to videotape the speeches (Morrill 2011).

Taylor complained that these cancellations were an obstruction to the right of free speech. However, nobody has to condone the ability or the right of hate groups to gather and incite racial and class hatred. Is it wrong to refuse to allow known terrorist groups to openly meet in public forums and advocate overthrowing our government and its laws? Public hotels are visited by individuals from a wide variety of ethnic groups, and they employ many "nonwhites." They have the right to protect these people from insult and injury by those who are openly advocating their mistreatment. I would hope that most public venues would not allow a Nazi rally in their facilities. The history of the Pioneer Fund and the American Renaissance Foundation shows that there is little difference between the ideals and goals of these organizations and those of racial hate groups that have caused misery throughout the centuries. Modern science now undermines all of their basic premises, and there is no reason to tolerate their hateful, dangerous, ancient, and outdated assertions. If just one racially motivated hate crime is prevented by depriving attendees of the stimulus these conferences provide for some of its more radical and deranged followers, then we have ample reason to close them down.

In January 2011, Congresswoman Gabrielle Giffords was shot, six other people were killed, and thirteen were wounded while attending a political rally in Arizona. Giffords is a moderate Democrat who had been narrowly reelected against a right-wing, Tea Party candidate a few months earlier. Giffords also was the first Jewish female to be elected to such a high position in Arizona. Among those killed was Judge John Roll, an ally of Giffords in dealing with the controversial issue of illegal immigration, a very hot topic in Arizona. The shooter was 22-year-old Jared Lee Loughner, a Tucson native. *Mein Kampf* was among his favorite books, and a Department of Homeland Security memo stated that Mr. Loughner was "possibly linked" to American Renaissance, which was presumably mentioned in some of his Internet postings; he allegedly posted links to American Renaissance.com on his blog and other websites. It appears that federal law enforcement officials pored over versions of a MySpace page that belonged to Loughner and over a YouTube video published just weeks before the shooting under an account linked to him. However, the MySpace page was removed minutes after the gunman was identified by officials (NC-ISAAC 2011; Jonsson 2011).

Jared Taylor denied any connection between Loughner and American Renaissance, and no connection was ever proven. However, when a group publishes a journal and holds conferences preaching pro-Nazi, anti-Semitic, xenophobic, and racial hatred and glorifies such hatemongers as Adolf Hitler, Jesse Helms, and David Duke, it should not be at all surprised when some people that are susceptible to this hatred (and love permissive gun laws) become active, racially motivated terrorists. History shows that if you justify, condone, and preach racial hatred, racial discrimination and racist-driven terrorism follows.

In February 2012, Stormfront advertised the next American Renaissance conference on its website (along with an ad for the latest David Duke video). Stormfront has been described as the Internet's first major hate site: The conference, entitled Defending the West, was scheduled to be held in March 2012 at an undisclosed location near Nashville, Tennessee. And indeed, the conference was held on March 16–18 on state property, at the Montgomery Bell Park Inn, which is in a state park. The location had been kept secret in anticipation of protests. Presumably, the event went smoothly; the American Renaissance website says that it took place "with the complete cooperation of the state authorities and without a demonstrator in sight" (Wolff 2014). Apparently, however, some did protest at the entrance to the park, out of the sight of participants.

Welcoming remarks were, as usual, given by Jared Taylor. Among others, Richard Lynn and J. Philippe Rushton were scheduled featured speakers, though it appears that Rushton did not attend. (Rushton died of cancer in early October 2012 [Terry 2012].) The first speaker before the small audience of 150 was Robert Weissberg, an emeritus professor of the University of Illinois at Urbana-Champaign. He stated that "racial nationalism is intuitive and written in our genes" and that "even children are conscious of race" (quoted in Wolff 2014). Henry Wolff (2014), an American Renaissance member, summarizes his presentation as follows:

> [Prof. Weissberg proposed] "A Politically Viable Alternative to White Nationalism" [instead of proposing the complete elimination of nonwhites]. . . . [He] argued that an "80 percent solution" would be one that enforced the "First-World" standards of excellence and hard work that attract and reward whites. He pointed out that there are still many "Whitopias" in America and that there are many ways to keep them white, such as zoning that requires large houses, and a cultural ambiance or classical music and refined demeanor that repels undesirables. This approach to maintaining whiteness has the advantage that people can make a living catering to whites in their enclaves. . . . In answer to questions about the adequacy of his "enclave" solution for poor whites who cannot afford to live in them, Prof. Weissberg expressed the hope that less financially successful whites could draw on their sturdy, warrior heritage to protect their own enclaves.

Another speaker, Donald Templer, a retired professor of psychology from Alliant International University, spoke on the "geographical distribution of intelligence." His presentation repeated age-old eugenics claims that sounded very much like Nazi proclamations. According to Henry Wolff (2014):

> Professor Templer spoke at length about the dysgenic effects of social policy. He noted that 70 percent of white women on welfare in the United States have IQs of less than 90, and that only 5 percent have IQs over 110. Voluntary, compensated sterilization for welfare recipients would be cost saving, and prevent many burdens on society.
>
> Professor Templer pointed out that prison inmates are of below average intelligence, and that permitting conjugal visits is dysgenic in that it allows prisoners to reproduce . . . with the result that society must support genetically disadvantaged children.

[Professor Templer stated that] the average IQ in Arab countries is 84, so it is foolish to expect these countries to establish democratic regimes. . . . Prof. Templer wondered whether the flowering of ancient Middle East cultures must have required a higher average intelligence than that of to-day's Arabs. Declines could have come about through mixture with African slaves, and the emphasis of Islam on rote memorization may have rewarded plodders rather than creative thinkers. Prof. Templer closed with a call for a recognition of reality rather than the current fetish of egalitarian fantasy.

American Renaissance glowingly describes Lynn's presentation, which evidenced a strong adherence to the ideology of Galton, Gobineau, and Madison Grant:

Richard Lynn, the author of many by-now standard works on eugenics, dysgenics, and race differences in intelligence, spoke about the world-wide problem of declining intelligence. He briefly summarized the field of eugenics since Francis Galton coined the term in 1883, noting that the West has now had six generations of dysgenic fertility that has led to a decline in genetic IQ of about one point per generation. . . . Prof. Lynn warned that there is a global dysgenic effect, because the high-IQ peoples of Europe and East Asia have sub-replacement fertility while sub-Saharan Africans have six to eight children. . . . Eugenics is the obvious solution. . . . "We in the West," as Prof. Lynn argued, "have become too nice." We do not have the will to stop dysgenic reproduction or dysgenic immigration, which leads Prof. Lynn to predict that "the torch of civilization will pass from the Europeans to the Chinese."

The website goes on:

As he always has, since the very first AR conference in 1994, attorney Sam G. Dickson brought the conference to a close. He began with a new retelling of the story of the emperor's new clothes. Even a child, he noted, can see the foolishness of multi-culturalism and forced mixing of the races. . . . Mr. Dickson closed with a description of what can be the only guarantor of our survival as a people and culture: the ethnostate. . . . Here "the health of the people will be the supreme law," and if, today, whites are the problem, in our own state whites would be the solution. Only in an ethnostate, he said, can there be true tolerance, because when people are all of one stock, naturally occurring diversity poses no threats and truly can be celebrated.

The report of the conference concludes that "with this invigorating benediction we returned to our homes, encouraged and inspired to fight on."

And fifteenth-century polygenics lives on, without any benefit of modern science. Racial bigotry, hatred, and intolerance are alive and well. As I read the proud description of the conference proceedings on the American Renaissance Foundation website, I was struck by how it is so reminiscent of the literature of nineteenth-century proponents of slavery, the early twentieth-century eugenicists, and the proponents of Nazism. The time warp is remarkable; to most rational people, it would be hard to believe, indeed shocking. One could almost read the description of the conference and think it was a joke, a satire. However, it was no joke; it was real. It is laughable, but it is not at all funny.

The American Immigration Control Foundation (AICF) and the Federation for American Immigration Reform (FAIR)

As I have noted, one of the major goals of the eugenics movement that came out of the First International Eugenics Congress of 1912 was limiting the immigration of various ethnic and racial groups. In 1922, Madison Grant, Charles Davenport, and Henry Fairfield Osborn established an advisory council of individuals to protect America "against indiscriminant immigration, criminal degenerates, and . . . race suicide" (Spiro 2009, 181). Madison Grant had been stimulated by his fear of the immigrants who were streaming into New York City at the turn of the twentieth century. This fear and hatred was a major stimulus of his writing of *The Passing of the Great Race* in 1916.

Madison Grant later became vice-president of the Immigration Restriction League. Harry Laughlin, another member of the big four of American eugenics, had a major hand in writing immigration laws and putting them into eugenic terms. Davenport kept Grant informed about eugenicists' research on immigration. Thus, as discussed earlier, the American eugenics movement was instrumental in the passing of the Immigration Act of 1924. The polygenic eugenicists inspired and influenced immigration laws, stimulated by economic and social factors and based on racially motivated fear, hatred, and prejudice. They ultimately had a direct influence on the death of millions of Europeans in the early decades of the twentieth century.

The Pioneer Fund and its new eugenics bigot brigade have once again become involved in immigration reform. Two of the major recipients of Pioneer Fund grants and funding from members of the Pioneer Fund in the past thirty years have been the American Immigration Control Foundation (AICF) and the Federation for American Immigration Reform (FAIR). The AICF has received less funding from the Pioneer Fund than FAIR, but the amounts it did receive were more significant to the group than they were to FAIR. This is because FAIR received substantial financial support from multimillionaire Cordelia Scaife May, a frequent donor to right-wing causes and a close friend of John Tanton, the founder and head of FAIR (Tucker 2002). FAIR also assisted the AICF behind the scenes (Crawford 1993).

In a mailing in 2000, the AICF listed immigrants as a major cause of high taxes, wasted welfare dollars, lost jobs, the high costs of education, and rising crime. It further claimed that immigrants drive up health care costs by getting free care and that they bring disease into the United States (Brugge 2010). The success of this anti-immigrant campaign is attributable to its ability to deflect anger about any problems in the economy, environment, or cultural situation of the United States onto the scapegoat of immigrants. The new eugenicists also can attract right-wing individuals and politicians by exploiting this popular issue. Those responsible for these publications claim that citizenship in the United States should be based on race and that ethnocentrism and xenophobia are part of human nature through natural selection. Nonwhites (which they often lump together as "browns") thus naturally create ethnic conflict, and the resulting mixed population is the main cause of America's decline (Elmer and Elmer 1988).

In a monograph the AICF published, John Vinson (1997), the president of the organization and editor of its journal, *Immigration Watch,* claimed that racial separation was part of God's plan and that multiculturalism violated biblical principles. Vinson believes that America is in danger, morally, spiritually, and economically, because certain racial groups and the descendants of certain earlier immigrants have attained influential positions in the fields of entertainment, media, education, business, and politics. The AICF makes no pretense about the fact that its views are predicated on eugenics (Tucker 2002). Over the years, the AICF has produced a number of pamphlets and books claiming that immigrants are the cause of crime, disease, an increase in welfare, and competition.

However, today the main focus of prejudice and hatred within anti-immigration organizations is Hispanics and Asians. In 1900, 85 percent of immigrants into the United States came from Europe and only 2.5 percent came from Latin America and Asia combined. By 1990, two-thirds of all immigrants were Latino and Asian (Lapham, Montgomery, and Niner

1993; Brugge 2010). These recent immigrants have tended to settle in five or six states, resulting in areas where people of color have become a majority, exacerbating racial tensions. The population of Hispanics in the United States is projected to reach 96 million by the mid-twenty-first century. The U.S. Census Bureau projects that by 2050, people of color will make up around half the U.S. population (Population Projection Program 2000). This, of course, is a very scary idea for racists and right-wing politicians, and it may be a major factor that drives some of the repressive policies of the right wing of the Republican Party. As Dan Stein, the current president of FAIR, stated on a 1996 television program produced by that organization: "How can we preserve America if it becomes 50 percent Latin American?" FAIR has been listed as a U.S. hate group by the Southern Poverty Law Center since 2007.

In the United States, the majority of wealth is owned by a very few individuals. As of 2010, the top 1 percent of households (the upper class) owned 35.4 percent of all private wealth; the next 19 percent owned 53.5 percent of the wealth. Thus, just 20 percent of the households owned a remarkable 89 percent of private wealth in the United States, leaving only 11 percent of the wealth for the bottom 80 percent of the population (generally speaking, those who work for wages or salaries). In terms of financial wealth (that is, total net worth minus the value of a home), the top 1 percent of households had an even greater share: 42.1 percent (Wolff 2010, 2012; Domhoff 2013). These figures became much more skewed with the housing bubble crash in the first decade of the twenty-first century. In fact, even by 1999, the top 1 percent of households in the United States owned more wealth than the entire bottom 95 percent (Collins, Leondar-Wright, and Sklar 1999). And the gap continues to widen. Between 1990 and 2005, the pay of CEOs increased around 300 percent, whereas the wages of production workers increased by only 4.3 percent (adjusted for inflation) (Domhoff 2013). Complicating these figures, in 2010, the average white household had almost twenty times as much total wealth as the average African American household and more than seventy times as much as the average Latino household. Furthermore, close to 100 percent of the wealth of African Americans and Hispanics was in the form of their principal residence (Domhoff 2013).

As Doug Brugge (2010) emphasizes:

Displaced workers, along with others who fear for their livelihood, are fertile ground in which to sow anti-immigrant sentiment, since angry and frustrated people often seek some target on which to blame their problems. The right wing has organized and manipulated such anger and resentment, turned it away from corporations, and directed it against the government, decrying high taxes and the inability of the state to solve problems such as social deterioration, homelessness, crime, and violence. In addition to the target of "failed liberal policies," immigrants make a convenient and tangible target for people's anger. Racial prejudice is often an encoded part of the message. . . . Right-wing populist themes are particularly effective at attracting working people disenchanted with the system.

In the mid-1990s, opposition to immigration became a major part of the Republican Party agenda. Immigrants, especially those who came from Mexico, became an enemy in the party's electoral strategy (Brugge 2010). In California, Proposition 187 was closely linked to the gubernatorial election of 1994. The initiative sought to establish a state-run citizenship screening system and prohibit illegal aliens from accessing health care, attending public schools, and participating in other social services. The proposition, which passed by a wide margin, mandated that teachers, doctors, social workers, and police check the immigration status of all persons seeking access to public education and health services from publicly funded agencies and denied these services to those in the United States without documentation. Fortunately, many of the provisions died as the result of court challenges (Southern Poverty Law Center 2012b), and a federal court in California ruled the law unconstitutional (McDonnell 1997).

FAIR endorsed the proposition and was linked to the issue by Alan Nelson, the author of Proposition 187. Nelson, who had been the director of the U.S. Immigration and Naturalization Service under President Reagan, had written anti-immigrant legislation in California for FAIR before writing Proposition 187. In fact, much of the work behind Proposition 187 had been done earlier in support of Proposition 63, a referendum to make English the official language of California. This was sponsored and mainly financed by U.S. English, an advocacy group led by FAIR. After the passage of Proposition 187, discrimination against Hispanics became rampant in California.

At the national level, after the 1996 national elections, the Republican-controlled Congress responded to the economic downturn and the heightened sentiment against immigrants. Rallying behind Newt Gingrich's "Contract with America" (shrinking the size of government, promoting lower taxes and greater entrepreneurial activity, and both tort reform and welfare reform; which I prefer to call "Contract on Americans"), Congress took up the issue of immigration and over the course of five months passed three laws that seriously diminished immigrant rights (Brugge 2010): (1) the Anti-Terrorism and Effective Death Penalty Act, enacted in April 1996, resulted in the indefinite detention of deportees from particular countries, the long-term detention of some asylum seekers, the deportation of legal immigrants for nonviolent offenses (that had sometimes taken place years ago), and a dramatic increase in the incarceration of immigrants; (2) the Welfare Reform Act, enacted in August 1996, which made citizenship an eligibility requirement for a wide range of federal benefits, allowed states to discriminate against noncitizen immigrants in their administration of the Temporary Assistance to Needy Families and Medicaid programs, and greatly reduced eligibility for Supplemental Security Income and food stamps; (3) the Illegal Immigration Reform and Immigrant Responsibility Act, enacted in September 1996, which sought to decrease "illegal" immigration. It increased border patrols, it increased penalties for crimes committed by an immigrant, it increased the earning requirements for individuals who are legally sponsoring new immigrants, and it required sponsors to sign legally binding affidavits of support. In the 2012 presidential campaign, immigration became a major topic in the Republican Party primary. Gingrich, who had run an advertisement calling Spanish "the language of the ghetto" during his reelection campaign, promised to control the border by January 1, 2012, though overall his stance on immigration was more liberal than that of the actual Republican Party presidential candidate, Mitt Romney (CNN 2012).

After the September 11, 2001, attacks on the Pentagon and the World Trade Center, the USA PATRIOT Act empowered the U.S. attorney general to detain and deport noncitizens with very little judicial review. President George W. Bush also signed an executive order that allowed secret military tribunals to try noncitizens, including U.S. residents, a law that did not meet constitutional standards. In December 2001, Attorney General Ashcroft told the Senate Judiciary Committee that in his view, citizens who criticize these measures are aiding terrorists and giving ammunition to America's enemies.

Shying away slightly from the overly repressive immigration laws, in 2007, President Bush backed a bipartisan bill for immigration reform, the Comprehensive Immigration Reform Act, that would have granted amnesty to undocumented immigrants. However, many of the anti-immigrant organizations that had been built up over the past twenty years (including

AICF and FAIR) led a drive that inundated Congress with e-mails, phone calls, and faxes to protest the bill. The campaign was so intense that it shut down the congressional phone system. Congress never voted on the bill. The majority of these organizations had been founded and funded by John Tanton (1934–), a retired Michigan ophthalmologist.

John Tanton, Founder of FAIR

For the past thirty years, John Tanton, the founder of FAIR, has been a major architect of the contemporary nativist movement. He has been at the heart of the white nationalist movement (Beirich 2008). FAIR, which was founded in 1979, is the most important contemporary organization fueling anti-immigration. It is mainly, but not exclusively, opposed to Hispanics and Catholics. FAIR is noted for its mix of bigotry and racist and eugenics attitudes. Many of its employees are members of white supremacist groups, and it has accepted over $1 million from the Pioneer Fund. Yet even given this widely recognized racist background, FAIR is still able to spread its propaganda in the popular media and have a relatively frequent voice in Congress. For example, in 2008, FAIR was quoted nearly 500 times in the mainstream media and its staff was featured several times on CNN and on other television news shows. On its website, FAIR claims to have testified in Congress on immigration bills "more than any other organization in America" (Southern Poverty Law Center 2009a). Since 2000 it has made dozens of appearances before Congress. On its current website, FAIR lists endorsements of a number of Republican congressmen.

As Henry Fernandez, a senior fellow and expert on immigration at the Center for American Progress, states: "The sad fact is that attempts to reform our immigration system are being sabotaged by organizations fueled by hate. Many anti-immigration leaders have backgrounds that should disqualify them from even participating in mainstream debate, yet the American press quotes them without ever noting their bizarre and often racist beliefs" (quoted in Southern Poverty Law Center 2009a). Although Tanton has claimed that he is not a racist, there are many indications that belie that claim, including evidence in Tanton's papers and letters, which are housed at the Bentley Historical Library at the University of Michigan (Beirich 2008).

Tanton has been the main personality behind FAIR and a number of other anti-Latino and white supremacist groups for decades, including U.S. Inc., The Social Contract Press, and FAIR's Research and Publications

Committee. As FAIR's annual report for 2004 claimed, Tanton's "visionary qualities have not waned one bit" (Southern Poverty Law Center 2012b) over the years. What is Tanton's vision? In 1986, in an internal document, Tanton warned colleagues and staff of a coming "Latin onslaught." He worried that given their high birth rate, Hispanics would take over political power by "simply being more fertile." He also questioned the morals and educability of Hispanics. Tanton had kept his racism under cover until a memo was leaked in 1988. After that, Arnold Schwarzenegger and Walter Cronkite, seemingly previously unaware of the racist attitudes of Tanton and his organization, resigned from U.S. English, another group Tanton headed. U.S. English Director Linda Chavez, a former Reagan administration official, a staunch conservative, and a strong opponent of affirmative action, also left the organization, stating that Tanton's remarks were "repugnant and not excusable" and were "anti-Catholic and anti-Hispanic" (Southern Poverty Law Center 2012b).

In 1995, Tanton's Social Contract Press republished, and the AICF distributed, a virulently racist French book, *The Camp of the Saints* by Jean Raspail. The book describes the takeover of France by "swarthy hordes" of immigrants who arrive on dilapidated ships from India. As William H. Tucker (2002, 191–192) notes:

> As if the author had systematically rooted through the thesaurus searching out every possible synonym for "repulsive," almost every page included a reminder of how vile was the cargo carried by these ships. Led by a "turd eater," they were "vermin" with "grotesque misshapen bodies," and "horrid twisted limbs," contaminated with "some gangrenous rot," "complete with [an] assortment of pus, scab, and chancre," a horde of "ignorant Ganges monsters, misery incarnate, absolute zeros," "a sweating, starving mass, stewing in urine and noxious gases," "sleeping in dung and debris," turning the ocean into "one big, festering sore," and creating a "horrible stench . . ." [These new arrivals were awaited in the West by their racial brethren, the] "kinky-haired, swarthy-skinned, long-despised phantoms; all the teeming ants toiling for the white man's comfort; all the swill men and sweepers, the troglodytes, the stinking drudges, the swivel-hipped menials . . . the lung-spewing hackers; all the numberless, nameless, tortured, tormented, indispensable mass" that performed the essential tasks too unpleasant for Westerners to do for themselves.

Yet it was neither these new immigrants nor their waiting brethren who were the true villains in Raspail's tome. The true villains were the white liberals who were "sucked in by all that brotherhood crap," who "preach universal mongrelization" and "empty out all our hospital beds so that cholera-ridden and leprous wretches could sprawl between our clean white sheets . . . and cram our brightest, cheeriest nurseries full of monster children."

Tanton wrote that he was "honored" to republish the novel and called it "prescient." He stated that "we are indebted to Jean Raspail for his insights into the human condition, and for being twenty years ahead of his time. History will judge him more kindly than have some of his contemporaries" (quoted in Southern Poverty Law Center 2009a). Linda Chavez called the volume "without doubt the most vehemently racist book I have ever read" (quoted in Tucker 2002, 193). In an afterword to the republished edition of *The Camp* (Raspail [1975] 1995), Raspail wrote that he feared that "the proliferation of other races doomed our race, my race, to extinction." (quoted in American Renaissance 1995). Tanton, mirroring Raspail's sentiments, voiced the opinion that once whites became racially conscious, there would be a "war of each against all" (quoted in Southern Poverty Law Center 2009a).

Letters housed at the Bentley Historical Library indicate that Tanton practically worshipped John B. Trevor Sr., a principal architect of the Immigration Restriction Act of 1924 and a distributor of pro-Nazi propaganda. Trevor Sr. had warned shrilly of the "diabolical Jewish control" of America (Beirich 2008; see Chapter 8). Tanton, who defends the Immigration Restriction Act, arranged for the Bentley Historical Library to house the papers of Trevor Sr. and his son, a longtime Pioneer Fund board member and a close friend of Tanton until his death in 2006. In a letter to Donald Collins (see below) in 2001, Tanton wrote that the work of Trevor should serve FAIR as "a guidepost to what we must follow again this time" (quoted in Southern Policy Law Center 2012b). In fact, one of the main goals of FAIR appears to be reversing the Immigration and Nationality Act, signed by President Johnson in 1965, which ended the racist quotas established by the 1924 act that limited immigration mainly to northern Europeans. FAIR president Dan Stein, along with other white nationalists, believes that the 1965 act is a disaster for Western civilization and Anglo-Saxon dominance and calls it "a key mistake in national policy" (quoted in Southern Poverty Law Center 2012b).

Tanton's letters also reveal his interest in eugenics. For example, he inquired in 1969 if there were still laws allowing forced sterilization in Michigan. Thirty years later, he worried about "less intelligent" people being allowed to have children and opined that modern medicine and social programs are eroding the human gene pool (Beirich 2008). In 1993, he wrote to fellow eugenicist Garrett Hardin: "I've come to the point of view that for European-American society to persist requires a European-American majority, and a clear one at that" (quoted in Beirich 2008). In a 1996 letter to Roy Beck, head of the anti-immigration group NumbersUSA, he wondered whether minorities, who currently make up 85 percent of schoolchildren in the lower grades, could run an advanced society.

Tanton's letters show that Jared Taylor was one of his closest friends. Shortly after Taylor began publishing *AR*, Tanton expressed approval of the opposition to affirmative action Taylor had expressed in his book *Paved with Good Intentions: The Failure of Race Relations in Contemporary America* (1992a). In 1993, Tanton wrote to Taylor and three of his *AR* colleagues, including Wayne Lutton (see below) and Samuel Francis, suggesting that they go after Stanley Fish (1993), the well-known literary critic and academic who had recently defended affirmative action in *The Atlantic*. Throughout the years Tanton has promoted Taylor and American Renaissance. In 1991, he wrote a letter to Pioneer Fund president Harry Weyher extolling *AR*'s efforts. In 1998, he wrote to a number of FAIR employees encouraging them to receive *AR* so they could keep track of "those *on our same side*" (Tanton's emphasis). Tanton also corresponded for years with Samuel Francis, until his death in 2005. Tanton was a houseguest of and corresponded with Samuel Dickson, a Georgia lawyer who has represented the Ku Klux Klan and who has written for and serves on the editorial board of Holocaust denial publications. As mentioned earlier in this chapter, Dickson has given the closing remarks at the American Renaissance conferences since their inception in 1994 (including the 2012 conference). Tanton's letters have expressed his anti-Semitism and his problems with the Roman Catholic Church and several Protestant denominations, especially the Lutheran Church, because they see themselves as universal and transcending national boundaries (Beirich 2008).

Tanton is surrounded by a group of employees and colleagues with similar racist views. Wayne Lutton is a longtime deputy of Tanton's who works at his Petoskey, Michigan, office and edits the journal *The Social Contract,* published by Tanton's press. Lutton is a white supremacist who holds leadership positions in a number of white nationalist hate groups and at a number of publications of such groups, including *American Renaissance*. Under

Lutton's editorship, *The Social Contract* has published articles by many white supremacists. The lead article for a special issue on "Europhobia" was written by John Vinson, president of the AICF. In the article, Vinson argues that multiculturalism is replacing "Euro-American culture" with "dysfunctional Third World cultures" (quoted in Southern Poverty Law Center 2009b). Tanton also contributed to this issue in an article that blamed unwarranted hatred and fear of whites on multiculturalists and immigrants (Southern Poverty Law Center 2009b). FAIR still advertises the journal on its website, and Tanton serves on its editorial board.

By 1994, FAIR had received around $1.2 million from the Pioneer Fund. However, after some negative publicity in 1994, FAIR stopped openly soliciting Pioneer Fund donations. This reluctance to receive funding from the Pioneer Fund was short lived. In 1997, Tanton met with Weyher and a number of FAIR board members to discuss possible fund-raising efforts for FAIR, and in 1998 Tanton wrote a letter to John Trevor Jr., a member of the Pioneer Fund board, thanking him for his handsome contribution to FAIR. In 2001, Tanton affirmed that he is quite "comfortable being in the company of other Pioneer Fund grantees" (quoted in Southern Poverty Law Center 2009a). FAIR is no longer reticent about the Pioneer Fund and defends it on its website (Southern Poverty Law Center 2011).

Other Tanton and FAIR Connections

One of Tanton's heroes is the late Garrett Hardin. Hardin, also a recipient of Pioneer Fund support, feared that freedom to breed would bring ruin to all. He argued that the world was being filled with the next generation of breeders, who needed to be stopped, and that providing aid to starving Africans only encouraged population growth (Spencer 1992). Earlier, Richard Lamm, the three-time governor of Colorado and a current member of FAIR's advisory board, had voiced similar concerns. Sounding very much like the deadly physicians of the Third Reich, he said that "terminally ill people have a duty to die and get out of the way" (quoted in Beirich 2007). Lamm also expressed fear of race wars in his 1985 novel, *Megatraumas: America at the Year 2000*. Tanton has also expressed pro-eugenics viewpoints, for example in a 1996 letter that explained how FAIR's new website emphasized how mankind used eugenic principles in breeding programs for plants and lower animals in order to condition the public about how eugenics might be used on humans. He stated: "We report ways [eugenics] is currently being done, but under the term genetics rather than eugenics" (quoted in Southern Poverty Law Center 2009b).

There are other connections between FAIR and racist organizations. Staff members of FAIR, such as Eastern Regional Coordinator Jim Stadenraus, have participated in anti-immigration conferences with Jared Taylor, the founder of American Renaissance. Television programming produced by FAIR has featured members of the Council of Conservative Citizens and has had members of this racist hate group on its payroll. The council is a direct descendent of the segregationist Mississippi Citizens' Councils, which described blacks as "a retrograde species of humanity" (quoted in Southern Poverty Law Center 2014). FAIR's television program *Borderline* featured interviews with *AR*'s Jared Taylor and Samuel Francis. Besides being a frequent contributor to *AR*, Francis served as editor of the newsletter of the Council of Conservative Citizens until his death in 2005.

Another member of FAIR's board of advisors and board of directors is Donald Collins. Collins frequently contributes to a website called VDARE .com, which publishes white supremacist and anti-Semitic diatribes. He also publishes in Roger Pearson's *Journal of Social, Political, and Economic Studies*. In his writings Collins frequently attacks the Catholic Church for being pro-immigration. Joe Guzzardi, a member of FAIR's board of advisors, is a former editor of VDARE.com. On this website, he frequently writes about how Latino immigrants wish to "reconquer" the United States and that this is one of the goals of their immigration. This conspiracy theory is proclaimed by numerous hate groups (see Appendix B).

Recently, FAIR has become very active in promoting anti-immigrant laws at the state and local levels. On April 23, 2010, Arizona governor Jan Brewer signed the Support Our Law Enforcement and Safe Neighborhoods Act, Arizona's newest and most controversial immigration law. It was written with the help of Kris Kobach (1966–), a Republican who is currently the Kansas secretary of state. Kobach works for FAIR's legal arm, the Immigration Reform Law Institute. The law forces police officers to detain individuals whom they suspect to be illegal immigrants and makes it a misdemeanor under state law for noncitizen immigrants to fail to carry their immigration papers. This law *compels* law enforcement to engage in racial profiling. The law was challenged in the federal courts in suits filed by the Justice Department, the American Civil Liberties Union, several individuals, and several Arizona cities. In June 2012, the U.S.

Supreme Court upheld the provision requiring immigration status checks during law enforcement stops, but it struck down three other provisions as violations of the Supremacy Clause of the U.S. Constitution (Barnes 2012). More recently, the United States District Court for the District of Arizona ruled that the Maricopa County Sheriff's Office, headed by Sheriff Joe Arpaio, has engaged in unconstitutional racial profiling, a legal decision that further limits the law (Southern Poverty Law Center 2012b). Earlier, Kobach helped pass anti-immigrant ordinances in Farmers Branch, Texas; Hazleton, Pennsylvania; and other cities. These laws seek to punish those who aid and abet "illegal aliens." The laws have proven a massive financial burden to the towns that pass them and in many cases have sparked racial strife and economic disorder (Southern Poverty Law Center 2012b).

In June 2011, Alabama passed an anti-immigration bill, the Beason-Hammon Alabama Taxpayer and Citizen Protection Act, which is even stricter than Arizona's law. One of its co-authors, state legislator Mickey Hammon, boasted that the bill "attacks every aspect of an illegal immigrant's life. It will make it difficult for them to live here so they will deport themselves" (quoted in *The Economist* 2012). Since then, many of the provisions of the law have been struck down by federal and appellate courts. These include a section that prohibited illegal immigrants from working; a section forbidding citizens from concealing, harboring, or transporting illegal immigrant workers; the section that made hiring or retaining an illegal immigrant actionable; the directive that required schools to determine the immigration status of their pupils; a section making it a crime for illegal immigrants to not have proper identification; and a section forbidding illegal immigrants from doing business with the state. The authors of the bill and the Alabama legislators hoped that it would lead to "self-deportation," and this has indeed happened. Reports estimate that thousands of legal and illegal Hispanics have left the state. This law created a climate in the state of Alabama that is similar to the one that led 60,000 Jews to leave Germany during World War II (Weiss 2010). After the law was passed, Alabama farmers complained that their crops were rotting in the fields because of a lack of farm workers, and building companies had rising costs because too few low-wage workers were available. An economist at the University of Alabama, Samuel Addy, estimated that the cost to the state—taking into account the decline in productivity, the increased cost of enforcing the law, and declines in aggregate consumer spending and tax revenue—was in the billions of dollars. This did not include the cost of defending the law in the courts (Addy 2011). The law is based on a model written by Kobach, tweaked to eliminate some of the problems of enforcement in the Arizona law. Kobach claims that with the country's toughest immigration law, Alabama "has done a great service to America" (Talbot 2011). On August 20, 2012, the United States Court of Appeals for the Eleventh Circuit partially invalidated the Beason-Hammon Alabama Taxpayer and Citizen Protection Act.

By September 2011, Indiana, Georgia, and South Carolina had passed somewhat similar measures and were facing legal action. However, by that time, Arizona-like bills had been defeated or had failed to progress in at least six states and the momentum seemed to be shifting away from such initiatives. Opposition from business leaders, legislators' concerns about the legal costs of defending these measures, and negative publicity appeared to be slowing these efforts. Even in Arizona, additional tough measures against illegal immigration were having a difficult time passing in the Arizona Senate (Riccardi 2011).

Kobach flippantly brags that he wrote the Alabama bill on his laptop computer during a hunting trip, while sitting in a turkey blind. There were few turkeys to be seen that day, allowing him time to write the bill, and he claimed that "a bad day for turkeys turned out to be a good day for constitutional law" (quoted in Talbot 2011). I would argue that it was a good day for turkeys and a bad day for constitutional law. Kobach regards tweaking immigration laws in response to legal action against them as a hobby and has worked on "improved" laws for Georgia, Indiana, Kansas, Missouri, Pennsylvania, South Carolina, and Texas. He has said, "Some politicians golf in their spare time. I spend mine defending American sovereignty" (quoted in Holman 2011).

In 1990, Kobach wrote a book opposing anti-apartheid efforts in South Africa (Kobach 1990). Before joining FAIR, he was U.S. Attorney General John Ashcroft's top immigration advisor. Before he was elected as secretary of state of Kansas, he was the chair of the Kansas Republican Party. After the 9/11 attacks, he headed Department of Justice efforts to tighten border security. He developed a program to monitor Arab and Muslim men that was later dropped after complaints about racial profiling and discrimination. He first got interested in immigration issues as a law student when he read about California's Proposition 187. Kobach has been placed on the Southern Poverty Law Center's "Profiles of 21 Nativist Leaders" watch list for his legal work on behalf of FAIR (Southern Poverty Law Center 2008). One of the principal supporters of Kobach's Arizona law was Kansas state senator Russell Pearce. In 2006, Pearce forwarded an e-mail to his supporters from the neo-Nazi National Alliance entitled "Who Rules America?" The article was critical of the media for supporting multiculturalism and racial equality and for presenting the Holocaust as a fact. Recently, Pearce was photographed embracing J. T. Ready, an Arizona member of the neo-Nazi National Socialist Movement (Beirich 2010). Kobach claims that his laws are not racially motivated and have no racial implications (Terbush 2012). To my mind this indicates that he is ignorant or unaware of the background and ideology of the people he keeps company with, he is extremely naïve about and ignorant of the history and implications of such laws, he is simply politically motivated, and/or he is lying. Benjamin Johnson, director of the American Immigration Council, says of Kobach: "I think he believes in what he is saying, but I also think he recognizes it as a way to make a name for himself on the national stage. . . . I can't judge what's in his heart, but I don't believe his ideas are right. I think he's irresponsible, and smart enough to know it" (quoted in Talbot 2011).

The Arizona and Alabama anti-immigration laws have been likened by many to Nazi laws that forced Jews to carry identification during the Third Reich. Cardinal Roger Mahony of Los Angeles, head of the largest Catholic diocese in the United States, said of the Arizona law (quoted in Watanabe 2010):

> The Arizona legislature just passed the country's most retrogressive, mean-spirited, and useless anti-immigrant law. . . . American people are fair-minded and respectful. I can't imagine Arizonans now reverting to German Nazi and Russian Communist techniques whereby people are required to turn one another in to the authorities on any suspicion of documentation. Are children supposed to call 911 because one parent does not have proper papers? Are family members and neighbors now supposed to spy on one another, create total distrust across neighborhoods and communities, and report people because of suspicions based upon appearance?

Republican member of Congress Connie Mack said of the bill (The Hill 2010), "This law of 'frontier justice'—where law enforcement officials are required to stop anyone based on 'reasonable suspicion' that they may be in the country illegally—is reminiscent of a time during World War II when the Gestapo in Germany stopped people on the street and asked for their papers without probable cause. . . . This is not the America I grew up in and believe in, and it's not the America I want my children to grow up in." Concerning the Alabama law, journalist Jean Damu (2012) writes: "The Nuremberg Laws . . . addressed economic and everyday social relations. For instance, entering into a contract with a Jew became illegal. Providing social services to Jews became illegal. Jews were relegated to their own schools. . . . The intent of all this naturally was to get the Jews to leave Germany. . . . Though the Nuremberg Laws were, we would like to think, far more extensive, invasive and racist than anything that could possibly be accepted anywhere in America in 2012, there is a disturbing overlap of key provisions of the laws; and the intent, to get the Jews in Germany and undocumented immigrants here, to 'deport themselves,' is the same." Damu then goes on to list key provisions of the Alabama law that underline the commonalities between the Alabama and Nuremburg laws.

Another FAIR initiative seeks to end the birthright citizenship provisions of the Fourteenth Amendment, a longtime goal of the group. This initiative was launched in January 2011 by the Immigration Reform Law Institute, FAIR's legal arm, working in partnership with State Legislators for Legal Immigration, a coalition of around seventy legislators from thirty-eight states that is dedicated to removing "economic attractions" for immigrants and eliminating their "unlawful invasion" into the United States. When the Fourteenth Amendment was ratified in 1868, it ensured that the children of slaves could not be denied citizenship. It now ensures that almost all children born in the United States are automatically granted citizenship (Southern Poverty Law Center 2012b). Kobach is one of the main legal minds behind this effort of State Legislators for Legal Immigration. The group often is described as the legislative arm of FAIR, and on its website it claims to have a "working partnership" with FAIR. It is the main instigator behind many FAIR initiatives, including a number of attempts in various states to pass laws modeled on Arizona's Support Our Law Enforcement and Safe Neighborhoods Act (Southern Poverty Law Center 2012b). State Legislators for Legal Immigration has written model legislation and has created an "online model legislation clearing house" that anti-immigration groups can use to develop state laws and generate grassroots support for state-level immigration enforcement restriction efforts (Southern Poverty Law Center 2011).

In January 2012, Republican Mitt Romney, later a candidate for president, announced that Kris Kobach had endorsed his campaign. Kobach had told FOX news reporter Neil Cavuto: "Mitt Romney stands apart from the others. He's the only one who's taken a strong across-the-border position on immigration" (quoted in Beadle 2012). And in a press release, he said: "We need a president who will finally put a stop to a problem that has plagued our country for generations. . . . Mitt Romney is the candidate who will finally secure the borders and put a stop to the magnets . . . that encourage illegal aliens to remain in our country unlawfully" (quoted in American Presidency Project n.d.). Romney, for his part, said: "I'm so proud to earn Kris's support. Kris has been a true leader on securing our borders and stopping the flow of illegal immigration into this country." He added: "We need more conservative leaders like Kris. . . . With Kris on the team, I look forward to working with him to take forceful steps to curtail illegal immigration and to support states like South Carolina and Arizona that are stepping forward to address this problem" (quoted in American Presidency Project n.d.). A political advisor to Romney, Eric Fehrnstrom, has stated that the candidate wanted to make immigrants' lives so unbearable they would be forced to leave (quoted in Beadle et al. 2012). The idea is to repress Hispanics so harshly that they will fear the daily curtailing of the civil rights of themselves and their families and friends so much that they will leave the state and the country of their own accord; what the advocates of these racially motivated laws call "self-deportation." In fact, during his presidential campaign, "self-deportation" became a political sound bite for Romney, like "raising taxes," and "Obamacare." As late as April 2013, Kris Kobach was still pushing self-deportation at the Senate Judiciary Committee's hearings on the current immigration bill (Gordon 2013).

FAIR runs many political advertisements for right-wing candidates, against candidates who do not supports its racist positions, and for right-wing causes, many related to immigration policy. Since the Supreme Court's 2010 ruling that corporations are legal persons, a decision that allows corporations to contribute to political campaigns without disclosure, it is difficult to know just how much and who the Pioneer Fund, FAIR, and other racist, neo-Nazi, and white supremacist groups are funding. Racism has not gone away in the United States, and with a black president in his second term, open, blatant racism has once again become very prominent in many spheres of American life. As Brugge (2010) has stated: "An atmosphere of fear, a competitive mentality and a sense of increasingly scarce resources create a fertile soil for anti-immigrant advocates who raise the alarm that newcomers will take your job, your home, and your culture, or worse threaten your sense of safety. . . . The decline in the economic position of the average American and the threat of violence are understandable motivators of fear. But to blame immigrants as the source of these problems is to scapegoat an easy, unpopular target and, at times, divert responsibility from more culpable parties."

Just as was the case in the late 1800s and early 1900s, unfortunately, the message that immigrants are the problem has been all too successful. And once again, this message is being propagated and led by racist eugenicists driven by ancient polygenic-like ideology. The Pioneer Fund has continued to follow the strategy Draper developed and the Satterfield plan of the mid-1960s (see Chapter 8). It has continued to influence policy by convincing the public that black and brown citizens have completely different natures than Western European whites do and that they are inferior to them and to demonstrate that black/brown inferiority is not attributable to environmental factors, mistreatment, or discrimination but to inherent biological shortcomings. It has provided pseudoscientific documentation of racial differences and has continued to award grants to educational institutions and individuals for "research projects" that fall in the category of pseudoscience. But most of its resources are devoted to disseminating virulent hate-based messages and literature to newspapers, periodicals, wire services, and radio and television stations. The Pioneer Fund also continues to subsidize its own periodicals and has provided funds for a number of other organizations and fronts, such as the AICF and FAIR, that distribute eugenicist, racist propaganda and attempt to influence opinions and policies in the United States. Now that the U.S. Supreme Court has ruled that the government may not ban political spending by corporations in elections (which enables groups to secretly spend as much as they want on local, state, and federal candidates and initiatives), there is no telling how important these racist hate groups may become in influencing America's future.

On February 10, 2012, before the presidential elections, many members of Congress, including Senate Majority Leader Mitch McConnell; presidential candidates Mitt Romney, Rick Santorum, and Newt Gingrich; and Marco Rubio, at the time one of the top candidates for running mate for Romney, addressed a meeting of the Conservative Political Action Conference (CPAC), an annual conference for conservative activists and officials. One of the scheduled events at this conference was listed in the program as follows:

9:50 Immigration—High Fences, Wide Gates: States vs. the Feds, the Rule
of Law & American Identity—Marriott Ballroom
    Alex Nowrasteh, Policy Analyst, Competitive Enterprise Institute
    The Honorable David Rivera (R-FL), United States Representative
    The Honorable Mario Diaz-Balart (R-FL), United States Representative
    The Honorable Kris Kobach, Secretary of State of Kansas
    Robert Vandervoort, Executive Director, ProEnglish
    Moderator: Niger Innis, National Spokesperson, Congress of Racial
Equality

As can be seen, Kris Kobach, Republican congressmen, and Niger Innis
(an extremely conservative political advisor to the failed Republican presidential candidate Herman Cain) shared the stage with the current director
of ProEnglish, Robert Vandervoort, a white nationalist from Illinois. Vandervoort has served as the head of a nativist organization, Chicagoland
Friends of American Renaissance, a part of Jared Taylor's racist organization. The group often holds events featuring numerous white nationalists.
Vandervoort also has made appearances at white nationalist events outside
Illinois. Both Kobach and Vandervoort have worked for organizations
founded by John Tanton of FAIR. As we have seen, Kobach serves as counsel for the Immigration Reform Law Institute. Vandervoort recently was
hired as the executive director of ProEnglish, the Tanton-founded English-
only project of his group U.S. Inc.
    Besides this session, Vandervoort also hosted the panel at the Conserva-
tive Political Action Conference entitled "The Failure of Multiculturalism:
How the Pursuit of Diversity Is Weakening the American Identity." Peter
Brimelow, founder and head of the White nationalist website VDARE,
participated in this panel. Brimelow reflected during the panel that after
"Obama's racial-socialist coup," he feared that the United States was
doomed to face a "minority occupation government." He called on the Republican Party to focus on becoming the party of white voters by attacking
"ethnic lobbies," affirmative action, bilingual education, and "taxpayer sub-
sidies to illegal aliens." Earlier in the session, Vandervoort delivered a ram-
bling presentation from Serge Trifkovic (a conservative commentator who
was unable to attend the conference). This paper focused on how the "cult
of non-white, non-male, non-heterosexual victimhood" and "multicultur-
alist indoctrination" was ruining the West and predicted that "the native
Western majorities will melt away." The speech concluded, "Europeans and

our trans-Atlantic cousins are literally endangered species. The facilitators of our destruction must be neutralized if we are to survive." Then, Rosalie Porter, current chair of the ProEnglish board, criticized the Civil Rights Act and the Voting Rights Act for giving too much political influence to minorities (American Renaissance website 2012).

In his own talk, Brimelow warned that multiculturalism and bilingualism were "diseases" that could wreck American society because they empower minorities and suppress traditional American (that is, white) groups. He called multiculturalism and bilingualism a "ferocious attack on the working class." During a panel discussion, Brimelow accused the Democratic Party of giving "up on the white working class" and of using immigration to "elect a new polity" by increasing the number of ethnic minorities. Another panel participant was fiercely anti-immigration congressman Steve King (R-IA) who discussed his bill to make English the official language of the United States. During the panel discussion, King endorsed Brimelow's writings. The views expressed by the members of this group and what they stand for went completely unchallenged by the Republican Party (Sanchez 2012).

It appears that with the blessing of many of the right-wing leaders of the Republican Party, the Pioneer Fund, its grant recipients, and its followers, the racist white supremacists, neo-Nazis, and modern day eugenicists, might once again be making inroads in U.S. society with their politics of racism, eugenics, and anti-immigration. The dream of having a politically active and publicly supported conservative movement that is focused on old-fashioned racism and on challenging cultural diversity may once again be coming to fruition (Burghart 2012).

Thus, though its thought is disseminated widely through a number of different organizations, the Pioneer Fund has been the main catalyst supporting an insidious racist plan that is focused on keeping the dreams of the eugenics movement alive. This plan, the Satterfield plan, which was initially conceived of in the 1960s, appears to have been quite successful throughout the years. Since Wickliffe Draper's death in 1972, Harry Weyher has used the Pioneer Fund to carry out the goals of this racist plan. From that time on, the fund has followed the strategy of the Satterfield plan fairly closely. It has supported "research" to provide evidence of racial differences, supported a number of front groups to distribute the results of this "science," and supported organizations engaged in promoting legislation and litigation focused on racist issues and on opposition to immigration. Such organizations as the Pioneer Fund, the American Renaissance Foundation, and FAIR have successfully kept age-old racist theories in the eyes and minds of the public and continue to pursue the goals of the eugenics movement of the early twentieth century. Colonel Draper has passed on his legacy to a new generation of bigots. So far, they have lost their attempt to turn the United States and beyond into a racist world, but they continue in their attempt to win the 500-year-old war of hatred and intolerance.

Conclusion

As we have seen, Europeans and Western European colonists of the United States first categorized "others" mainly with two fairly unchanging paradigms: polygenism and monogenism. Polygenecists (or pre-Adamites) believed that non-Western Europeans were not created by God but were on the earth before Adam and that physical characteristics and complex behaviors were biologically fixed and immutable. No environmental conditions could improve their lot. Monogenecists believed that all humans were created by God but that "others" had degenerated from the original ideal because they lived in less-than-ideal environmental conditions (either a bad climate and/or uncivilized social conditions). To monogenecists, these poor creatures could eventually be "saved" by introducing them to Western European civilization. Within both paradigms, Western Europeans were considered superior to other peoples or races.

Until Darwin's time, polygenism was supported "scientifically" by a Humeian/Kantian ethnocentric inductive philosophy that assumed that Western European civilization was superior and could be created and achieved only by Western Europeans. Also, they believed that certain physical and behavioral traits (such as skull size and shape and mental capacity) were superior in Western Europeans and had been created by God in that way and were thus biologically fixed and immutable. The "science" of the monogenecists was bolstered in the early nineteenth century by Lamarck's theory of the inheritance of acquired characteristics, the idea that physical traits and complex behaviors could be changed from one generation to the next by changing environmental conditions.

At the beginning of the twentieth century, however, Lamarck's theory was discredited and Mendelian genetics was rediscovered (Moore 2001).

302

Laboratory experiments by August Weismann and others had discredited Lamarck's theories and monogenism was left with no scientific backing; environment could not overcome heredity. Biological determinism and polygenic racism were considered the only "scientifically" valid explanations for the differences among human populations and between "unfit" individuals and those with superior heredity. Eugenics was born and became the dominant paradigm; it became conventional wisdom both scientifically and among the public in Western Europe and the United States in the early 1900s. The polygenecists considered monogenism to be dead: there was no valid environmental explanation for human variation. Some individuals and peoples were simply biologically inferior to (certain) Western Europeans, and they carried and transmitted genetically fixed, inferior physical and behavioral traits.

Eugenics and scientific racism reached its peak in the United States in the mid-1920s. In 1924, Congress passed the Immigration Restriction Act and Virginia passed a sterilization law and a law preventing the marriage of people of two races. In the same year, as follow-ups to Madison Grant's (1916) *Passing of the Great Race,* many Grantian eugenicists published similar volumes, including Henry Pratt Fairchild, Ellsworth Huntington, Vernon Kellogg, Edward Ross, and Lothrop Stoddard. Eugenics courses became popular in high schools and colleges throughout the United States. American families entered "fitter family" contests and attended popular lectures by leading eugenicists, who by the 1920s had prominent positions in academe, business, and politics. The American Museum of Natural History displayed dioramas to educate the public about Osborn's and Grant's eugenic views on race. The eugenics organizations (the Galton Society, the ERO, the ERA, and the AES) were thriving and were well connected to scientific organizations, private and government funding agencies, educational institutions, and politics and politicians. Eugenicists controlled the National Research Council. Scientific racism was rampant in American society (Black 2003; Brace 2005; Spiro 2009). Nazism was on the rise in Europe.

However, the seeds of destruction of eugenics and scientific racism had already been planted. Franz Boas's two volumes, *Changes in Bodily Form* and *The Mind of Primitive Man,* both published in 1911, had undermined the "science" of eugenics. Boas showed that physical traits, such as skull shape, could change with environmental influences in a single generation. He also presented a more plausible explanation for the social and behavioral differences among peoples around the world. He developed the scientific, anthropological concept of "culture." According to this theory, differences among the peoples of the world were the product of their social histories and were determined by an interaction between their social and natural environments over generations. For the first time in history, a coherent theory of human variation explained differences in terms other than assumptions about the superiority of one group over the other and even questioned the biological validity of races among humans. The scientific evidence amassed for Boas's cultural explanation was much more convincing than the myths that had been passed down through the previous five centuries. A new scientific paradigm began to replace the conventional wisdom of the past 500 years.

We can see a direct linear relationship of racist ideology through the centuries, and we can follow a historical pattern of racial prejudice and hatred. This prejudice is fueled by the same underlying theory that human variation can be easily put into simple categories (racial classifications), that these categories are inherently distinct, that they can be ranked in ways that make some "races" superior to others, and that many of the most important human characteristics (such as intelligence, criminality, aggression, even the ability to navigate ships) are biologically determined and racially variable and cannot (or can only superficially) be influenced by environmental factors. This ancient ideology can be traced from its beginnings during the Spanish Inquisition and the pre-Adamite scientific justification of racism of the fifteenth century, through the initial period of European colonization, into the period of slavery in the eighteenth and nineteenth centuries, and into the eugenics movement and Nazism of the nineteenth and twentieth centuries.

This same ideology can currently be seen in a new "scientific racism" of the twenty-first century. We might have thought that it ended with the extremities of Nazism at the end of World War II, but we were wrong. Pre-Adamite-like biological determinism is still with us, and all of us who believe in human dignity, freedom, and justice must continue to fight against racial prejudice and those who spread hatred based on the idea that differences among humans exist. We must teach our children of the real wonders of human variation and enjoy the wonderful ethnic variation we find among us and around the world. We must teach tolerance and love, not the bigotry and hatred of the new bigot brigade as they spread their ancient and outdated myths of race and racism.

Science has shown that intelligence is not a unitary, simply measured, genetically based phenomenon, that IQ scores and intelligence (whatever that may be) are greatly influenced by environment and culture, and that there are no biological races among humans and there never have been. It is difficult to correlate a measure of IQ, based on a simple concept of a unitary, biologically based "intelligence," with another false concept of biologically based "races" when neither of these phenomena exist, no matter how tightly many bigoted, bad scientists might still cling to this age-old, culturally based, hateful ideology. The idea of a unitary purely genetically based intelligence and of biologically distinct races among humans is as outdated scientifically as the ideas that the earth is flat or that it was created in 4004 B.C.

Race is not a biological reality among humans; there are no human biological races. What humans have designated as races are based on nonexistent differences among peoples. People are more similar to one another biologically and genetically as a whole than they are to any of the classifications that racists have devised. Once the history of racism is understood, however, it is easy to see how we have been led to think of race as reality. It is a part of our cultural history, a part of our everyday reality. We are raised to believe that biological races actually exist. What I have attempted to show here is the long history of racial thinking in Western Europe and in the United States. This history is easily followed, through our literature and philosophy, for the past 500 years. An appreciation and understanding of the history of racism and race should help us understand where we are today and where we need to move in the future. Variation among humans is real. However, it is not based on any inborn heredity related to a racial history. It derives from our individual history, our individual genetics, and our culture. Many of us continue to be influenced by the illegitimate goals of racists and racism. Racist history is part of our background, of our culture. However, the idea that race is a biological reality is a fallacy with a long and complex history. This is a history that we must all be aware of in our political, economic, and cultural interactions in the future. Biologically valid races are not real, but cultural racism is, and we must understand how this cultural reality affects our everyday interactions. We must continue to fight race and racism with our knowledge and assessment of the history of racist thinking.

The reelection of Barack Obama and his inauguration on Martin Luther King Day demonstrated that a great deal has changed in the United States in recent years. It was only fifty years ago that there were marches in the streets to protest the notion of giving civil rights to African Americans and Medgar Evers was murdered by a member of the White Citizens'

Council of Mississippi. I think that I could correctly say that a black president could not have been elected even twenty years ago, and few would
have thought it possible in 2008 and again in 2012. Yet racism is still with
us. In the introduction to this book I mentioned journalist Guy Harrison's
(2010) surprise that he did not know that anthropologists had rejected the
biological concept of race sixty years ago and that nobody had taught him
about this until he took a course in anthropology in college. How can this
be when there is so much scientific evidence against biological racism?

Recently, in the wake of a jury's decision that the killer of a seventeen-
year-old black teenager, Trayvon Martin, was not guilty due to Florida's
"stand your ground" laws, Barack Obama declared that "Trayvon Martin
could have been me 35 years ago." Obama noted that "the African American community is looking at this issue through a set of experiences and a
history that doesn't go away."

> There are very few African American men who haven't had the experi-
> ence of walking across the street and hearing the locks click on the doors of
> cars. . . . There are very few African Americans who haven't had the experi-
> ence of getting on an elevator and a woman clutching her purse nervously
> and holding her breath until she had a chance to get off. . . .
>
> The African American community is also knowledgeable that there is a
> history of racial disparities in the application of our criminal laws. . . . The
> poverty and dysfunction that we see in those communities can be traced to
> a very difficult history. . . .
>
> And that all contributes I think to a sense that if a white male teen was
> involved in the same kind of scenario, that, from top to bottom, both the
> outcome and the aftermath might have been different. . . .
>
> I don't want us to lose sight that things are getting better. Each successive
> generation seems to be making progress in changing attitudes when it
> comes to race. It doesn't mean we're in a post-racial society. It doesn't mean
> that racism is eliminated. . . .
>
> But we should also have confidence that kids these days, I think, have
> more sense than we did back then, and certainly more than our parents did
> or our grandparents did; and that along this long, difficult journey, we're
> becoming a more perfect union—not a perfect union, but a more perfect
> union (Obama 2013).

I do think that we are improving. We have learned a great deal and we
need to teach our children what we have learned. Historically, race and
race concepts in the West were driven by an informal, mutually reinforc-

ing consortium of intellectuals, politicians, and financial backers. Currently, there are new and sinister consortia around us that are cleverly cloaking the motives behind the rhetoric of modern racist "intellectuals" and politicians and their financial supporters. Racism is still alive in the United States and the West. Only education about the real nature of human differences and about the history of the concept of racism will help us escape from these continuing cycles of ignorance, hatred, and fear. We need to be alert to the agendas of new racist alliances of intellectuals, politicians, and businesspeople in the United States and elsewhere. Prejudices about race are created by these new agendas. We must remember that although race does not represent a biological reality, the cultural reality of race is real. Although people are different, the main differences are due to the realities of their upbringing, to their culture, not to unclear biological and unchanging inherited differences. *Biologically,* Homo sapiens *is one race*. It is only by recognizing this fact and understanding its history that we might one day have a society in which all people are treated with dignity, equality, and kindness regardless of their ethnicity or culture.

Appendix A
The Eugenics Movement, 1890s–1940s

1891	Nathaniel Southgate Shaler writes *History of Man in America*
1903	Charles Davenport founds the Station for Experimental Evolution
1907	Indiana passes a sterilization law
1910	Charles Davenport founds Eugenics Record Office with Harry H. Laughlin as superintendent
1911	Charles Davenport writes *Heredity in Relation to Eugenics*
1912	First International Eugenics Congress is held
	Henry H. Goddard writes *The Kallikak Family*
1913	Davenport and Laughlin co-found the Eugenics Research Association
1914–1919	World War I
1916	Madison Grant writes *The Passing of the Great Race*
1917	Robert Yerkes, Henry Goddard, Lewis Terman, Carl Brigham administer intelligence tests for the U.S. Army
1918	Madison Grant, Charles Davenport, and Henry Fairfield Osborn co-found the Galton Society
1922	Madison Grant and others co-found the Eugenics Committee of the United States of America
1923	Carl Brigham writes *A Study of American Intelligence*
1923	Second International Eugenics Congress is held
1924	Virginia passes the Racial Integrity Act
	Congress passes the Immigration Restriction Act

1927	The U.S. Supreme Court rules, in *Buck v. Bell*, that compulsory sterilization of the "unfit" is legal
1932	Third International Eugenics Congress is held
1933	Adolf Hitler is named chancellor of Germany
1939–1945	World War II

Appendix B
The Pioneer Fund

1937	Five-member board of directors is appointed: Wickliffe Draper (founder), Harry Laughlin (president), Frederick Henry Osborn, Malcolm Donald, and John Marshall Harlan
1937–1941	Distributes the English-subtitled German film *Erbkrank (The Hereditary Defective)*
	Increases the income of families of officers of the United States Army Air Corps
1939–1966	Gives both financial and strategic planning aid to Earnest Sevier Cox and Mississippi senator Theodore Gilmore Bilbo for their efforts to get Congress to pass a black repatriation bill
1958	Draper replaces Frederick Osborn with Harry Frederick Weyher Jr. as president and appoints John B. Trevor Jr. as treasurer
1959–1979	Funds the International Association for the Advancement of Ethnology and Eugenics
1959–present	Funds the *Mankind Quarterly*; original Executive committee members include psychologist Henry Garrett, zoologist Robert E. Kuttner, political scientist A. James Gregor, and Donald A. Swan
1959–1963	Carlton Putnam founds the National Putnam Letters Committee
1961	Funds the publication and distribution of Carlton Putnam's *Race and Reason*
1963–1964	Grantee R. Travis Osborne and Pioneer Fund president Henry Garrett testify for the defendants in *Stell v. Savannah-Chatham*

County Board of Education in Savannah, Georgia, and in *Evers v. Dwyer* in Jackson, Mississippi, in an attempt to influence the lower courts to give rulings that would undermine *Brown v. Board of Education*

1964 Sovereignty Commission and the Citizens' Council of Mississippi work with the Council School Foundation in Mississippi to set up a private school system for whites

1964 Gives money to the state of Mississippi to implement the Satterfield plan for implementing a racist agenda, but after three civil rights workers are killed, Mississippi governor returns the money to Harry Weyher Jr.

1968–1976 Funds William Shockley and the Foundation for Research and Education on Eugenics and Dysgenics

1969–1978 Funds the Shockley-Jensen team after Arthur Jensen publishes his *Harvard Educational Review* article on race

1973–1999 Funds Arthur Jensen and his Institute for the Study of Educational Differences at the University of California, Berkeley

1973–1996 Funds Roger Pearson and his Institute for the Study of Man

1978 Roger Pearson acquires and becomes editor of *Mankind Quarterly*

1979–2000 Gives funds to support Thomas Bouchard's Minnesota Twin Family Study

1980s Funds the Foundation for Human Understanding for the publication of Jensen's *Straight Talk about Mental Tests* (1981) and Stanley Burnham's *America's Bimodal Crisis: Black Intelligence in White Society* (1985)

1981 Gives a grant to Pearson so he can purchase and house the library of Donald A. Swan

1980s–present Funds members of the "new bigot brigade": psychologist Hans J. Eysenck, sociologist Robert A. Gordon, educational psychologist Linda S. Gottfredson, geneticist Glayde Whitney, professor of philosophy Michael Levin, professor of psychology J. Philippe Rushton (president of the Pioneer Fund 2002–2012), and professor emeritus of psychology Richard Lynn (current co-president of the fund)

1983–present Funds the American Immigration Control Foundation

1984–present Funds John Tanton's Federation for American Immigration Reform

1991	Jared Taylor founds the American Renaissance Foundation and *American Renaissance* with strong links to the Pioneer Fund
1994	Richard Herrnstein and Charles Murray publish *The Bell Curve: Intelligence and Structure in American Life,* which draws on the pseudoscientific research of Pioneer Fund member Richard Lynn
2010–2011	Arizona and Alabama pass anti-immigrant laws written by Kris Kobach, a lawyer for Federation for American Immigration Reform

References

Addy, S. 2011. *The New Alabama Immigration Law: A Preliminary Macroeconomic Assessment.* Tuscaloosa, AL: Center for Business and Economic Research, University of Alabama.

Agassiz, L. 1850a. The Geographic Distribution of Animals. *Christian Examiner and Religious Miscellany* 48:181–204.

———. 1850b. The Diversity of Origins of the Human Races. *Christian Examiner and Religious Miscellany* 49:110–145.

———. 1854. Sketch of the Natural Provinces of the Animal World and Their Relation to Types of Man. In G. R. Nott and J. C. Gliddon, eds., *Types of Mankind,* lviii–lxxvi. Philadelphia: Lippincott, Grambo & Co.

Aichel, O., and O. von Vorwort Verschuer. 1934. Vorwort. *Zeitschrift für Morphologie und Anthropologie* 34: v–vi.

Allen, G. E. 1978. *Thomas Hunt Morgan: The Man and His Science.* Princeton, NJ: Princeton University Press.

———. 1986. The Eugenics Record Office at Cold Spring Harbor, 1910–1940. *Osiris* 2:225–264.

———. 2001a. The Biological Basis of Crime: An Historical and Methodological Study. *History and Sociology of the Physical and Biological Sciences* 31:183–222.

———. 2001b. Is a New Eugenics Afoot? *Science.* 294:59–71.

———. 2004. Heredity, Development, and Evolution at the Carnegie Institution of Washington. In J. Maienschein, N. Glitz, and G. E. Allen, eds., *Centennial History of the Carnegie Institution of Washington.* Cambridge: Press Syndicate of the University of Cambridge.

———. 2011. Eugenics and Modern Biology: Critiques of Eugenics, 1910–1945. *Annals of Human Genetics* 75:314–325.

———. 2013. "Culling the Herd": Eugenics and the Conservation Movement in the United States, 1900–1940. *Journal of the History of Biology* 46:31–72.

Amato G., and J. Gatesy. 1994. PCR Assays of Variable Nucleotide Sites for Identification of Conservation Units. In B. Schierwater, B. Streit, G. P. Wagner, and R. DeSalle, eds., *Molecular Ecology and Evolution: Approaches and Applications*, 215–226. Basel: Birkhäser Verlag.

American Association of Physical Anthropologists. 1996. AAPA Statement on Biological Aspects of Race. *American Journal of Physical Anthropology* 101:569–570.

American Presidency Project. n.d. Press Release: Mitt Romney Announces Support of Kansas Secretary of State Kris Kobach. January 11, 2012. Retrieved from http://www.presidency.ucsb.edu/ws/?pid=99028.

American Renaissance. 2014a. Conferences. American Renaissance website. Retrieved from http://www.amren.com/archives/conferences/.

———. 2014b. Print Back Issues. American Renaissance website. Retrieved from http://www.amren.com/archives/back-issues/.

———. 2014. The Morality of Survival, Part I. Retrieved from http://www.amren.com/archives/back-issues/july-1995/.

Anderson, S., and J. L. Anderson. 1986. *Inside the League: The Shocking Exposé of How Terrorists, Nazis, and Latin American Death Squads Have Influenced the World Anticommunist League.* New York: Dodd, Mead.

Anti-Defamation League. 2011. Jared Taylor/American Renaissance. Retrieved from http://www.adl.org/main_Extremism/jared_taylor_eia_entry.htm?Multi_page_sections=sHeading_7.

———. 2012. Jared Taylor/American Renaissance: Recent Activity. Retrieved from http://www.adl.org/Learn/Ext_US/jared_taylor/activity.asp?LEARN_Cat=Extremism&LEARN_SubCat=Extremism_in_America&xpicked=2&item=taylor.

Baehre, R. 2008. Early Anthropological Discourse on the Inuit and the Influence of Virchow on Boas. *Études/Inuit/Studies* 32:13–34.

Bahar, A., and W. Kugel. 2001. *Der Reichtagsbrand: Wie Geschichte gemacht wird.* Berlin: Edition Q.

Barash, D. P. 1995. Book review: Race, Evolution, and Behavior. *Animal Behaviour* 49:1131–1133.

Barbujani, G., A. Magagni, E. Minch, and L. L. Cavalli-Sforza. 1997. An Apportionment of Human DNA Diversity. *Proceedings of the National Academy of Sciences* 94:4516–4519.

Barkan, E. 1992. *The Retreat of Scientific Racism.* Cambridge: Cambridge University Press.

Barker, M. 2010. Foundations and Anthropology in the United States. Dissident Voice, September 15. Retrieved from http://dissidentvoice.org/2010/09/foundations-and-anthropology-in-the-united-state.

Barnes, R. 2012. Supreme Court Rejects Much of Arizona Immigration Law. *Washington Post,* June 25.

Baur, E. 1911. *Einführung in die experimentelle Vererbungslehre.* Berlin: Verlag von Gebrüder Borntraeger.

———. 1924. Berlin: From Our Regular Correspondent. *Journal of the American Medical Association* 21:1709–7110.

Baur, E., E. Fischer, and F. Lenz. (1921) 1931. *Menschliche Erblichkeitslehre und Rassen hygiene.* Munich: J. F. Lehmann.

Beadle, A. P. 2012. On MLK Day, Romney Campaigning with Anti-Immigrant Official Tied to Hate Groups. Thinkprogress Justice, January 16. Retrieved from http://thinkprogress.org/justice/2012/01/16/404357/romney-campaigning-with-anti-immigrant-official-with-ties-to-hate-groups-on-martin-luther-king-day/.

Beadle, A. P., M. Diamond, and I. Millhiser. 2012. Kris Kobach, Author of Anti-Immigrant State Laws, Backs Mitt Romney in GOP Race. Thinkprogress Justice, January 11. Retrieved from http://thinkprogress.org/justice/2012/01/11/402550/kris-kobach-author-of-anti-immigrant-state-laws-backs-mitt-romney-in-gop-race/.

Beals, K. L., C. L. Smith, and S. M. Dodd. 1984. Brain Size, Cranial Morphology, Climate, and Time Machines. *Current Anthropology* 25:301–330.

Beebee, H. 2011. David Hume. In D. Pritchard and S. Bernecker, eds., *The Routledge Companion to Epistemology,* 730–740. New York: Routledge.

Begon, M., and M. Mortimer. 1981. *Population Ecology: A Unified Study of Animals and Plants.* New York: Sunderland Sinauer Associated.

Beirich, H. 2007. Federation for American Immigration Reform's Hate Filled Track Record. *Intelligence Report* 128 (Winter). Retrieved from http://www.splcenter.org/get-informed/intelligence-report/browse-all-issues/2007/winter/the-teflon-nativists.

———. 2008. The Tanton Files: FAIR Founder's Racism Revealed. Hatewatch, September 19. Retrieved from http://www.splcenter.org/blog/2008/09/19/tanton-fair-founder-racism/.

———. 2010. Hate Group Lawyer Drafted Arizona's Anti-Immigration Law. Hatewatch, April 28. Retrieved from http://www.splcenter.org/blog/2010/04/28/hate-group-lawyer-drafted-arizona-anti-immigrant-law/.

Benedict, R. 1934. *Patterns of Culture.* Boston: Houghton Mifflin.

Benedict, R., and G. Wetfish. 1946. *The Races of Mankind.* Public Affairs Pamphlet no. 85. New York: Public Affairs Committee, Inc.

Benson, C. 1995. Ireland's "Low" IQ: A Critique. In R. Jacoby and N. Glauberman, eds., *The Bell Curve Debate: History, Documents, Opinions,* 222–233. New York: Times Books.

Benson, P. 2011. Altruism and Cooperation among Humans: The Ethnographic Evidence: Introduction Part III. In R. W. Sussman and C. R. Cloninger, eds., *Origins of Altruism and Cooperation,* 195–202. New York: Springer.

Berkowitz, A. 1999. Our Genes, Ourselves? In R. W. Sussman, ed., *The Biological Basis of Human Behavior,* 360–370. Upper Saddle River, NJ: Prentice Hall.

Berlet, C. 2003. Into the Mainstream. *Intelligence Report* 10 (Summer). Retrieved from http://www.splcenter.org/get-informed/intelligence-report /browse-all-issues/2003/summer/into-the-mainstream?page=0,1.

Biddiss, M. D. 1970. *Father of Racist Ideology: The Social and Political Thought of Count Gobineau.* London: Weidenfeld and Nicolson.

Billig, M. 1979. *Psychology, Racism, and Fascism.* Birmingham: A. F. and R. Publications.

Binet, A., and T. Simon. 1905. Méthodes nouvelles pour le diagnostic du niveau intellectuel des anormaux. *L'Année psychologique* 11:191–244.

———. 1908. Le development de l'intelligence chez les enfants. *L'Année psychologique* 14:1–94.

Black, E. 2001. *IBM and the Holocaust: The Strategic Alliance between Nazi Germany and America's Most Powerful Corporation.* New York: Crown Publishers.

———. 2003. *War against the Weak: Eugenics and America's Campaign to Create a Master Race.* New York: Four Walls Eight Windows.

———. 2009. American Corporate Complicity Created Undeniable Nazi Nexus. *The Cutting Edge,* March 9. Retrieved from http://www.thecutting edgenews.com/index.php?article=11168.

Blackman, D. A. 1999. Silent Partner: How the South's Fight to Uphold Segregation Was Funded Up North. Slavery by Another Name, June 11. http://www.slaverybyanothername.com/other-writings/silent-partner-how -the-souths-fight-to-uphold-segregation-was-funded-up-north/.

Blumenbach, J. F. 1795. *On the Natural Variety of Mankind.* Translated and edited by Thomas Bendyshe. In *The Anthropological Treatises of Johann Friedrich Blumenbach.* London: Longman, Green, Longman, Roberts, & Green.

Boas, F. 1887a. Anthropology in the American and British Associations for the Advancement of Science. *Science* 10 (July–December):231–232.

———. 1887b. Letter. *Science* 9 (June 17):587–589.

———. (1889) 1940. *Die Ziele der Ethnologie (The Aims of Ethnology).* Reprinted in F. Boas, *Race, Language, and Culture,* 626–638. Toronto: Collier-Macmillan Canada.

———. (1894a) 1940. The Half-Blood Indian: An Anthropometric Study. *Popular Science Monthly* 45:761–770. Reprinted in F. Boas, *Race, Language, and Culture,* 138–148. Toronto: Collier-Macmillan Canada.

———. (1894b) 1974. Human Facility as Determined by Race. *Proceedings of the American Association for the Advancement of Science* 43:301–327.

Reprinted in G. W. Stocking Jr., *The Shaping of American Anthropology, 1883–1811: A Franz Boas Reader,* 221–242. New York: Basic Books.

———. (1896) 1940. The Limitations of the Comparative Method in Anthropology. *Science* 4:901–908. Reprinted in F. Boas, *Race, Language, and Culture,* 270–280. Toronto: Collier-Macmillan Canada.

———. 1904a. Some Traits of Primitive Culture. *Journal of American Folklore* 17:243–254.

———. 1904b. The History of Anthropology. *Science* n.s. 20 (October 21):513–524.

———. 1911a. *Changes in Bodily Form of Descendants of Immigrants.* Reports of the Immigration Commission, vol. 38. Washington, DC: Government Printing Office.

———. 1911b. *The Mind of Primitive Man.* New York: Macmillan.

———. 1916. Eugenics. *Scientific Monthly* 3:471–478.

———. 1917. Inventing a Great Race: Review of *Passing of the Great Race* by Madison Grant. *New Republic* 9:305–307.

———. 1919. Scientists as Spies. *Nation* 109:797.

———. 1921. The Problem with the American Negro. *Yale Review* 10:384–395.

———. 1928. *Anthropology and Modern Life.* New York: Dover Publications.

———. 1931. Race and Progress. *Science* 74:1–8.

———. (1932) 1940. The Aims of Anthropological Research. *Science* 76:605–613. Reprinted in F. Boas, *Race, Language, and Culture,* 243–259. Toronto: Collier-Macmillan Canada.

———. (1936) 1940. History and Science in Anthropology: A Reply. *American Anthropologist* 38:137–141. Reprinted in F. Boas, *Race, Language, and Culture,* 305–311. Toronto: Collier-Macmillan Canada.

———. (1938) 1974. The Background of My Earlier Thinking. In G. W. Stocking Jr., ed., *The Shaping of American Anthropology, 1883–1911: A Franz Boas Reader,* 41–42. New York: Basic Books.

———. 1940. *Race, Language, and Culture.* Toronto: Collier-Macmillan Canada.

———. 1945. *Race and Democratic Society,* New York, J. J. Augustin.

Boas, F., A. Hrdlička, and A. M. Tozzer. 1919. *My Dear Professor Hale:* At the Meeting of the American Anthropological Association. *American Anthropologist* 21:216–218.

Bok, Hilary. 2010. Baron de Montesquieu, Charles-Louis de Secondat. In *Stanford Encyclopedia of Philosophy.* Retrieved from http://plato.stanford.edu/entries/montesquieu.

Bouchard, T. J. Jr. 1984. Twins Reared Together and Apart: What They Tell Us about Human Diversity. In S. Fox, ed., *Individuality and Determinism: Chemical and Biological Bases,* 147–184. New York: Plenum.

———. 1994. Genes, Environment, and Personality. *Science* 264:1700–1701.

Boyce, M. S. 1984 Restitution of *r-* and *K*-Selection as a Model of Density-Dependent Natural Selection. *Annual Review of Ecology and Systematics* 15:427–447.

Brace, C. L. 1995. *The Stages of Human Evolution.* 5th ed. Englewood Cliffs, NJ: Prentice Hall.

———. 1996. Racialism and Racist Agendas: Review of *Race, Evolution, and Behavior: A Life History Perspective,* by Philippe Rushton. *American Anthropologist* 98:176–177.

———. 1999. An Anthropological Perspective on "Race" and Intelligence: The Non-Clinal Nature of Human Cognitive Capabilities. *Journal of Anthropological Research* 55:245–264.

———. 2005. *"Race" Is a Four-Letter Word: The Genesis of the Concept.* New York: Oxford University Press.

Brace, C. L., G. R. Gamble, and J. T. Bond. 1971. Introduction to Jensenism. In C. L. Brace, G. R. Gamble, and J. T. Bond, eds., *Race and Intelligence.* Anthropological Studies no. 8. Washington, DC: American Anthropological Association.

Brigham, C. C. 1923. *A Study of American Intelligence.* Princeton, NJ: Princeton University Press.

———. 1930. Intelligence Tests of Immigrant Groups. *Psychological Review* 37:158–165.

British Medical Journal. 1912. First International Eugenics Conference. *British Medical Journal* 2 (2692): 253–255.

Brown, D. E. 1991. *Human Universals.* Philadelphia: Temple University Press.

Brugge, D. 2010. Pulling up the Ladder: The Anti-Immigration Backlash. Political Research Associates. Retrieved from http://www.publiceye.org/magazine/v09n2/immigran.html.

Bruinius, H. 2006. *Better for All the World: The Secret History of Forced Sterilization and America's Quest for Racial Purity.* New York: Alfred A. Knopf.

Buffon, comte de. 1785. *Histoire naturelle, générale and particuliere. Histoire naturelle,* tome V. Paris: Chez Sanson & Compagnie.

Bulmer, M. 2003. *Francis Galton: Pioneer of Heredity and Biometry.* Baltimore, MD: Johns Hopkins University Press.

Burghart, D. 2012. Kansas Secretary of State to Share Stage with White Nationalist at CPAC. Institute for Research and Education on Human Rights, February 9. Retrieved from http://www.irehr.org/issue-areas/race-racism-and-white-nationalism/item/396-kansas-secretary-of-state-to-share-stage-with-white-nationalist-at-cpac.

Burnham, S. 1985. *America's Bimodal Crisis: Black Intelligence in White Society.* Athens, GA: Social Sciences Press.

——. 1993. *America's Bimodal Crisis: Black Intelligence in White Society.* 2nd ed. Athens, GA: Foundation for Human Understanding.

Cain, O. P., and C. H. Vanderwolf. 1989. A Critique of Rushton on Race, Brain-Size Intelligence. *Personality and Individual Differences* 11:777–784.

Cartmill, M. 1998. The Status of the Race Concept in Physical Anthropology. *American Anthropologist* 100:651–660.

Cernovsky, Z. Z. 1993. Race and Brain Weight: A Note on J. P. Rushton's Conclusions. *Psychological Reports* 66:337–338.

Chamberlain, H. S. (1899) 1910. *Foundations of the Nineteenth Century.* 2 vols. Translated by J. Lees. London: John Lane, Broadly Head.

——. (1905) 1914. *Immanuel Kant. Die Persönlichkeit als Einführung in das Werk.* Munich: F. Bruckmann. Published in English as *Immanuel Kant: A Study and Comparison with Goethe, Leonardo da Vinci, Bruno, Plato and Descartes.* Translated by Baron A. B. F.-M. Redesdale. London: John Lane.

Charney, E. 2013. Still Chasing Ghosts: A New Genetic Methodology Will Not Find the "Missing Heritability." *Independent Scientific News,* Sept. 19. Retrieved from http://www.independentsciencenews.org/health/still-chasing -ghosts-a-new-genetic-methodology-will-not-find-the-missing-heritability/.

Chase, A. 1977. *The Legacy of Malthus: The Social Costs of the New Scientific Racism.* New York: Alfred A. Knopf.

Clifford, D. 2004. Lamarck (1744–1829). The Victorian Web: History, Literature, and Culture in the Age of Victoria. Retrieved from http://www.victorianweb .org/science/lamarck1.html.

Cloud, J. 2010. Why DNA Isn't Your Destiny. *Time Magazine,* January 6. Retrieved from http://www.time.com/time/magazine/article/0,9171 ,1952313,00.html.

CNN. 2012. Florida Debate January 26, 2012. YouTube video. Retrieved from http://www.youtube.com/watch?v=dj10nUvBQWk.

Cohen, J. 2007: *Christ Killers: The Jews and the Passion from the Bible to the Big Screen.* New York: Oxford University Press.

Cole, D. 1999. *Franz Boas: The Early Years, 1858–1906.* Vancouver: Douglas & McIntyre.

Collins, C., B. Leondar-Wright, and H. Sklar. 1999. *Shifting Fortunes: The Perils of the Growing American Wealth Gap.* Boston: United for a Fair Economy.

Coon, C. S. 1962. *The Origin of Races.* New York: A. A. Knopf.

Cornwell, J. 2003. *Hitler's Scientists: Science, War and the Devil's Pact.* New York: Penguin.

Count, E. W. 1950. *This Is Race: An Anthology Selected from the International Literature on the Races of Man.* New York: Henry Schuman.

Covert Action. 1986. The Checkered Careers of James Angleton and Roger Pearson. *Covert Action* 25 (Winter):27–38.

Cox, E. S. 1923. *White America.* Charlottesville, VA: University of Virginia Press.

———. 1951. *Teutonic Unity.* Richmond, VA: Published by the author.

Craig, E. 1987. *The Mind of God and the Works of Man.* Oxford: Clarendon Press.

Cravens, H. 1978. *The Triumph of Evolution: American Scientists and the Heredity-Environment Controversy, 1900–1944.* Philadelphia: University of Pennsylvania Press.

———. 2009. Race, IQ, and Politics in Twentieth-Century America. In P. Farber and H. Cravens, ed., *Race and Science: Scientific Challenges to Racism in Modern America,* 152–184. Corvallis, OR: Oregon State University Press.

Crawford, J. 1993. *Hold Your Tongue: Bilingualism and the Politics of English Only.* Boston: Addison-Wesley.

Crew, F. A. E., C. D. Darlington, J. B. S. Haldane, C. Harland, L. T. Hogben, J. S. Huxley, H. J. Muller, J. Needham, G. P. Child, P. C. Koller, P. R. David, W. Landauer, G. Dahlberg, H. H. Plough, T. Dobzhansky, B. Price, R. A. Emerson, J. Schultz, C. Gordon, A. G. Steinberg, J. Hammond, C. H. Waddington, and C. L. Huskins. 1939. Social Biology and Population Improvement. *Nature* 144:521–522.

Damu, J. 2012. Arizona-Alabama Immigration Laws Recall 1935 Germany. People's World, January 26. Retrieved from http://www.peoplesworld.org /alabama-arizona-immigration-laws-recall-1935-germany/.

Darnell, R. 1971. The Professionalization of American Anthropology: A Case Study in the Sociology of Knowledge. *Social Science Information* 10:83–103.

Darwin, C. (1859) 1860. *The Origin of Species by Means of Natural Selection.* 6th ed. London: A. L. Burt.

———. (1871) 1874. *The Descent of Man and Selection in Relation to Sex.* 2nd ed. Chicago: The Henneberry Company.

Davenport, C. B. 1910. *Eugenics: The Science of Human Improvement by Better Breeding.* New York: Henry Holt & Company.

———. 1911. *Heredity in Relation to Eugenics.* New York: Henry Holt & Company.

———. 1912. The Inheritance of Physical and Mental Traits of Man and Their Application to Eugenics. In W. E. Castle, J. M. Coulter, C. B. Davenport, E. M. East, and W. L. Tower, eds., *Heredity and Eugenics,* 269–288. Chicago: University of Chicago Press.

———. 1913. *State Laws Limiting Marriage Selection Examined in the Light of Eugenics.* Eugenics Record Office Bulletin no. 9. Cold Spring Harbor, NY: Eugenics Record Office.

———. 1920. Twins in the Census. *Eugenical News* 4:16.

———. 1928. Race Crossing in Jamaica. *Scientific Monthly* 27 (September):225–228.

Davenport, C. B., and A. G. Love. 1921. *The Medical Department of the United States Army in the World War.* Vol. 15, *Statistics.* Part I. Washington, DC: Government Printing Office.

Davenport, C. B., and H. Laughlin. 1915. *How to Make a Eugenical Family Study.* Eugenics Record Office Bulletin no. 13. Cold Spring Harbor, NY: Eugenics Record Office.

———. 1917. The Great Nordic Race. *Eugenical News* 2 (2):10–11.

Davenport, C. B., and M. Steggerda. 1929. *Race Crossing in Jamaica.* Publication 395. Washington, DC: Carnegie Institution of America.

Dawkins, R. 1976. *The Selfish Gene.* New York: Oxford University Press.

Degler, C. N. 1991. *In Search of Human Nature: The Decline and Revival of Darwinism in American Social Thought.* New York: Oxford University Press.

Demerec, M., ed. 1951. *Origin and Evolution of Man.* Cold Spring Harbor Symposia on Quantitative Biology, vol. 15. Cold Spring Harbor, NY: Biology Laboratory.

Dobzhansky, T. 1950. Evolution in the Tropics. *American Scientist* 38:209–221.

Domhoff, G. W. 2013. Wealth, Income, and Power. Who Rules America? Retrieved from http://whorulesamerica.net/power/wealth.html.

Dorsey, G. A. 1928. *Race and Civilization.* New York: Longmans, Green & Company.

Dugdale, R. L. (1877) 1891. *The Jukes: A Study of Crime, Pauperism, Disease, and Heredity.* 5th ed. New York: G.P. Putnam's Sons.

Duke, D. 1998. *My Awakening: A Path to Racial Understanding.* Mandeville, LA: Free Speech Press.

East, E. M., and D. F. Donald. 1919. *Inbreeding and Outbreeding: Their Genetic and Sociological Significance.* Philadelphia: Lippincott.

The Economist. 2012. Caught in the Net: Alabama's Immigration Law Is Proving Too Strict and Too Costly. *The Economist,* January 12. Retrieved from http://www.economist.com/node/21543541.

Edgar, H. J. H., and K. L. Hunley, eds. 2009. Race Reconciled? How Biological Anthropologists View Race. Special issue, *American Journal of Physical Anthropology* 139:1–107.

Edsall, T. B., and D. A. Vise. 1985. CBS Fight a Litmus Test for Conservatives: Helms Group Faces Legal Hurdles in Ideological Takeover Bid. *Washington Post,* March 31. Retrieved from http://www.ferris.edu/isar/Institut/pioneer/helms.htm.

Ehrenreich, E. 2007. Otmar von Verschuer and the "Scientific" Legitimization of Nazi Anti-Jewish Policy. *Holocaust and Genocide Studies* 21:55–72.

Eisenberg, L. 1995. The Social Construction of the Human Brain. *American Journal of Psychiatry* 152:1563–1575.

Elden, S. 2011. *Reintroducing Kant's Geography*. Albany: State University of New York Press.

Ellwood, C. A. 1906. The Color Line. *American Journal of Sociology* 11:572.

Elmer, G. A., and E. E. Elmer. 1988. *Ethnic Conflict Abroad: Clues to America's Future*. Monterey, VA: American Immigration Control Foundation.

Engs, R. C. 2005. *The Eugenics Movement: An Encyclopedia*. Westport, CT: Greenwood Press.

Epstein, E. J. 1983. Ruling the World of Money. *Harper's Magazine* (November):25–32.

Erickson, P. A. 1986. The Anthropology of Josiah Clark Nott. *Kroeber Anthropological Society Papers* 65–66:103–120.

Estabrook, A. H. 1916. *The Jukes in 1915*. Washington, DC: Carnegie Institution of Washington.

Eysenck, H. J. 1971. *The IQ Argument: Race, Intelligence, and Education*. New York: Library Press.

———. 1985. *Decline and Fall of the Freudian Empire*. Washington, D.C.: Scott-Townsend Publishers.

Eze, E. C. 1995. The Color of Reason: The Idea of "Race" in Kant's Anthropology. In K. M. Faull, ed., *Anthropology and the German Enlightenment: Perspectives on Humanity*, 200–241. Lewisburg, PA: Bucknell University Press.

Farber, S. L. 1981. *Identical Twins Reared Apart: A Reanalysis*. New York: Basic Books.

Farber, P. L. 2009. Changes in Scientific Opinion on Race Mixing: The Impact of the Modern Synthesis. In P. Farber and H. Cravens, ed., *Race and Science: Scientific Challenges to Racism in Modern America*, 130–151. Corvallis: Oregon State University Press.

———. 2011. *Mixing Races: From Scientific Racism to Modern Evolutionary Ideas*. Baltimore, MD: Johns Hopkins University Press.

Fischer, E. 1913. *Die Rehobother Bastards und das Bastardierungsproblem beim Menschen: Anthropologische und Ethnographiesche Studien am Rehobother Bastardvolk in Deutsch-Südwest-Afrika*. Jena: G. Fischer.

———. 1914. Racial Hybridization. *Journal of Heredity* 10:465–467.

———. 1937. Vorwort. In M. Grant, *Die Eroberung eines Kontinents oder die Verbreitung der Rassen in Amerika*. Translated by C. C. Bryant. Berlin: Alfred Metzner.

———. 1959. *Begegnungen mit Toten: Aus den Erinnerungen eines Anatomen*. Freiburg: H. F. Schulz.

Fish, J. M. 1999. Why Psychologists Should Learn Some Anthropology. In R. W. Sussman, ed., *The Biological Basis of Human Behavior,* 197–198. Upper Saddle River, NJ: Prentice Hall.

———. 2002. *Race and Intelligence: Separating Science from Myth.* Mahwah, NJ: Lawrence Erlbaum Associates.

———. 2013. *The Myth of Race.* Lexington, KY: Argo Navis.

Fish, S. 1993. Reverse Racism, or How the Pot Got to Call the Kettle Black. *The Atlantic,* November 1. Retrieved from http://www.theatlantic.com /magazine/archive/1993/11/reverse-racism-or-how-the-pot-got-to-call-the -kettle-black/304638/.

Fitzgerald, F. S. 1925. *The Great Gatsby.* New York: Charles Scribner's Sons.

Fletcher, R. 1891. The New School of Criminal Anthropology. *American Anthropologist* 4:201–236.

Francis, S. 1995. Prospects for Racial and Cultural Survival. *American Renaissance* 6:1–6.

———. 1998, December. Race and American Identity (Part I): To Claim That We Are a "Universal Nation" Is to Deny the Past. *American Renaissance* 9 (2):1–4. Retrieved from http://www.unz.org/Pub/AmRenaissance-1998dec -00001.

———. 1999, January. Race and American Identity (Part II): Americans Have Long Taken Racial Nationalism for Granted. *American Renaissance* 10 (1):5. Retrieved from http://www.unz.org/Pub/AmRenaissance-1999jan -00005.

Frick W. 1934. German Population and Race Politics. *Eugenical News* 19 (2):33–38.

Fry, D. P. 2006. *The Human Potential for Peace: An Anthropological Challenge to Assumptions about War and Violence.* New York: Oxford University Press.

———. 2013. *War, Peace, and Human Nature: The Convergence of Evolutionary and Cultural Views.* Oxford: Oxford University Press.

Fuentes, A. 2012. *Race, Monogamy, and Other Lies They Told You: Busting Myths about Human Nature.* Berkeley, CA: University of California Press.

Galton, F. 1865. Hereditary Talent and Character. *Macmillan's Magazine* 12:157–166, 318–227.

———. 1869. *Hereditary Genius: An Inquiry into Its Laws and Consequences.* London: Macmillan.

———. 1875. The History of Twins, as a Criterion of the Relative Powers of Nature and Nurture. *Fraser's Magazine* 12:566–576.

———. 1883. *Inquiries into Human Faculty and Its Development.* London: Macmillan.

Garis, R. L. 1927. *Immigration Restriction: A Study of the Opposition to and Regulation of Immigration into the United States.* New York: Macmillan.

Garrett, H. E., and H. Bonner. 1961. *General Psychology.* 2nd rev. ed. New York: American Book Company.

Geertz, C. 1973. *The Interpretation of Cultures: Selected Essays by Clifford Geertz.* New York: Basic Books.

George, W. C. 1962. *The Biology of the Race Problem.* New York: Putnam National Letters Committee.

Gillette, A. 2011. *Eugenics and the Nature-Nurture Debate in the Twentieth Century.* New York, Palgrave Macmillan.

Gilman, J. 1924. Statistics and the Immigration Problem. *American Journal of Sociology* 30:29–48.

Gledhill, J. 2008. Anthropology and Espionage (2). *Media/Anthropology,* November 29. Retrieved from http://johnpostill.com/2008/11/29/anthropology-and-espionage-2/.

Gobineau, J.-A. Comte de 1853–1855. *Essai sur l'ine'galité des races humaines.* 4 vols. Paris: Didot Frères.

———. 1856. *The Moral and Intellectual Diversity of Races.* Philadelphia: Lippincott, Grambo & Co.

Goddard, H. H. 1912. *The Kallikak Family: A Study in the Heredity of Feeblemindedness.* New York, Macmillan.

———. 1913. The Binet Tests in Relation to Immigration. *Journal of Psycho-Asthenics* 18:105–107.

———. (1914) 1973. *Mental Illness and Social Policy.* New York: Arno Press Inc.

———. 1919. *Psychology of the Normal and Subnormal.* New York: Dodd, Mead & Co.

———. 1920. *Human Efficiency and Levels of Intelligence.* Princeton, NJ: Princeton University Press.

———. 1927. Who Is a Moron? *Scientific Monthly* 24:41–46.

Gordon, I. 2013. Big Surprise: Kris Kobach Still Believes in Self-Deportation. *Mother Jones,* April 22. Retrieved from http://www.motherjones.com/kris-kobach-self-deportation-senate-immigration-bill.

Gossett, T. F. 1965. *Race: The History of an Idea in America.* New York: Schocken Books.

Gottfredson, L. J. 1997. Why g Matters: The Complexity of Everyday Life. *Intelligence* 24:79–132.

Gould, S. J. 1977a. *Ever since Darwin.* New York: W. W. Norton.

———. 1977b. *Ontogeny and Phylogeny.* Cambridge, MA: Belknap Press.

———. 1994. The Curveball. *New Yorker,* November 28, 139–149.

———. 1995. Mismeasure by Any Measure. In R. Jacoby, and N. Glauberman, eds., *The Bell Curve Debate: History, Documents, Opinions,* 3–13. New York: Times Books.

———. 1996. *The Mismeasure of Man*. Rev. and exp. ed. New York: W. W. Norton.

Grant, M. 1916. *Passing of the Great Race*. New York: Charles Scribner's Sons.

———. 1918. *Passing of the Great Race: Or the Racial Basis of European History*. Rev. ed. New York: Charles Scribner's Sons.

———. 1919. Discussion of Article on Democracy and Heredity. *Journal of Heredity* 10:165.

———. 1925. America for the Americans. *Forum* 74:346–355.

———. 1928. Further Notes on the Racial Elements of European History. *Eugenical News* 13;120.

———. 1933. *The Conquest of a Continent, or the Expansion of Races in America*. New York: C. Scribner's Sons.

Graves, J. L. Jr. 2001. *The Emperor's New Clothes: Biological Theories of Race at the Millennium*. New Brunswick, NJ: Rutgers University Press.

———. 2002a. What a Tangled Web He Weaves. *Anthropological Theory* 2:131–154.

———. 2002b. The Misuse of Life History Theory: J. P. Rushton and the Pseudoscience of Racial Hierarchy. In J. M. Fish, ed., *Race and Intelligence: Separating Science from Myth*, 57–94. Mahwah, NJ: Lawrence Erlbaum Associates.

Gravlee, C. C., R. H. Bernard, and W. R. Leonard. 2003a. Heredity, Environment, and Cranial Form: A Reanalysis of Boas's Immigration Data. *American Anthropologist* 105:125–138.

———. 2003b. Boas's Changes in Bodily Form: The Immigrant Study, Cranial Plasticity, and Boas's Physical Anthropology. *American Anthropologist* 105:326–332.

Greene, S. 1994. Letter to the Editor. *New York Times Book Review,* November 13, 75.

Gregory, W. K. 1919. The Galton Society for the Study of the Origin and Evolution of Man. *Science* 49:267–268.

Grimm, J. 1848. *Geschichte der deutschen Sprache*. Leipzig: Weidmannsche buchhandlung.

Groves, C. 2008. *Extended Family: Long Lost Cousins. A Personal Look at the History of Primatology*. Arlington, VA: Conservation International.

Guisepi, R. A. 2007. Asian Americans. International World History Project. Retrieved from http://history-world.org/asian_americans.htm.

Günther, H. F. K. (1922) 1930. *Rassenkunde des deutschen Volkes*. 15th ed. München: J. F. Lehmann.

———. 1927. *The Racial Elements of European History*. Translated by G. C. Wheeler. London: Methuen & Co.

Guyer, P. (1998) 2004. Kant, Immanuel. In E. Craig, ed., *Routledge Encyclopedia of Philosophy*. London: Routledge. Retrieved from http://www.rep.routledge .com/article/DB047.

Hachee, M. R. 2011. Kant, Race, and Reason. Retrieved from https://www.msu .edu/~hacheema/kant2.htm?iframe=true, August 3, 2013.

Hacking, I. 1995. The Looping Effect of Human Kinds. In D. Sperber, ed., *Causal Cognition: An Interdisciplinary Approach*, 351–383. Oxford: Oxford University Press.

Haeckel, E. 1874. *Anthropologenie: Keimes- und Stammes-Geschichte des Manschen*. Leipzig: Engelmann.

———. 1892. *The History of Creation*. Translated by E. R. Lankester. London: Kegan Paul, Trench, Trubner & Co.

———. (1899) 1900. *The Riddle of the Universe: At the Close of the Nineteenth Century*. Translated by J. McCabe. New York: Harper & Brothers.

Hall, E. R. 1960. Zoological Subspecies of Man. *Mankind Quarterly* 1:118. Retrieved from http://www.amfirstbooks.com/IntroPages/ToolBarTopics /Articles/Other_Topics/Environ_vs_Genetic/ZoologicalSubspeciesOfMan .html.

Hammond, J. H. 1981. Josiah C. Nott. In D. G. Faust, ed., *The Ideology of Slavery: Proslavery Thought in the Antebellum South*, 206–238. Baton Rouge: Louisiana State University Press.

Hanke, L. 1949. *The Spanish Struggle for Justice in the Conquest of America*. Philadelphia: University of Pennsylvania Press.

Hankins, F. H. 1926. *The Racial Basis of Civilization: A Critique of the Nordic Doctrine*. New York: Knopf.

Hardwick, S. W. 2002. *Mythic Galveston: Reinventing America's Third Coast*. Baltimore, MD: Johns Hopkins University Press.

Harrison, F. V., ed. 1998. Contemporary Issues Forum: Race and Racism. *American Anthropologist* 100:601–864.

Harrison, G. P. 2010. *Race and Reality: What Everyone Should Know about Our Biological Diversity*. Amherst, NY: Prometheus Books.

Hays, H. R. 1964. *From Ape to Angel: An Informal History of Social Anthropology*. New York: Capricorn Books Edition.

Hearnshaw, L. 1979. *Cyril Burt: Psychologist*. Ithaca, NY: Cornell University Press.

Helman, C. G. 1994. *Culture, Health and Illness*. Oxford: Butterworth-Heinemann.

Heron, D. 1913. *Mendelism and the Problem of Mental Defect I. A Criticism of Recent American Work*. Questions of the Day and the Fray no. 7. London: Dulau.

Herrnstein, R. J., and C. Murray. 1994. *The Bell Curve: Intelligence and Class Structure in American Life*. New York: Free Press.

Herskovits, M. J. 1928. *The American Negro: A Study in Racial Crossing.* New York: Alfred A. Knopf.

———. 1930. *The Anthropometry of the American Negro.* New York: Columbia University Press.

Higham, C. 1983. *Trading with the Enemy: The Nazi-American Money Plot 1933–1949.* New York: Barnes & Noble Books.

The Hill. 2010. Mack (R) Compares Ariz. Law to Nazi Germany. Briefing Room; The Hill's Political Blog, April 29. Retrieved from http://thehill.com /blogs/blog-briefing-room/news/95123-mack-r-compares-ariz-law-to-nazi -germany.

Hirobe, I. 2001. *Japanese Pride, American Prejudice: Modifying the Exclusion Clause of the 1924 Immigration Act.* Palo Alto, CA: Stanford University Press.

Hitler, A. (1925) 1939. *Mein Kampf.* Translated by James Murphy. London: Hurst and Blackett.

Holden, C. 1980. Identical Twins Reared Apart. *Science* 207:1323–1328.

Hollander, E. P. 1993. Obituary: Otto Klineberg (1899–1992). *American Psychologist* 48:909–910.

Holman, R. 2011. Kobach Wrote Alabama Law in a Turkey Blind. *Wichita Eagle,* June 20. Retrieved from http://blogs.kansas.com/weblog/category /immigration/page/3/.

Horgan, J. 1993. Eugenics Revisited. *Scientific American* 268:122–131.

Horowitz, I. L. 1995. The Rushton File. In R. Jacoby and N. Glauberman, eds., *The Bell Curve Debate: History, Documents, Opinions,* 179–200. New York: Times Books.

Horsman, R. 1987. *Josiah Nott of Mobile: Southerner, Physician, and Racial Theorist.* Baton Rouge: Louisiana State University Press.

Hrdlička, A. 1919. *Physical Anthropology: Its Scope and Its Aims; Its History and Present Status in the United States.* Philadelphia: Wistar Institution of Anatomy and Biology.

Human Rights Library. 2000. *United States, Initial Report to the Committee on the Elimination of Racial Discrimination (September 2000).* University of Minnesota. Retrieved from http://www1.umn.edu/humanrts/usdocs /cerdinitial.html.

Hume, D. (1739–1740) 1978. *A Treatise of Human Nature.* 2nd ed., revised and edited by P. H. Nidditch. Oxford: Clarendon Press.

ISAR. 1998. Pioneer Fund, Certificate of Incorporation (1937). Retrieved from http://www.ferris.edu/isar/Institut/pioneer/pfund.htm.

Jablonski, N. G. 2006. *Skin: A Natural History.* Berkeley: University of California Press.

———. 2012. *Living Color: The Biological and Social Meaning of Skin Color.* Berkeley: University of California Press.

Jackson, J. P. Jr., and A. S. Winston. 2009. The Last Repatriationist: The Career of Ernest Sevier Cox. In P. Farber and H. Cravens, eds., *Race and Science: Scientific Challenges to Racism in Modern America,* 58–80. Corvallis: Oregon State University Press.

Jacoby, R., and N. Glauberman, eds. 1995. *The Bell Curve Debate: History, Documents, Opinions.* New York: Times Books.

James, W. 1881. *The Principles of Psychology.* 2 vols. Cambridge, MA: Harvard University Press.

———. 1904. Herbert Spencer. *Atlantic Monthly* 94 (1):98–108.

Jennings, H. 1932. Eugenics. In E. Seligman and A. Johnson, eds., *Encyclopedia of the Social Sciences.* New York: Macmillan.

Jensen, A. R. 1967. The Culturally Disadvantaged: Psychological and Educational Aspects. *Educational Research* 10:4–20.

———. 1969. How Much Can We Boost IQ and Scholastic Achievement? *Harvard Educational Review* 38:1–123.

———. 1981. *Straight Talk about Mental Tests.* New York: Free Press.

———. 1995. Psychological Research on Race Differences. *American Psychologist* 50:41–42.

———. 1998. *The g Factor: The Science of Mental Ability.* Westport, CT: Praeger.

———. 2006. *Clocking the Mind: Mental Chronometry and Individual Differences.* Oxford: Elsevier.

Jewish Virtual Library. 2013. Nuremberg Trial Defendants: Wilhelm Frick: 1877–1946. The American-Israeli Cooperative Enterprise. Retrieved from http://www.jewishvirtuallibrary.org/jsource/Holocaust/Frick.html.

Jonkers, P. A. E. 2008, September 2. Roger Pearson. Retrieved from http://moversandshakersofthesmom.blogspot.com/2008/09/roger-pearson.html.

Jonsson, P. 2011. American Renaissance: Was Jared Lee Loughner Tied to Anti-Immigrant Group? *Christian Science Monitor,* January 9. Retrieved from http://www.csmonitor.com/layout/set/print/content/view/print/355125.

Joseph, J. 2001. Separated Twins and the Genetics of Personality Differences: A Critique. *American Journal of Psychology* 114 (1): 1–30.

———. 2004. *The Gene Illusion: Genetic Research in Psychiatry and Psychology under the Microscope.* New York: Algora.

———. 2010. Genetic Research in Psychology and Psychiatry: A Critical Overview. In K. E. Hood, C. T. Halpern, G. Greenberg, and R. M. Lerner, eds., *Handbook of Developmental Science, Behavior and Genetics,* 557–625. New York: Wiley-Blackwell.

Jurmain, R., L. Kilgore, W. Trevathan, and R. L. Ciochon. 2014. *Introduction to Physical Anthropology.* Belmont, CA: Wadsworth, Cengage Learning.

Kamen, H. 1998. *The Spanish Inquisition: A Historical Revision*. New Haven, CT: Yale University Press.

Kamin, L. J. 1974. *The Science and Politics of IQ*. Potomac, NJ: Lawrence Erlbaum Associates.

———. 1995. Lies, Damned Lies, and Statistics. In R. Jacoby and N. Glauberman, eds., *The Bell Curve Debate: History, Documents, Opinions*, 179–200. New York: Times Books.

Kant, I. (1764) 1965. *Observations on the Feeling of the Beauty and the Sublime*. Translated by J. R. Goldthwait. Berkeley: University of California Press.

———. (1775) 1950. On the Different Races of Man. In E. W. Count, *This Is Race*, 16–24. New York: Shuman.

———. (1798) 1974. *Anthropology from a Pragmatic Point of View*. Translated by Mary J. Gregor. The Hague: Martinus Nijhoff.

———. 1802. *Physische Geographie*. Königsberg: Bey Göbbels und Unzer.

Kellogg, V. 1922. *Human Life as the Biologist Sees It*. New York: H. Holt and Company.

Kelsey, C. 1913. Review of Boas's *The Mind of Primitive Man. Annals of the American Academy of Political and Social Sciences* 46:203–204.

———. 1916. *The Physical Basis of Society*. New York: Appleton.

Kenrick, K. L. 1914. The Case against Eugenics. *British Review* 64–81.

King, L. L. 1973. The *Traveling Carnival of Racism. New Times,* December 28, 36.

Klineberg, O. 1928. An Experimental Study of Speed and Other Factors in "Racial" Differences. *Archives of Psychology* 15:1–111.

———. 1931. A Study of Psychological Differences between "Racial" and National Groups in Europe. *Archives of Psychology* 132:1–58.

———. 1935a. *Negro Intelligence and Selective Migration*. New York: Columbia University Press.

———. 1935b. *Race Differences*. New York: Harper-Collins.

———, ed. 1944. *Characteristics of the American Negro*. New York: Harper and Brothers.

Kobach, K. W. 1990. *Political Capital: The Motives, Tactics, and Goals of Politicized Businesses in South Africa*. Ann Arbor, MI: University of Michigan Press.

Köpping, K.-P. 2005. *Adolf Bastian and the Psychic Unity of Mankind: The Foundations of Anthropology in Nineteenth-Century Germany*. Münster: Lit Verlag.

Kroeber, A. L. 1915. Eighteen Professions. *American Anthropologist* 17:283–288.

———. 1916. Inheritance by Magic. *American Anthropologist* 18:26–27.

———. 1917. The Superorganic. *American Anthropologist* 19:163–213.

———. 1923. *Anthropology.* New York: Harcourt, Brace & Co.

Kühl, S. 1994. *The Nazi Connection: Eugenics, American Racism, and German National Socialism.* New York: Oxford University Press.

Kuhn, T. 1962. *The Structure of Scientific Revolutions.* Chicago: University of Chicago Press.

Kurtagic, A. 2011. Interview with Richard Lynn. Mermod and Mermod Publishing Group, September 18. Retrieved from http://www.wermodan dwermod.com/newsitems/news160920111400.html.

Kuttner, R. E., ed. 1967. *Race and Modern Science.* New York: Social Science Press.

La Peyrère, I. 1656. *A Theological Systeme upon That Presupposition That Men Were Before Adam.* London.

Lagnado, L. M., and S. C. Dekel. 1991. *Children of the Flames: Dr. Josef Mengele and the Untold Story of the Twins of Auschwitz.* New York: William Morrow and Company.

Lamarck, J.-B. 1778. *Flore françoise.* Paris: l'Imprimerie Royale.

———. 1809. *Philosophie zoologique.* Paris: Librairie F. Savy.

Lamm, R. 1985. *Megatraumas: America at the Year 2000.* Boston: Houghton Mifflin.

Lane, C. 1995. Tainted Sources. In R. Jacoby and N. Glauberman, eds., *The Bell Curve Debate: History, Documents, Opinions,* 125–139. New York: Times Books.

Langford, E. [R. Pearson]. 1966. Editorial. *New Patriot* 8 (June).

Lapham, S. J., P. Montgomery, and D. Niner. 1993. *We the American . . . Foreign Born.* Washington, DC: U.S. Department of Commerce, Economics and Statistics Administration, Bureau of the Census.

Larson, E. J. 1995. *Sex, Race, and Science: Eugenics in the Deep South.* Baltimore, MD: Johns Hopkins University Press.

Laughlin, H. H. 1914a. *Report of the Committee to Study and to Report on the Best Practical Means of Cutting off the Defective Germ-Plasm in the American Population. II. The Legal, Legislative, and Administrative Aspects of Sterilization.* Eugenics Record Office Bulletin no. 10a. Cold Spring Harbor, NY: Eugenics Record Office.

———. 1914b. *Report of the Committee to Study and to Report on the Best Practical Means of Cutting off the Defective Germ-Plasm in the American Population. I. Scope of the Committee's Work.* Eugenics Record Office, Bulletin no. 10b. Cold Spring Harbor, NY: Eugenics Record Office.

———. 1922. *Eugenical Sterilization in the United States.* Chicago: Psychopathic Laboratory of the Municipal Court of Chicago.

———. 1924. *Europe as an Emigrant-Exporting Continent and the United States as an Immigrant-Receiving Nation. Hearings before the Committee* on *Immigration and Naturalization, House of Representatives, Sixty-Eighth Congress, First Session, March 8, 1924.* Washington, DC: Government Printing Office.

———. 1928. Walter A. Plecker to Harry H. Laughlin, November 22. Harry H. Laughlin Papers, Pickler Memorial Library, Truman State University, Kirksville, MO.

———. 1939. *Immigration and Conquest: A Report of the Special Committee on Immigration and Naturalization of the Chamber of Commerce of the State of New York.* New York: Chamber of Commerce of the State of New York.

Lerner, R. M. 1992. *Final Solutions: Biology, Prejudice, and Genocide.* University Park: Pennsylvania State University Press.

Levin, M. 1995. The Evolution of Racial Differences in Morality. *American Renaissance* 6 (April):3, 5.

Lewontin, R. C. 2005. Confusions about Human Races. Is Race "Real"? Retrieved from http://raceandgenomics.ssrc.org/Lewontin/.

Lichtenstein, G. 1977. Fund Backs Controversial Study of "Racial Betterment." *New York Times,* December 11.

Lieberman, L. 2001. How "Caucasoids" Got Such Big Brains and How They Shrunk: From Morton to Rushton. *Current Anthropology* 42:69–95.

Lifton, R. J. 1986. *The Nazi Doctors: Medical Killing and the Psychology of Genocide.* New York: Basic Books.

Linklater, M. 1995. The Curious Laird of Nigg. In R. Jacoby and N. Glauberman, eds., *The Bell Curve Debate: History, Documents, Opinions,* 140–143. New York: Times Books.

Linnaeus, C. 1758. *Systema naturæ per regna tria naturæ, secundum classes, ordines, genera, species, cum characteribus, differentiis, synonymis.* Tomus I. Editio decimal. Stokholm: Holmiæ (Salvius).

Little, M. A. 2010. Franz Boas's Place in American Physical Anthropology and Its Institutions. In M. A. Little and K. A. R. Kennedy, eds., *Histories of American Physical Anthropology in the Twentieth Century,* 55–85. Lanham, MD: Lexington Books.

Little, M. A., and K. A. R. Kennedy, eds. 2010. *Histories of American Physical Anthropology in the Twentieth Century.* Lanham, MD: Lexington Books.

Littlefield, A., L. Lieberman, and L. T. Reynolds. 1982. Redefining Race: The Potential Demise of a Concept in Physical Anthropology. *Current Anthropology* 23:641–654.

Livingstone, D. N. 1987. *Nathaniel Southgate Shaler and the Culture of American Science.* Tuscaloosa: University of Alabama Press.

Livingstone, F. B. 1962. On the Nonexistence of Human Races. *Current Anthropology* 3:279.

Locke, John. (1690) 1980. *Second Treatise of Government*. Reprinted in *John Locke Second Treatise of Government*. C. B. McPherson, ed., Indianapolis: Hackett Publishing Co.

Lombardo, P. A. 2002. "The American Breed": Nazi Eugenics and the Origins of the Pioneer Fund. *Albany Law Review* 65:743–830.

———. 2008. *Three Generations, No Imbeciles: Eugenics, the Supreme Court, and Buck V. Bell*. Baltimore, MD: Johns Hopkins University Press.

Longhurst, J. E. 1964. *The Age of Torquemada*. Sandoval, NM: Coronado Press.

Lowie, R. H. 1929. *Are We Civilized? Human Culture in Perspective*. New York: Harcourt, Brace and Company.

Ludmerer, K. 1972. *Genetics and American Society*. Baltimore, MD: Johns Hopkins University Press.

Lynn, R. 1972. *Personality and National Character*. Oxford: Pergamon.

———. 1978. Ethnic and Racial Differences in Intelligence, International Comparisons. In R. T. Osborne, C. E. Noble, and N. Weyl, eds., *Human Variation: The Biopsychology of Age, Race, and Sex*, 261–286. New York: Academic Press.

———. 1987. The Intelligence of the Mongoloids: A Psychometric, Evolutionary and Neurological Theory. *Personality and Individual Differences* 8:813–844.

———. 1991a. The Evolution of Racial Differences in Intelligence. *Mankind Quarterly* 32:99–121.

———. 1991b. Civilization and the Quality of Population. *Journal of Social, Political and Economic Studies* 1:121–123.

———. 1994a. Some Reinterpretations of the Minnesota Transracial Adoption Study. *Intelligence* 19:21–27.

———. 1994c. Research That Was to Prove Jensen Wrong Proves Him Right. *American Renaissance* 5 (March):4.

———. 2001. *Eugenics: A Reassessment*. Westport, CT: Greenwood.

———. 2004. A Review of "A New Morality from Science: Beyondism." *Irish Journal of Psychology* 2.

———. 2011. *Dysgenics: Genetic Deterioration in Modern Populations*. 2nd rev. ed. Ulster: Ulster Institute for Social Research.

Lynn, R., and T. Vanhanen. 2002. *IQ and the Wealth of Nations*. Westport, CT: Praeger/Greenwood.

MacArthur, R., and E. O. Wilson. 1967. *The Theory of Island Biogeography*. Princeton, NJ: Princeton University Press.

Mack, M. 2003. *German Idealism and the Jew: The Inner Anti-Semitism of Philosophy and German Jewish Responses*. Chicago: University of Chicago Press.

MacKenzie, D. A. 1981. *Statistics in Britain, 1865–1930: The Social Construction of Scientific Knowledge.* Edinburgh: Edinburgh University Press.

Malthus T. R. (1798) 2008. *An Essay on the Principle of Population.* Oxford: Oxford World's Classics.

Marks, J. 1995. *Human Biodiversity: Genes, Race, and History.* New York: Aldine du Gruyter.

———. 2002. Folk Heredity. In J. M. Fish, *Race and Intelligence: Separating Science from Myth,* 95–112. Mahwah, NJ: Lawrence Erlbaum Associates.

———. 2010a. Why Were the First Anthropologists Creationists? *Evolutionary Anthropology* 19:222–226.

———. 2010b. The Two 20th Century Crises of Racial Anthropology. In M. A. Little and K. A. R. Kennedy, eds., *Histories of American Physical Anthropology in the Twentieth Century,* 187–206. Lanham, MD: Lexington Books.

———. 2012. Why Be against Darwin? Creationism, Racism, and the Roots of Anthropology. *Yearbook of Physical Anthropology* 55:95–104.

Martin, J. S. 1950. *All Honorable Men.* Boston: Little, Brown.

Masters, W. H., and V. E. Johnson. 1966. *Human Sexual Response.* Toronto: Bantam Books.

May, J. A. 1970. *Kant's Concept of Geography and Its Relation to Recent Geographical Thought.* Toronto: University of Toronto Press.

McCulloch, R. 1995. The Preservation Imperative: Why Separation Is Necessary for Survival. *American Renaissance* 6:5–6. Retrieved from http://www.amren.com/archives/back-issues/august-1995/.

McDonnell, P. J. 1997. Prop. 187 Found Unconstitutional by Federal Judge. *Los Angeles Times,* November 15.

McDougall, W. 1909. *An Introduction to Social Psychology.* London: Methuen & Co.

———. 1923. *Is America Safe for Democracy?* New York: Charles Scribner's Sons.

———. 1925. *Ethics and Some Modern World Problems.* London: Methuen.

Mead, M. 1928. *Coming of Age in Samoa.* New York: Morrow.

Meaney, M. J. 2010. Epigenetics and the Biological Definition of Gene × Environment Interactions. *Child Development* 81:41–79.

Medawar, P. B. 1975. Review of *Francis Galton: The Life of and Work of a Victorian Genius.* By D. W. Forrest. *Times Literary Supplement,* January 24, 83.

Mercer, J. 1994. A Fascination with Genes: Pioneer Fund Is at Center of Debate over Research on Race and Intelligence. *Journal of Higher Education* 28 (December 7):28.

Michael, G. 2008. *Willis Carto and the American Far Right.* Gainesville: University Press of Florida.

Miller, A. 1995. Professors of Hate. In R. Jacoby and N. Glauberman, eds., *The Bell Curve Debate: History, Documents, Opinions,* 162–178. New York: Times Books.

Miller, P. 2012. A Thing or Two about Twins. *National Geographic* 221:38–65.

Miller, S., and S. A. Ogilvie. 2006. *Refuge Denied: The* St. Louis *Passengers and the Holocaust.* Madison: University of Wisconsin Press.

Mills, C. W. 1997. *The Racial Contract.* Ithaca, NY: Cornell University Press.

Milner, R. 2009. *Darwin's Universe: Evolution from A to Z.* Berkeley: University of California Press.

Molnar, S. 2006. *Human Variation: Races, Types, and Ethnic Groups.* 6th ed. Englewood Cliffs, NJ: Prentice Hall.

———. 2010. *Human Variation: Races, Types, and Ethnic Groups.* 8th ed. Englewood Cliffs, NJ: Prentice Hall.

Montagu, A. 1942. *Man's Most Dangerous Myth: The Fallacy of Race.* New York: Columbia University Press.

———. 1964. *The Concept of Race.* London: Collier-Macmillan.

———. 1997. *Man's Most Dangerous Myth: The Fallacy of Race.* 6th ed. Walnut Creek, CA: AltaMira Press.

Montesquieu, C. L. de Secondat. 1748. *De l'Esprit des Loix, ou du rapport que les lois doivent avoir avec la constitution de chaque gouvernement, mœurs, climat, religion, commerce, etc. (sic); à quoi l'auteur a ajouté des recherches sur les lois romaines touchant les successions, sur les lois françaises et sur les lois féodales, s.d.* Genève: Barrillot & Fils.

Moore, R. 2001. The "Rediscovery" of Mendel's work. *Bioscene* 27:13–24.

Morgan, T. H. 1925. *Evolution and Genetics.* 2nd ed. Princeton, NJ: Princeton University Press.

———. 1932. *Scientific Basis of Evolution.* New York: W. W. Norton.

Morrill, J. 2011. White Nationalist Group to Hold Annual Conference in Charlotte. *Charlotte Observer,* January 20. Retrieved from http://www.mcclatchydc.com/2011/01/20/107141/white-nationalist-group-to-hold.html#storylink=cpy.

Morton, S. G. 1839. *Crania Americana: Or, a Comparative View of Various Aboriginal Nations of North and South America; to Which Is Prefixed an Essay on the Varieties of the Human Species.* Philadelphia: J. Dodson.

———. 1844. *Crania Aegyptiaca: Or, Observations on Egyptian Ethnography Derived from Anatomy.* Philadelphia: John Pennington.

Mukhopadhyay, C., R. Henze, and Y. Moses. 2014. *How Real Is Race? A Sourcebook on Biology, Culture, and Race.* Lanham, MD: AltaMira Press.

Muller, H. J. 1934. Dominance of Economics over Eugenics. In H. F. Perkins and H. H. Laughlin, eds., *A Decade of Progress in Eugenics,* 138–144. Baltimore, MD: Williams & Wilkins.

Müller-Hill, B. 1998. *Murderous Science: Elimination by Scientific Selection of Jews, Gypsies, and Others in Germany, 1933–1945*. Translated by G. R. Fraser. Plainview, NY: Cold Spring Harbor Press.

Murphy, C. 2012. *God's Jury: The Inquisition and the Making of the Modern World*. New York: Houghton Mifflin Harcourt.

Nash, G. 1999. *Forbidden Love: The Secret History of Mixed-Race America*. New York: Henry Holt and Company.

Nature. 1912. The First International Eugenics Congress. *Nature* 89 (August 1): 558–561.

NBC Nightly News. 2011. Transhumanism: The Obvious Offspring of Eugenics and Genocide. Info Wars, NBC Nightly News. YouTube video, uploaded December 20. Retrieved from http://www.youtube.com/watch?v=AApMqG7ADsg.

NC-ISAAC (North Carolina's Information Sharing and Analysis Center). 2011, January 18. New Century Foundation, American Renaissance, and Jared Lee Loughner. Special Information Bulletin 29. Retrieved from http://info.publicintelligence.net/NC-ISAAC-Loughner.pdf.

Nei, M., and A. K. Roychoudhury. 1974. Evolutionary Relationships of Human Populations on a Global Scale. *Molecular Biology and Evolution* 10:927–943.

Neugebauer, C. M. 1990. The Racism of Hegel and Kant. In H. Odera Oruka, ed., *Sage Philosophy: Indigenous Thinkers and Modern Debate on African Philosophy*, 259–272. Leiden: E. J. Brill.

New York Times. 1912. First Eugenics Congress. *New York Times*, July 25.

———. 1932. The Week in Science: Eugenists and Geneticists at Odds. *New York Times*, August 28.

Nordenskiöld, E. 1928. *The History of Biology: A Survey*. New York: Alfred A. Knopf.

Nott, J. C. 1844. *Two Lectures, on the Natural History of the Caucasian and Negro Races*. Mobile, AL: Dade and Thompson.

———. 1856. Appendix. In Comte de Gobineau, *The Moral and Intellectual Diversity of Races*. Philadelphia: Lippincott, Grambo & Co.

Nott, J. C., and G. R. Gliddon. 1854. *Types of Mankind*. Philadelphia: Lippincott, Grambo & Co.

Oakesmith, J. 1919. *Race and Nationality*. London: Heinemann.

Obama, B. 2013. Full Text: President Obama's Remarks on Trayvon Martin. *National Journal*, July 19. Retrieved from http://www.nationaljournal.com/whitehouse/full-text-president-obama-s-remarks-on-trayvon-martin-20130719.

Odum, H. W. 1910. *Social and Mental Traits of the Negro. Research into the Conditions of the Negro Race in Southern Towns. A Study of Race Traits, Tendencies, and Prospects*. New York: Columbia University Press.

———. 1913. Negro Children in the Public Schools of Philadelphia. *Annals of the American Academy of Political and Social Sciences* 49:186–208.

Okuefuna, D. 2007. *Racism: A History.* Three-part documentary series, aired on BBC, March 22, March 28, and April 4.

Orsucci, A. 1998. Ariani, indogermani, stirpi mediterranee: aspetti del dibattito sulle razze europee (1870–1914). *Cromohs* 3:1–9. Retrieved from http://www.cromohs.unifi.it/3_98/orsucci.html.

Ortner, D. J. 2010. Aleš Hrdlička and the Founding of the *American Journal of Physical Anthropology:* 1918. In M. A. Little and K. A. R. Kennedy, eds., *Histories of American Physical Anthropology in the Twentieth Century,* 87–104. Lanham, MD: Lexington Books.

Osborn, H. F. 1918. Preface. In M. Grant, *Passing of the Great Race: Or the Racial Basis of European History,* xi–xiii. Rev. ed. New York: Charles Scribner's Sons.

———. 1926. The Evolution of Human Races. *Natural History* 26:3–13.

Osborne, R. T. 1962. School Achievement of White and Negro Children of the Same Mental and Chronological Ages. *Mankind Quarterly* 2:26–29.

Oshinsky, D. 1996. *"Worse Than Slavery": Parchman Farm and the Ordeal of Jim Crow Justice.* New York: Free Press.

Panofsky, A. L. 2005. The Gene for Trouble? The Social Roots for Controversy in Behavior Genetics. Retrieved from https://files.nyu.edu/alp219/public/Panofsky%20Gene%20for%20Trouble%20Writing%20sample.pdf.

Papavasiliou, C. 1999. Interview with Dr. Robert Gordon. Creative Consciousness Evolution. Retrieved from http://www.euvolution.com/euvolution/interview05.html.

Patterson, Q. 2001. The Root of Conflict in Jamaica. *New York Times,* July 23, A17.

Paul, D. B. 1995. *Controlling Human Heredity: 1865 to the Present.* Atlantic Highlands, NJ: Humanities Press.

Pauwels, J. R. 2003. Profits *über Alles!* American Corporations and Hitler. *Labour/Le Travail* 51 (Spring):223–249. Retrieved from http://www.intelltheory.com/burt.shtml.

Pearl, R. 1927. The Biology of Superiority. *American Mercury,* November, 257–266.

Pearson, K. 1925. Foreword. *Ann Eugen* 1:1–4.

Pearson, R. 1966. *Eugenics and Race.* Colchester, UK: Clair Publications.

———. 1969. The Indo-European Trustee Family System: A Comparative Study in Basic Social Organization. PhD diss., University College London.

———. 1995. The Concept of Heredity in Western Thought: Part II—The Myth of Biological Egalitarianism. *Mankind Quarterly* 35:346.

———. 1996. *Heredity and Humanity: Race, Eugenics and Modern Science.* Washington, DC: Scott-Townsend.

Pendergrast, M. 1993. *For God, Country, and Coca-Cola: The Unauthorized History of the Great American Soft Drink and the Company That Makes It.* New York: Charles Scribner's Sons.

Pianka, E. R. 1970. On *r*- and *K*-Selection. *American Naturalist* 104:592–597.

Pickens, D. K. 1968. *Eugenics and the Progressives.* Nashville, TN: Vanderbilt University Press.

Pim, J. E. 2010. *Nonkilling Societies.* Honolulu: Center for Global Nonkilling.

Ploetz, A. 1895. *Grunlinien einer Rassenhygiene.* Berlin: S. Fischer.

Plucker, J. 2012. The Cyril Burt Affair. Human Intelligence. Retrieved from http://www.indiana.edu/~intell/burtaffair.shtml.

Poliakov, L. 1971. *The Aryan Myth: A History of Racist and Nationalist Ideas in Europe.* London: Sussex University Press.

Pool, J., and S. Pool. 1979. *Who Financed Hitler: The Secret Funding of Hitler's Rise to Power, 1919–1933.* New York: Dial Press.

Popkin, R. H. 1973. The Marrano Theology of Isaac Peyrère. *Studi Internazionali Di Filosofia* 5:97–126.

———. (1974) 1983. The Philosophical Basis of Modern Racism. In C. Walton and J. P. Anton, eds., *Philosophy and the Civilizing Arts: Essays Presented to Herbert W. Schneider on His 80th Birthday,* 126–165. Athens: Ohio University Press. Reprinted in R. H. Popkin, *The High Road to Pyrrhonism,* 79–102. Indianapolis: Hackett Publishing Company.

———. 1976. The Pre-Adamite Theory in the Renaissance. In E. P. Mahoney, ed., *Philosophy and Humanism: Renaissance Essays in Honor of Paul Oskar Kristelle,* 50–69. Leiden: E. J. Brill.

Population Projection Program. 2000. *Projections of the Resident Population by Age, Sex, Race and Hispanic Origin: 1999 to 2100.* Washington, DC: Population Division, U.S. Census Bureau.

Porter, D. 1999. Eugenics and the Sterilization Debate in Sweden and Britain before World War. *Scandinavian Journal of History* 24:145–162.

Price, A. 2001. *The Last Year of the Luftwaffe: May 1944–May 1945.* London: Greenhill.

Price, D. 2000. Anthropologists as Spies. *Nature,* November 2. Retrieved from http://www.thenation.com/article/anthropologists-spies#.

Proctor, R. N. 1988. *Racial Hygiene: Medicine under the Nazis.* Cambridge: Harvard University Press.

Punnett, R. C. 1909. *Mendelism.* New York: Wilshire.

Putnam, C. 1961. *Race and Reason: A Yankee View.* Cookeville, TN: New Century Books.

Raspail, J. (1975) 1995. *The Camp of the Saints.* Translated by Norman Shapiro. Petoskey, MI: The Social Contract Press.

Read, D. W. 2012. *How Culture Makes Us Human.* Walnut Creek, CA: Left Coast Press.

Rectenwald, M. 2008. *Darwin's Ancestors: The Evolution of Evolution.* The Victorian Web: History, Literature, and Culture in the Age of Victoria. Retrieved from http://www.victorianweb.org/science/darwin/rectenwald .html.

Reed, A. Jr. 1995. Intellectual Brown Shirts. In R. Jacoby and N. Glauberman, eds., *The Bell Curve Debate: History, Documents, Opinions,* 263–268. New York: Times Books.

Relethford, J. H. 2004. Boas and Beyond: Migration and Craniometric Variation. *American Journal of Human Biology* 16:79–386.

———. 2013. *The Human Species: An Introduction to Biological Anthropology.* New York: McGraw-Hill.

Riccardi, N. 2011. On Immigration, Momentum Shifts away from Arizona. *Los Angeles Times,* March 6.

Richardson, K., and J. Joseph. 2011. Misleading Treatments: Attempts to Validate the EEA in Twin Research. Comment on "Twin Research: Misperceptions." *Psychology Today,* August 29. Retrieved from http://www .psychologytoday.com/blog/twofold/201108/twin-research-misperceptions /comments.

Richmond Times-Dispatch. 1924. Racial Integrity. *Richmond Times-Dispatch,* February 18.

Riddle, O. 1947. Biographical Memoir of Charles Benedict Davenport 1866–1944. *National Academy of Sciences Biographical Memoirs* 25: 73–110.

Ripley, W. Z. 1899. *The Races of Europe: A Sociological Study.* New York: D. Appleton & Company.

Roddy, Dennis. 2005, January 23. Jared Taylor, a Racist in the Guise of "Expert." *Pittsburgh Post-Gazette,* January 23. Retrieved from http://www .post-gazette.com/pg/05023/446341.stm.

Roff, D. 1992. *The Evolution of Life Histories: Theory and Analysis.* London: Routledge, Chapman and Hall.

Roosevelt, T. 1913. T. Roosevelt to C. Davenport, January 3. Reprinted at DNA Learning Center. Retrieved from http://www.dnalc.org/view/11219-T -Roosevelt-letter-to-C-Davenport-about-degenerates-reproducing-.html.

Rose, R. J. 1982. "Separated Twins: Data and Their Limits." *Science* 215: 959–960.

Rosenberg, J. 2013. Voyage of the *St. Louis.* About.com 20th Century History. Retrieved from http://history1900s.about.com/od/holocaust/a/stlouis.htm.

Rosenthal, S. J. 1995. The Pioneer Fund Financier of Fascist Research. *American Behavioral Scientist* 39:44–61.

Rüdin, E. 1930. Hereditary Transmission of Mental Diseases. *Eugenical News* 15: 171–174.

———. 1933. Eugenic Sterilization: An Urgent Need. *Birth Control Review* 27 (4):102–104.

———. 1938. Honor of Prof. Dr. Alfred Ploetz. *ARGB* 32:473–474.

Rushton, J. P. 1985. Differential K Theory: The Sociobiology of Individual and Group Differences. *Personality and Individual Differences* 6:441–452.

———. 1988. Race Differences in Behaviour: A Review and Evolutionary Analysis. *Personality and Individual Differences* 9:1009–1024.

———. 1995. *Race, Evolution, and Behavior: A Life History Perspective.* New Brunswick, NJ: Transaction Publishers.

———. 1999. *Race, Evolution, and Behavior: A Life History Perspective.* Special abridged ed. New Brunswick, NJ: Transaction Publishers.

———. 2011, March 9. Race and IQ—Dr. John Philippe Rushton. Disclose TV, March 9. Retrieved from http://www.disclose.tv/action/viewvideo/68820 /Race_and_IQ___Dr_John_Philippe_Rushton/.

Rushton, J. P., and A. F. Bogaert. 1988. Race versus Social Class Differences in Sexual Behavior: A Follow-Up Test of the *r*/*K* Dimension. *Journal of Research in Personality* 22:259–272.

Rushton, J. P., and A. R. Jensen. 2005. Thirty Years of Research on Group Differences in Cognitive Ability. *Psychology, Public Policy, and the Law* 11:235–294.

———. 2006. The Totality of Available Evidence Shows Race-IQ Gap Still Remains. *Psychological Science* 17:921–922.

———. 2010. The Rise and Fall of the Flynn Effect as a Reason to Expect a Narrowing of the Black–White IQ Gap. *Intelligence* 38:213–219.

Samelson, F. 1979. Putting Psychology on the Map: Ideology and Intelligence Testing. In A. R. Buss, ed., *Psychology in Social Context,* 103–168. New York: Irvington Publishers.

Sanchez, M. 2012. Commentary: Why Didn't GOP Question CPAC Panelist's Alleged White Supremacist Ties? *Kansas City Star,* February 13. Retrieved from http://www.mcclatchydc.com/2012/02/13/138717/commentary-why -didnt-gop-question.html#storylink=cpy.

Sautman, B. 1995. Theories of East Asian Superiority. In R. Jacoby and N. Glauberman, eds., *The Bell Curve Debate: History, Documents, Opinions,* 201–221. New York: Times Books.

Schiller, M. 1995. Separation: Is There an Alternative? *American Renaissance* 6:1, 3–6. Retrieved from http://www.scribd.com/doc/45644375/199502 -American-Renaissance.

Science Encyclopedia. 2013. Eugenics—Criticisms of Eugenics. Retrieved from http://science.jrank.org/pages/9250/Eugenics-Criticisms-Eugenics.html.

Sedgewick, J. 1995. Inside the Pioneer Fund. In R. Jacoby and N. Glauberman, eds., *The Bell Curve Debate: History, Documents, Opinions,* 144–161. New York: Times Books.

Selden, S. 1999. *Inheriting Shame: The Story of Eugenics in America.* New York: Teachers College Press.

Shaler, N. S. 1884. The Negro Problem. *Atlantic Monthly,* November, 697–698.

———. 1891. *Nature and Man in America.* New York: C. Scribner's Sons.

Shipman, P. (1994) 2002. *The Evolution of Racism: Human Differences and the Use and Abuse of Science.* Cambridge: Harvard University Press.

Shockley, W. 1972. Dysgenics, Geneticity, Raceology: A Challenge to the Intellectual Responsibility of Educators. *Phi Delta Kappan* 53:297–307.

Shurkin, J. 2006. *Broken Genius: The Rise and Fall of William Shockley, Creator of the Electronic Age.* London: Macmillan.

Silver, J. W. 1984. *Running Scared: Silver in Mississippi.* Jackson: University of Mississippi Press.

Slotkin, J. S. 1965. *Readings in Early Anthropology.* Chicago: Methuen.

Smedley, A. 1999. *Race in North America: Origin and Evolution of a World-view.* 2nd ed. Boulder, CO: Westview.

Smedley, A., and B. D. Smedley. 2012. *Race in North America: Origin and Evolution of a Worldview.* 4th ed. Boulder, CO: Westview.

Smith, H. M., D. Chiszar, and R. R. Montanucci. 1997. Subspecies and Classification. *Herpetological Review* 28:13–16.

Southern Poverty Law Center. 2008. Profiles of 21 Nativist Leaders. *Intelligence Report* 129 (Spring). Retrieved from http://www.splcenter.org/get-informed/intelligence-report/browse-all-issues/2008/spring/the-nativists.

———. 2009a. FAIR: The Action Arm. Southern Poverty Law Center. Retrieved from http://www.splcenter.org/publications/the-nativist-lobby-three-faces-of-intolerance/fair-the-action-arm.

———. 2009b. John Tanton and the Nativist Movement. Southern Poverty Law Center. Retrieved from http://www.splcenter.org/publications/the-nativist-lobby-three-faces-of-intolerance/john-tanton-and-the-nativist-movement.

———. 2011. The Partner: Behind the Legislators. Southern Poverty Law Center. Retrieved from http://www.splcenter.org/get-informed/publications/attacking-the-constitution-slli-and-the-anti-immigrant-movement/the-partner.

———. 2012a. Jared Taylor. Southern Poverty Law Center. Retrieved from http://www.splcenter.org/get-informed/intelligence-files/profiles/jared-taylor.

———. 2012b. Federation for American Immigration Reform. Southern Poverty Law Center. Retrieved from http://www.splcenter.org/get-informed/intelligence-files/groups/federation-for-american-immigration-reform-fair.

———. 2014. Council of Conservative Citizens. Retrieved from http://www.splcenter.org/get-informed/intelligence-files/groups/council-of-conservative-citizens.

Sparks, C. S., and R. L. Jantz. 2002. A Reassessment of Human Cranial Plasticity: Boas Revisited. *Proceedings of the National Academy of Sciences* 99:14636–14639.

Spencer, C. 1992. Interview with Garrett Hardin. *Omni* 14:55–63.

Spencer, F. 1968. Hrdlička, Aleš. In D. L. Sills, ed., *International Encyclopedia of the Social Sciences*, 273–308. New York: Macmillan.

Spencer, H. 1864. *Principles of Biology.* 2 vols. London: Williams and Norgate.

Spier, L. 1959. Some Central Elements in the Legacy. In W. R. Goldschmidt, ed., *The Anthropology of Franz Boas*, 41–51. New York: American Anthropological Association.

Spiro, J. P. 2009. *Defending the Master Race: Conservation, Eugenics, and the Legacy of Madison Grant.* Burlington: University of Vermont Press.

Stamm, J. 2009. Hitler, Socialism, and the Racial Agenda. Part II, Sir Francis Galton and Eugenics. *The Epoch Times,* March 6. Retrieved from http:// www.theepochtimes.com/n2/opinion/hitler-socialism-racial-agenda-part-ii -13209.html.

Stansfield, W. D. 2005. The Bell Family Legacies. *The Journal of Heredity* 96:1–3.

Stearns, S. C. 1976. Life-History Tactics: A Review of the Ideas. *Quarterly Review of Biology* 51:3–47.

———. 1977. Evolution of Life-History Traits—Critique of Theory and a Review of Data. *Annual Review of Ecology and Systematics* 8:145–171.

———. 1983. The Influence of Size and Phylogeny on Patterns of Covariation in the Life-History Traits of Mammals. *Oikos* 41:173–187.

———. 1992. *The Evolution of Life Histories.* Oxford: Oxford University Press.

Stein, L. 1950. *The Racial Thinking of Richard Wagner.* New York: Philosophical Library.

Sternberg, R. J. 2007. Critical Thinking in Psychology Is Really Critical. In R. Sternberg, H. Roediger III, and D. Halpern, eds., *Critical Thinking in Psychology*, 289–296. New York: Cambridge University Press.

Stocking, G. W. Jr. 1968. *Race, Culture, and Evolution: Essays in the History of Anthropology.* New York: The Free Press.

———. 1992. *The Ethnographer's Magic and Other Essays in the History of Anthropology.* Madison: University of Wisconsin Press.

———, ed. 1974. *The Shaping of American Anthropology, 1883–1911: A Franz Boas Reader.* New York: Basic Books.

Stoddard, L. 1920. *The Rising Tide of Color: Against White World-Supremacy.* New York: Charles Scribner's Sons.

———. 1923. *The Revolt against Civilization: The Menace of the Under Man.* New York: Charles Scribner's Sons.

————. 1940. *Into the Darkness: Nazi Germany Today.* New York: Duell, Sloan & Pierce.

Stokes, W. E. D. 1917. *The Right to Be Well Born: or, Horse Breeding in Its Relation to Eugenics.* New York: C. J. O'Brien.

Stoskepf, A. 1999. The Forgotten History of Eugenics. *Rethinking Schools* 13 (3). Retrieved from http://www.rethinkingschools.org/archive/13_03/eugenic.shtml.

Sussman, R. W. 1998. No Forum for Racist Propaganda. *Anthropology Newsletter* 39:2.

————. 1999. The Nature of Human Universals. In R. W. Sussman, ed., *The Biological Basis of Human Behavior: A Critical Review,* 2nd ed., 246–252. Upper Saddle River, NJ: Prentice Hall.

————. 2010. Human Nature and Human Culture. *American Anthropologist* 112:514–515.

————. 2011. A Brief History of Primate Field Studies: Revised. In C. J. Campbell, A. Fuentes, K. C. Mackinnon, S. K. Bearder, and R. S. M. Stumpf, eds., *Primates in Perspective,* 2nd ed., 6–11. New York: Oxford University Press.

Sussman, R. W., and C. R. Cloninger, eds. 2011. *Origins of Altruism and Cooperation.* New York: Springer.

Sussman, R. W., and J. Marshack. 2010. Are Humans Inherently Killers? Global Nonkilling Working Papers #1 2010:7–28. Honolulu: Center for Global Nonkilling.

Sutton, A. 1976. *Wall Street and the Rise of Hitler.* Seal Beach, CA: '76 Press.

Suzuki, D. 1995. Correlation as Causation. In R. Jacoby and N. Glauberman, eds., *The Bell Curve Debate: History, Documents, Opinions,* 280–282. New York: Times Books.

Swan, D. A. 1954. Likes Facism. *Exposé* 34:4.

Szathmáry, E. J. E. 2010. Founding of the American Association of Physical Anthropologists: 1930. In M. A. Little and K. A. R. Kennedy, eds., *Histories of American Physical Anthropology in the Twentieth Century,* 127–139. Lanham, MD: Lexington Books.

Talbot, G. 2011. Kris Kobach: The Kansas Lawyer behind Alabama's Immigration Law. AL.com, October 16. Retrieved from http://blog.al.com/live/2011/10/kris_kobach_the_kansas_lawyer_1.html.

Tattersall, I., and R. DeSalle. 2011. *Race? Debunking a Scientific Myth.* College Station: Texas A&M University Press.

Taylor, H. F. 1980. *The IQ Game: A Methodological Inquiry into the Heredity-Environment Controversy.* New Brunswick, NJ: Rutgers University Press.

Taylor, J. 1983. *Shadows of the Rising Sun: A Critical View of the "Japanese Miracle."* New York: Morrow.

————. 1992a. *Paved with Good Intentions: The Failure of Race Relations in Contemporary America*. New York: Carroll & Graf.

————. 1992b. A Conversation with Arthur Jensen. American Renaissance, August and September. Retrieved from http://www.amren.com/news/2011/12/a_conversation/.

————. 1997. Why Race Matters. *American Renaissance* 10 (October):6.

————. 1998. *The Real American Dilemma: Race, Immigration, and the Future of America*. Oakton, VA: New Century Foundation.

————. 1999. The Racial Revolution. *American Renaissance* 10 (March):1–6.

————. 2011. *White Identity: Racial Consciousness in the 21st Century*. Oakton, VA: New Century Foundation.

Templeton. A. R. 1983. The Evolution of Life Histories under Pleiotropic Constraints and *K*-selection. In H. I. Freedman and C. Strobeck, eds., *Population Biology*, 64–71. Berlin: Springer-Verlag.

————. 1998. Human Races: A Genetic and Evolutionary Perspective. *American Anthropologist* 100:632–650.

————. 2002. The Genetic and Evolutionary Significance of Human Races. In J. M. Fish, *Race and Intelligence: Separating Science from Myth*, 31–56. Mahwah, NJ: Lawrence Erlbaum Associates.

————. 2003. Human Races in the Context of Recent Human Evolution: A Molecular Genetic Perspective. In A. H. Goodman, D. Heath, and M. S. Lindee, eds., *Genetic Nature/Culture: Anthropology and Science beyond the Two-Culture Divide*, 234–257. Berkeley: University of California Press.

————. 2007. Genetics and Recent Human Evolution. *Evolution* 61:1507–1519.

————. 2013. Biological Races in Humans. *Studies in History and Philosophy of Science Part C: Studies in History and Philosophy of Biological and Biomedical Sciences* 44:262–271.

Terbush, J. 2012. Chris Hays to AZ Immigration architect: convince me this isn't about racism. July 1, 2012. Retrieved from http://www.rawstory.com/rs/2012/07/01/chris-hayes-to-az-immigration-law-architect-convince-me-this-isnt-about-racism/?utm_source=feedburner&utm_medium=feed&utm_campaign=Feed%3A+TheRawStory+%28The+Raw+.

Terman, L. M. 1916. *The Measurement of Intelligence: An Explanation of and a Complete Guide for the Use of the Stanford Revision and Extension of the Binet-Simon Intelligence Scale*. Boston: Houghton Mifflin.

————. 1922. Were We Born That Way? *World's Work* 44:655–660.

Terry, D. 2012. Leading Race "Scientist" Dies in Canada. Southern Poverty Law Center. Reprinted at Salon.com, October 6. Retrieved from http://www.salon.com/2012/10/06/leading_race_scientist_dies_in_canada/singleton/.

Thomas, W. I. 1912. Race Psychology: Standpoint and Questionnaire, with Particular Reference to the Immigrant and the Negro. *American Journal of Sociology* 17:725–775.

Tobias, P. V. 1970. Brain-Size, Grey Matter and Race—Fact or Fiction? *American Journal of Physical Anthropology* 32:3–25.

Tucker, W. H. 1997. Re-Reconsidering Burt: Beyond a Reasonable Doubt. *Journal of the History of the Behavioral Sciences* 33:145–162.

———. 2002. *The Funding of Scientific Racism: Wickliffe Draper and the Pioneer Fund.* Champaign: University of Illinois Press.

UNESCO. 1950. Statement by Experts on Race Problems. *Man* 50:138–139.

———. 1952. *The Race Question in Modern Science. The Concept of Race: Results of an Enquiry.* Paris: UNESCO.

———. 1961. *The Race Question in Modern Science: Race and Science.* New York: Columbia University Press.

United Jewish Appeal of Toronto. 2003. "Enlightened" Immanuel Kant Racist. Reprinted from *National Post.* Retrieved from http://www.jewishtoronto .com/page.aspx?id=46607&print=1.

United States Holocaust Memorial Museum. 2012. The Voyage of the *St. Louis.* Holocaust Encyclopedia. Retrieved from http://www.ushmm.org /wlc/en/article.php?ModuleId=10005267.

Urbani, B., and A. Viloria. 2008. Ameranthropoides loysi *Montandon 1929: The History of a Primatological Fraud.* Buenos Aires: LibrosEnRed.

Van de Pitte, F. P. 1971. *Kant as Philosophical Anthropologist.* The Hague: Martinus Nijhoff.

Van Wagenen, B. 1912. Preliminary Report of the Committee of the Eugenic Section of the American Breeders' Association to Study and to Report on the Best Practical Means for Cutting off the Defective Germ-Plasm in the Human Population. In *Problems in Eugenics: Papers Communicated to the First Eugenics Conference,* 460–479. Adelphi, London: Eugenics Education Society. Retrieved from https://archive.org/details/problem sineugeni00inte.

Vinson, J. C. 1997. *Immigration and Nation: A Biblical View.* Monterey, VA: American Immigration Control Foundation.

Virchow, R. 1880. Ausserordentliche Zusammenkunft im Zoologischen Garten am 7 November 1880. Eskimos von Labrador. *Zeitschrift für Ethnologie* 12: 253–274.

von Verschuer, O. 1934. Introduction. In E. Fischer, O. Aichel, and O. vonVer-schuer, eds., *Festband, Eugen Fischer zum 60. Geburtstage: Zeitschrift für Morphologie und Anthropologie, Bd 34.* Stuttgart: Schweizerbart.

———. 1938. *The Racial Biology of the Jews.* Forschungen zur Judenfrage (Studies on the Jewish Problem), Volume III. Hamburg: Hanseatische

Verlagsanstalt. Translated by Charles Weber. Retrieved from http://www
.stormfront.org/forum/t43804/.

———. 1941. *Leitfaden der Rassenhygiene*. Stuttgart: Georg Thieme.

Washburn, S. L. 1984. Review of *A History of Physical Anthropology: 1930–1980*. *Human Biology* 56:393–410.

Watanabe, T. 2010. Cardinal Mahony Criticizes Arizona Immigration Bill. *Los Angeles Times*, April 20. Retrieved from http://articles.latimes.com/2010/apr/20/local/la-me-0420-mahony-immigration-20100420.

Watson, J. B. 1924. *Behaviorism*. New York: W. W. Norton.

Weindling, P. 1988. From Philanthropy to International Science Policy: Rockefeller Funding of Biomedical Sciences in Germany, 1920–1940. In N. A. Rupke, ed., *Science, Policy and the Public Good: Essays in Honor of Margaret Gowing*, 119–140. New York: Macmillan Press.

———. 1989. *Health, Race, and German Politics between National Unification and Nazism, 1870–1945*. Cambridge: Cambridge University Press.

Weiner, M. 1995. *The Global Migration Crisis: Challenges to States and Human Rights*. New York: Harper Collins.

Weinstein, A. 1932. Heredity. In E. Seligman and A. Johnson, eds., *Encyclopedia of the Social Sciences*. New York: Macmillan.

Weinstein, D. 2012. Herbert Spencer. In Edward N. Zalta, ed., *The Stanford Encyclopedia of Philosophy,* Fall 2012 edition. Retrieved from http://plato.stanford.edu/archives/fall2012/entries/spencer/.

Weiss, K. M. and A. V. Buchanan. 2009. *The Mermaid's Tale: Four Billion Years of Cooperation in the Making of Living Things*. Cambridge: Harvard University Press.

Weiss, S. F. 1990. The Race Hygiene Movement in Germany, 1904–1945. In M. B. Adams, ed., *The Wellborn Science: Eugenics in Germany, France, Brazil, and Russia*, 8–68. Oxford: Oxford University Press.

———. 2010. *The Nazi Symbiosis: Human Genetics and Politics in the Third Reich*. Chicago: University of Chicago Press.

Weizmann, F. 2001. Review of *Race, Evolution, and Behavior: A Life History Perspective*. *Canadian Psychology* 42:339–441.

Weizmann, F., N. I. Weiner, D. L. Wiesenthal, and M. Ziegler. (1990) 1999. Differential K Theory and Racial Hierarchies. *Canadian Psychology* 31:1–13. Reprinted in R. W. Sussman, ed., *The Biological Basis of Human Behavior: A Critical Review,* 204–214. Upper Saddle River, NJ: Prentice Hall.

Weizmann, F., N. I. Weiner, D. L. Wiesenthal, and M. Ziegler. 1991. Discussion: Eggs, Eggplants, and Eggheads: A Rejoinder to Rushton. *Canadian Psychology* 32:43–50.

White, Charles. 1799. *An Account of the Regular Gradation in Man, and in Different Animals and Vegetables.* London: C. Dilly.

Whitney, G. 1995. Ideology and Censorship in Behavior Genetics. *Mankind Quarterly* 35:327. Reprinted at Prometheism: The 21st Century Cult of Prometheus. Retrieved from http://www.prometheism.net/ideology/.

———. 1998. Foreword. In D. Duke, *My Awakening: A Path to Racial Understanding.* Covington, LA: Free Speech Press.

———. 2002. Subversion of Science: How Psychology Lost Darwin. *Journal of Historical Review* 21:20–30.

Williams, B. J. 1973. *Evolution and Human Origins: An Introduction to Physical Anthropology.* New York: Harper & Row.

Wilson, E. O. 1975. *Sociobiology: The New Synthesis.* Cambridge: Harvard University Press.

———. 1998. *Concilience: The Unity of Knowledge.* New York: Random House.

Winston, A. S. 1996. The Context of Correctness: A Comment on Rushton. *Journal on Social Stress and the Homeless* 5:231–250.

———. 2013. Shared Eugenic Visions: Raymond B. Cattell and Roger Pearson. Biographies: Institute for the Study of Academic Racism. Retrieved from www.ferris.edu/isar/bios/cattell/HPPB/visions.htm.

Wissler, C. 1923. *Man and Culture.* New York: Thomas Y. Crowell.

Wistrich, R. (1982) 1984. *Who's Who in Nazi Germany.* New York: Bonanza Books.

Wolff, E. N. 2010. *Recent Trends in Household Wealth in the United States: Rising Debt and the Middle-Class Squeeze—An Update to 2007.* Working Paper no. 589. Annandale-on-Hudson, NY: The Levy Economics Institute of Bard College.

———. 2012. *The Asset Price Meltdown and the Wealth of the Middle Class.* New York: New York University Press.

Wolff, H. 2014. AmRen Conference Held in Tennessee. American Renaissance. http://www.amren.com/features/2012/03/amren-conference-held-in-tennessee/.

Woodworth, R. W. 1910. Racial Differences in Mental Traits. *Science* 31:178–181.

———. 1939. *Selected Papers of R. W. Woodworth.* New York: Henry Holt.

Wrangham, R. W. 1996. *Demonic Males: Apes and the Origins of Human Violence.* Boston: Houghton Mifflin.

X, Jacobus. 1896. *Untrodden Fields of Anthropology: Observations on the Esoteric Manners and Customs of Semi-Civilized Peoples; Being a Record of Thirty Years' Experience in Asia, Africa and America.* Paris: Libraire de Bibliophiles.

Yates, F. 1992. *The Art of Memory.* London: Pimlico.

Yerkes, R. M. 1923. Testing of the Human Mind. *Atlantic Monthly* 131:358–370.

———. 1941. Manpower and Military Effectiveness: The Case for Military Engineering. *Journal of Consulting Psychology* 5:205–209.

———, ed. 1921. *Psychological Examining in the United States Army.* Washington, DC: Government Printing Office.

Zenderland, L. 1998. *Measuring Minds: Henry Herbert Goddard and the Origins of American Intelligence Testing.* Cambridge: Cambridge University Press.

Zuckerman, M., and N. Brody. 1988. Oysters, Rabbits and People: A Critique of "Race Differences in Behavior" by J. P. Rushton. *Personality and Individual Differences* 9:1025–1033.

Acknowledgments

I would like to thank the many people from many walks of life who read the book and provided comments on it: Julia Katris, Christopher Shaffer, and a number of my colleagues, including Garland Allen and Ian Tattersall. My academically accomplished family, including my wife, Linda, my sister, Sylvia, and my daughter, Diana, offered helpful feedback and support. Alan Templeton provided input on some chapters. I also received comments from three anonymous reviewers. I would like to thank Jennifer Moore and Micah Zeller at the Washington University Library for their assistance in obtaining some of my figures.

Finally, I thank Michael G. Fisher, my editor at Harvard University Press, for his tremendous assistance in this project and Lauren K. Esdaile of Harvard University Press. Kate Babbitt did an excellent job in copyediting. Edward Wade of Westchester Publishing Services served as an exceptional production editor. Gerald Early and a fellowship in residence at the Center for the Humanities at Washington University in 2010 provided funding for the beginning of this project, and I thank all of the people who provided comments and criticisms during many of my presentations on the topic.

Index